MW00461012

COMPUTATIONAL GEOMETRY FOR DESIGN AND MANUFACTURE

Series: Mathematics and its Applications

MATHEMATICS & AND ITS APPLICATIONS

Series Editor: Professor G. M. Bell,

Chelsea College, University of London

Mathematics and its applications are now awe-inspiring in their scope, variety and depth. Not only is there rapid growth in pure mathematics and its applications to the traditional fields of the physical sciences, engineering and statistics, but new fields of application are emerging in biology, ecology and social organisation. The user of mathematics must assimilate subtle new techniques and also learn to handle the great power of the computer efficiently and economically.

The need of clear, concise and authoritative texts is thus greater than ever and our series will endeavour to supply this need. It aims to be comprehensive and yet flexible. Works surveying recent research will introduce new areas and up-to-date mathematical methods. Undergraduate texts on established topics will stimulate student interest by including applications relevant at the present day. The series will also include selected volumes of lecture notes which will enable certain important topics to be presented earlier than would otherwise be possible.

In all these ways it is hoped to render a valuable service to those who learn, teach, develop and use mathematics.

MATHEMATICAL MODELS IN SOCIAL LIFE AND MANAGEMENT SCIENCES
DAVID BURGHES and ALISTAIR D. WOOD, Cranfield Institute of Technology

MODERN INTRODUCTION TO CLASSICAL MECHANICS AND CONTROL
DAVID BURGHES, Cranfield Institute of Technology and
ANGELA DOWNS, University of Sheffield

VECTOR & TENSOR METHODS
FRANK CHORLTON, University of Aston, Birmingham

LECTURE NOTES ON QUEUEING SYSTEMS
BRIAN CONOLLY, Chelsea College, London University

MATHEMATICS FOR THE BIOSCIENCES
G. EASON, C. W. COLES and G. GETTINBY, University of Strathclyde

HANDBOOK OF HYPERGEOMETRIC INTEGRALS: Theory, Applications, Tables, Computer Programs
HAROLD EXTON, The Polytechnic, Preston

COMPUTATIONAL GEOMETRY FOR DESIGN AND MANUFACTURE
IVOR D. FAUX and MICHAEL J. PRATT, Cranfield Institute of Technology

APPLIED LINEAR ALGEBRA
R. J. GOULT, Cranfield Institute of Technology

GENERALISED FUNCTIONS: Theory, Applications
R. F. HOSKINS, Cranfield Institute of Technology

MECHANICS OF CONTINUOUS MEDIA
S. C. HUNTER, University of Sheffield

USING COMPUTERS
BRIAN MEEK and SIMON FAIRTHORNE, Queen Elizabeth College, University of London

ENVIRONMENTAL AERODYNAMICS
R. S. SCORER, Imperial College of Science and Technology, University of London

LIQUIDS AND THEIR PROPERTIES: A Survey for Scientists and Technologists
H. N. V. TEMPERLEY, University College of Swansea, University of Wales and
H. D. TREVENA, University of Wales, Aberystwyth

COMPUTATIONAL GEOMETRY FOR DESIGN AND MANUFACTURE

I. D. FAUX, B.Sc., Ph.D.,

and

M. J. PRATT, M.A., M.Sc.,

Department of Mathematics,
Cranfield Institute of Technology

ELLIS HORWOOD LIMITED
Publishers Chichester

Halsted Press: a division of
JOHN WILEY & SONS
New York - Chichester - Brisbane - Toronto

[redacted]

The publisher's colophon is reproduced from James Gillison's drawing of the ancient Market Cross, Chichester

First published in 1979 by
ELLIS HORWOOD LIMITED
Market Cross House, Cooper Street, Chichester, West Sussex, England
REPRINTED 1980

Distributors:

Australia, New Zealand, South-east Asia:
Jacaranda-Wiley Ltd., Jacaranda Press,
JOHN WILEY & SONS INC.,
G.P.O. Box 859, Brisbane, Queensland 40001, Australia.
Canada:
JOHN WILEY & SONS CANADA LIMITED
22 Worcester Road, Rexdale, Ontario, Canada.
Europe, Africa:
JOHN WILEY & SONS LIMITED
Baffins Lane, Chichester, West Sussex, England.
North and South America and the rest of the world:
HALSTED PRESS, a division of
JOHN WILEY & SONS
605 Third Avenue, New York, N.Y. 10016, U.S.A.

British Library Cataloguing in Publication Data

Faux, I. D.
Computational geometry for design and manufacture – (Mathematics and its applications).
1. Geometry 2. Design
I. Title II. Pratt, M. J. III. Series
516'.002'474 QA445 78-40637

ISBN 0–85312–114–1 (Ellis Horwood Ltd., Publishers)
ISBN 0–470–26473–X (Halsted Press) Clothbound
ISBN 0–470–27069–1 (Halsted Press) Paperback

Typeset in Press Roman by Ellis Horwood Limited, Chichester, Sussex.

Printed in the U.S.A. by Eastern Graphics, Inc., Old Saybrook, Connecticut.

Table of Contents

Chapter 5 Curve and Surface Design

Chapter 6 Composite Curves and Splines

Authors' Preface

The purpose of this book is to outline the principal developments during the last thirty years or so in the area which Forrest (1971) has called **computational geometry** – 'the computer representation, analysis and synthesis of shape information'. In our opinion the purely geometrical origins of this subject have been to some extent obscured in recent years, and we have endeavoured to redress the balance here by presenting it from a geometrical rather than an analytical point of view. To this end we start by reviewing those aspects of various branches of geometry which are relevant to our main theme, and extensive use is made of this material later in the book in our treatment of various types of curve and surface representations. We also provide Appendices dealing with matrices, determinants and basic numerical analysis, which are intended to make the book reasonably self-contained for the reader who is familiar with elementary calculus to the level taught in most undergraduate engineering courses.

The material for the book started life as a set of notes for a course on applied geometry given from time to time at Cranfield to practising engineers and industrial mathematicians who are actively concerned with computer-aided design (CAD) or manufacture (CAM). The book represents a considerable amplification of the original notes, but is aimed principally at the same people. We hope that some academics will find it a useful introduction to the subject, however, despite the fact that its style may not appeal to pure mathematicians.

We have necessarily been selective in our choice of subject matter. Because we are primarily concerned with the broad mathematical principles of the methods described, we do not go into details of their practical implementation, but the fairly numerous references given should enable readers to seek out these details if they require them. We have drawn extensively on the published work of others, and have done our best to give credit where it is due.

We should like to express our thanks to Dr. K. R. Butterfield for helpful discussions and for making available to us work from his Ph.D. thesis. We are also grateful to Mr. S. S. Gould and Miss E. McLellan of the Department of

Electronic Systems Design, Cranfield, for providing information relating to the numerical control aspects of the book.

It is a pleasure to record our particular thanks to Mrs. Iris Harrison for her continued patience and her thoroughness in the production of a difficult typescript. We must also thank our publisher, Ellis Horwood, for his forbearance in awaiting the completion of the book.

IVOR FAUX

Cranfield, January 1978. MICHAEL PRATT

Introduction

Although its origins can be traced back earlier, numerical geometry first assumed practical importance during the Second World War when production pressures, particularly in the aircraft industry, stimulated the development of new methods of design. Hitherto the design process had been mainly graphical, using the techniques given in books on technical descriptive geometry such as that by Wellman (1957). The new methods were based on analytical curves, conics in particular (Liming, 1944). They avoided much of the painstaking manual draughting work formerly undertaken, and introduced in its place a considerable element of computation, which led to the widespread use of mechanical and electro-mechanical calculators. The design stage now required less time than formerly, and a much higher degree of dimensional integrity was attained than had been possible by graphical means.

With the advent of the electronic computer more ambitious techniques were developed, mostly departing from the spirit of the traditional graphical methods. Previously, surfaces had been represented by the construction of a number of longitudinal curves to blend a set of previously defined cross-sections, a process known as **lofting.** Many of the new methods treat longitudinal and cross-sectional curves in an equal manner, however, and regard them as dividing the surface into an assembly of curvilinear quadrilateral **patches.** Each of these patches may be completely specified by means of a mathematical formula. This constitutes a major advance over the older methods in which it was not possible to define the surface itself, but only a system of lines lying on the surface.

One of the earliest patch systems was that of Ferguson (1963), which deviates still further from traditional practice in using parametric rather than cartesian coordinates in its curve and surface definitions. This has since become standard usage, for a number of reasons. First, it enables twisted curves in three dimensions to be given a simple mathematical representation; these would formerly have been defined in terms of their projections onto two mutually perpendicular coordinate planes. Secondly, it avoids certain problems which can arise in representing closed curves and curves with vertical tangents in a fixed coordinate system. Thirdly, and perhaps most importantly, it enables coordinate transformations such as translation and rotation to be performed very simply. In other words, the use of parametric techniques frees us from dependence on any particular system of coordinates. As pointed out by Forrest (1972c), shape is independent of frame of reference, and so the parametric approach may be regarded as an entirely natural development. Its use would be impossible without the aid of computers, however.

At the same time as systems of this kind were being devised, automatic draughting machines, graphical display units and numerically controlled machine tools were also becoming available. Parametric curve and surface descriptions proved to be well suited for use with these. The graphics displays require the provision of coordinate transformations, projections, perspectives and so on, all of which are most simply handled in parametric terms. They enable us to display a mathematically defined three-dimensional object on a graphics screen as it would appear from any chosen viewpoint.

Once a mathematical representation of the shape of an object has been computed, it is natural to store the representation in the computer. The advantages of doing so are manifold, and some of them are listed below:

(a) the shape of the object is held within the computer in terms of purely numerical information. Problems corresponding to those previously caused by paper shrinkage or draughting aberrations do not arise;
(b) the computer can readily calculate such geometrical properties of the shape as volumes, cross-sectional curves or cross-sectional areas;
(c) conversion between different units of measurement is easily achieved within the computer;
(d) information concerning the shape may be output visually,via draughting machines or graphics terminals, or numerically. In particular it is now possible to produce numerical control tapes which enable the shape to be machined automatically, or to provide information in a form suitable for immediate processing by a structural analysis program. This suggests the possibility of integrating the whole of the manufacturing process, from design and analysis through to production, using the computer as an intermediary.

Since the first surface-defining systems became operational, a number of important mathematical developments have occurred. We indicate some of them briefly here; they are dealt with more fully later in the book. First, the properties of spline curves have been much exploited. These are approximate mathematical analogues of the flexible metal or wooden splines traditionally used by draughtsmen to draw curves fitting a given set of points. Accordingly they are smooth in nature, and an easy extension of their theory leads on to correspondingly smooth spline surfaces. In the last few years it has been shown how any spline curve or surface can be expressed in terms of fundamental splines or B-splines (Curry and Schoenberg, 1966), which have the property of being only locally non-zero. This formulation enables local modifications to be made to a shape during the design process without the need for its numerical representation to be recomputed from scratch each time.

Secondly, a very general theory of surface patches was evolved by Coons (1964), who showed how four arbitrary boundary curves can be blended into a single smooth patch and demonstrated how inter-patch continuity of gradient and curvature may be achieved. This work unified much of what had gone before, but its very generality made it difficult for the mathematical layman to understand.

By this time methods of surface description had become far removed from their origins in the geometrical methods of manual draughting. But the intended users of most surface-defining systems are the latter-day equivalents of draughtsmen and loftsmen who, while they have a good intuitive grasp of geometrical ideas, cannot be expected to have an advanced knowledge of the mathematics of surface patches. This raises the problem of how to make the considerable freedom offered by modern methods of patch definition available to the operator in terms which he can easily understand. One answer is to make the system completely automatic, demanding as information only the positions of points through which the surface is desired to pass, and removing all other freedom from the operator. This kind of system is known as a **surface fitting** system; it performs what a numerical analyst would refer to as two-dimensional interpolation.

A third major mathematical development was the introduction by Bézier (1971) of his UNISURF system, which went against this trend towards total automation. UNISURF is founded on an ingenious reformulation of the mathematics on which Ferguson's original system was based. It enables sections of curves and surfaces to be freely designed by an operator who needs no advanced mathematical training and who works entirely in terms of elementary geometrical concepts. This was the first practical **surface design** system.

Basically, UNISURF operates as follows. The designer defines an open polygon, made up of straight lines, which is displayed on a graphics screen. The system responds with a smooth curve which approximates the polygon. By making modifications to his polygon, the designer can modify the curve in a fairly predictable way until it satisfies whatever criteria he wishes to impose on it, whether aesthetic or otherwise. This is an example of an **interactive** process, amounting to a dialogue between the designer and the computer. Surfaces are defined analogously, in terms of an open polyhedron. While UNISURF works in terms of simple polynomial functions, a similar system has recently been proposed by Gordon and Riesenfeld (1974) which makes use of the B-splines referred to earlier and exploits their properties in providing the facility for purely local modification of curves and surfaces.

A problem which is central to the whole area of curve fitting and design is that no clear definition exists of what constitutes a 'fair' curve. Any traditional draughtsman or loftsman would have little hesitation in judging whether a given curve was fair or not, though he would find it difficult to explain the reasons for his decision. However he would generally be at variance with his colleagues if asked to draw the 'fairest' curve through a given set of points. Experience with computational curve and surface-defining systems has shown that fairness involves more than mere continuity of gradient and of curvature. Obviously these properties are desirable, and they are not difficult to attain, but they do not in themselves guarantee acceptable results. This is because a curve can be very smooth in the mathematical sense but nonetheless contain a great many undulations; the function sin x, which possesses continuous derivatives of all orders, is a simple example. Although most practical systems to date have been based upon polynomials, which can also possess an embarrassing number of maxima and minima, an increasing interest is now being shown in more exotic but less oscillatory functions for curve synthesis. One example is the **spline in tension** (Schweikert, 1966); another is the nonlinear spline used by the AUTOKON system (Mehlum, 1969).

This brief survey of our subject should serve to show that the techniques it employs are based on a number of mathematical disciplines which are usually treated separately. The purpose of the early part of this book (Chapters 1 to 4) is to draw together those parts of classical analytic, algebraic and differential geometry which have direct relevance to our main topic, and to review the vector algebra which has become the accepted language of numerical geometry. This material is used to unify and, we hope, to clarify, the geometrical aspects of the various techniques for the fitting and design of curves and surfaces which are dealt with in Chapters 5, 6, 7 and 8. Chapter 9 discusses some specialised numerical techniques related to surface design and manufacture.

In order to make the book self-contained we provide Appendices which summarise elementary matrix algebra, determinants, and some basic numerical methods. Concerning this last subject it is perhaps worth making the point that many of its standard methods may be applied more or less directly to the type of curve fitting we are interested in. The main difference is that in curve fitting in the usual mathematical sense we are trying to approximate a function, and can clearly define our notions of accuracy and hence state acceptable tolerances. In numerical geometry we are trying to represent not a function but a shape, and our criteria of acceptability are much more elusive, as already observed.

Finally, we should mention one method of shape description with which this book does not deal, namely the building up of computer representations of complex shapes from representations of simpler components. We do not underestimate the importance of this type of approach, but we are here primarily concerned with the problem of describing individual components rather than assemblies of components. The reader should refer, for instance, to Braid (1973) or Woo (1977a) for details of successful systems of this kind.

Chapter 1

Plane co-ordinate geometry

1.1 A REVIEW OF THE FUNDAMENTAL IDEAS

1.1.1 Cartesian co-ordinates in the plane

The simplest plane co-ordinate system is the familiar **Cartesian co-ordinate system.** Two perpendicular lines are drawn which form the **co-ordinate axes,** and their intersection is known as the **origin of co-ordinates** O. On each axis, we define one side of the origin to be the **positive axis** and the other side to be the **negative axis.**

Referring to Figure 1.1, we may choose the positive x- and y-axes to be the lines Ox and Oy respectively. It is usual to choose these positive axes in such a way that rotation from Ox to Oy is anticlockwise.

The **co-ordinates** of any point P in the plane are defined by drawing lines through P parallel to the Ox and Oy axes to intersect them at Y and X respectively. The co-ordinates x and y of the point P are the lengths OX and OY, as shown in Figure 1.2. However, if X or Y lie on the negative axis, the co-ordinate concerned is taken as *minus* the length OX or OY.

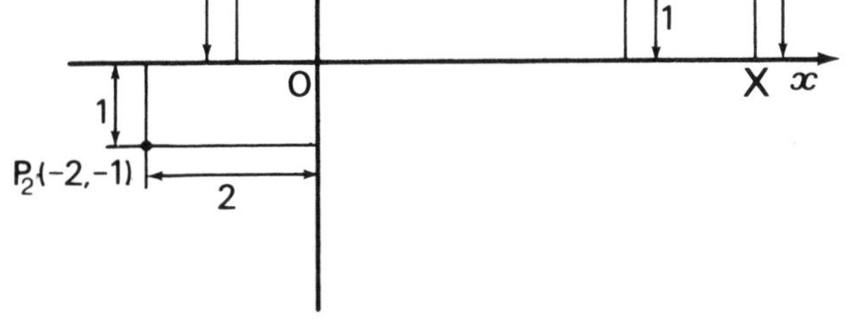

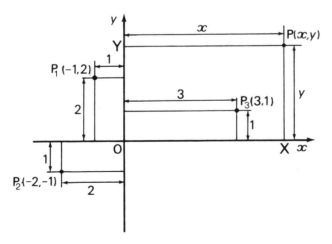

Figure 1.1

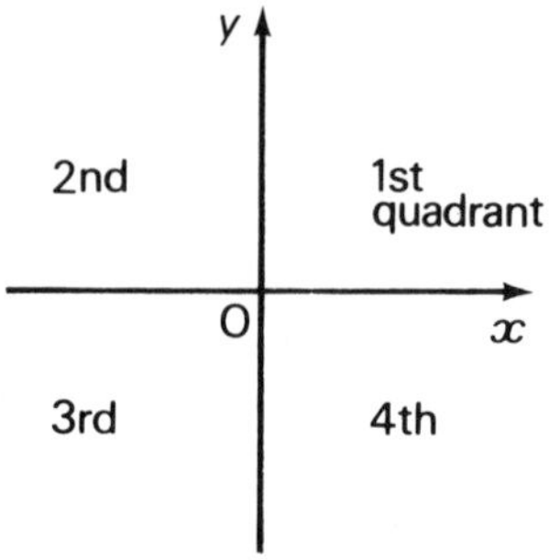

Figure 1.2

The co-ordinates are usually bracketed together in the order (x,y), and we then refer to 'the point $P(x,y)$'. In Figure 1.1, the three points $P_1(-1,2)$, $P_2(-2,-1)$ and $P_3(3,1)$ are shown, illustrating the use of positive and negative co-ordinates.

The axes divide the plane into four **quadrants** and the conventional names of these are given in Figure 1.2.

Cartesian co-ordinate geometry uses such co-ordinates to describe the relationship between points, lines and curves. For a full and rigorous treatment of plane co-ordinate geometry, the reader is referred to the texts listed in the bibliography. We will assume that the reader is familiar with the basic methods of plane co-ordinate geometry, and we shall simply review some of the results without proof in order to emphasize some of those features which are important from a computational point of view.

1.1.2 Equations of a straight line

The most familiar equation of a straight line is

$$y = mx + c, \tag{1.1}$$

in which m is the slope, and c the intercept on the y-axis (see Figure 1.3). This is an **explicit equation** for y, giving a direct prescription enabling us to compute y at any value of x. There is, however, one drawback: vertical lines such as $x = 1$ cannot be accommodated by such an equation.

Expressed in terms of two points (x_1,y_1) and (x_2,y_2) lying on the line, the explicit equation (1.1) becomes

$$y = \frac{y_2 - y_1}{x_2 - x_1} x + \frac{y_1x_2 - y_2x_1}{x_2 - x_1} \cdot \tag{1.2}$$

This equation may be expressed more symmetrically as

$$(x_2 - x_1)(y - y_1) = (y_2 - y_1)(x - x_1). \tag{1.3}$$

The equation is now in **implicit form**, and y can only be evaluated for given values of x by solving this linear equation, effectively by reproducing equation (1.2). However, the implicit formula does allow for vertical lines: if $x_2 = x_1$, and $y_2 \neq y_1$, we obtain the equation of the vertical line $x = x_1$.

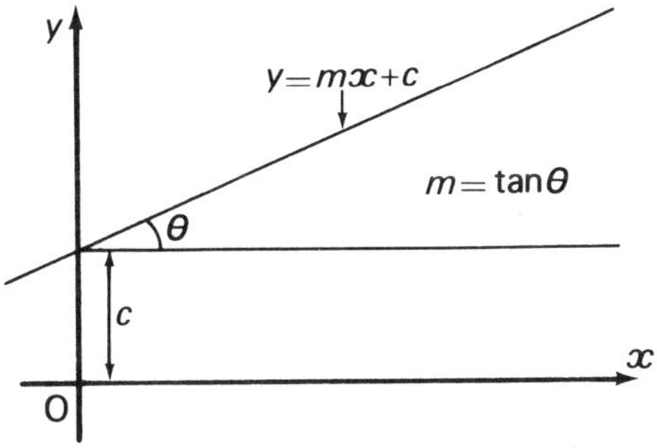

Figure 1.3

Whilst the problem of vertical lines is easily dealt with when solving a problem by hand, it is a nuisance when programming geometrical problems for a computer. Moreover, we must avoid lines which are nearly vertical as these may lead to overflow or rounding error problems. The implicit form avoids the need for special provision for such lines. In its general form, this is written as

$$ax + by + c = 0. \tag{1.4}$$

A vertical line is simply a line for which $b = 0$. One feature should be noted which is characteristic of all implicit equations: the coefficients are not uniquely defined, because the equation is satisfied when any multiples λa, λb and λc are used in place of a, b and c.

In order to obtain a unique description of any given line, the coefficients can be scaled so that $a^2 + b^2 = 1$ and $c < 0$. The two dimensional versions of the APT part programming language incorporate scaling of this kind.

1.1.3 Equations of plane curves

The simplest way to define a plane curve is to use the **explicit form** $y = f(x)$, where $f(x)$ is a prescribed function of x, enabling us to tabulate and plot the function in the familiar way (see Figure 1.4). In doing so, we make assumptions about the nature of the curve between tabulated points. The question of interpolation between known values is taken up in more detail in Appendix A5.

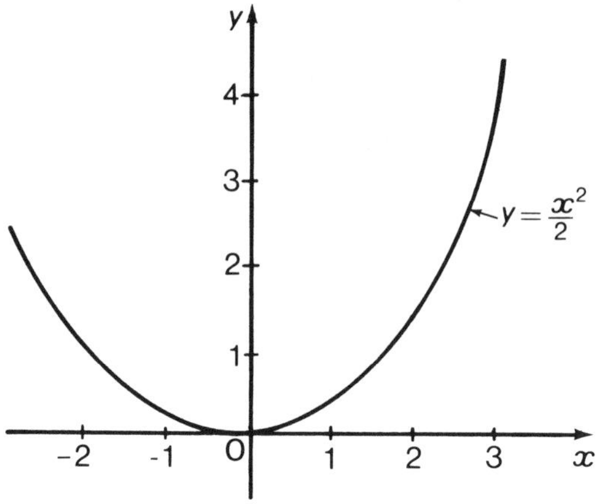

Figure 1.4

The explicit form is satisfactory when the function is single-valued and the curve has no vertical tangents. However, this precludes many curves of practical importance such as circles, ellipses and the other conic sections.

The equation $x^2 + y^2 - r^2 = 0$ is an **implicit equation** for the circle shown in Figure 1.5. The value of y is not described directly as a function of x. If we require an explicit equation, the circle must be divided into two segments, with $y = +\sqrt{(r^2 - x^2)}$ for the upper half and $y = -\sqrt{(r^2 - x^2)}$ for the lower half. This kind of segmentation again creates cases which are a nuisance in computer programs.

We will write the general implicit equation of a curve in the form $f(x,y) = 0$, where $f(x,y)$ is a prescribed function of x and y. We can determine whether or not a point (x,y) lies on the curve, but we cannot directly calculate points on the curve unless the equation can be reduced to an explicit equation for x or y.

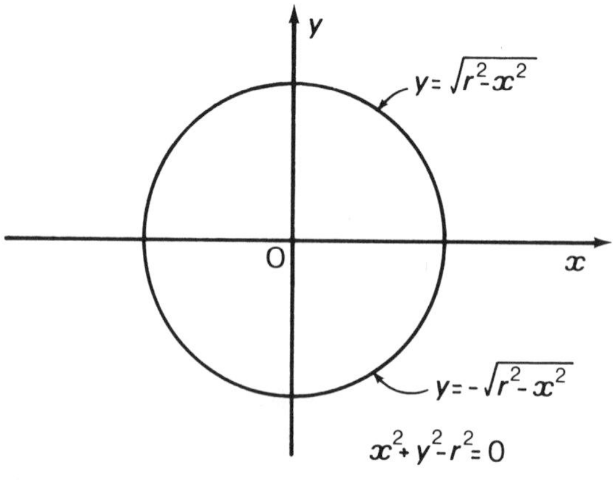

Figure 1.5

If $f(x,y)$ is a continuous function of x and y in the vicinity of the curve, then the value of $f(x,y)$ can be used as a measure of proximity to the curve in automatic search procedures designed to plot the curve or to move a machine tool along it. In practice, we usually require a continuous tangent, which is ensured by the existence and continuity of the partial derivatives $\dfrac{\partial f}{\partial x}$ and $\dfrac{\partial f}{\partial y}$ provided both are not zero at the same point. If continuous curvature is also required, it is sufficient for the second partial derivatives to be continuous in addition.

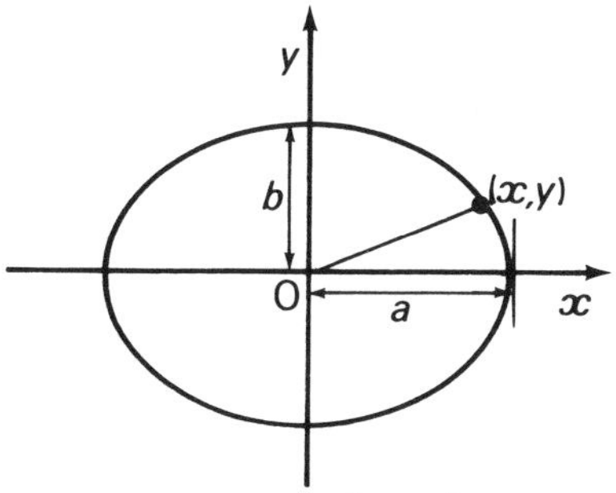

Figure 1.6: The Ellipse $\dfrac{x^2}{a^2} + \dfrac{y^2}{b^2} - 1 = 0$

The most commonly used implicit equations are those of the conic sections. The familiar equations given in Figures 1.6 to 1.8 describe these conic sections in their standard position and orientation.

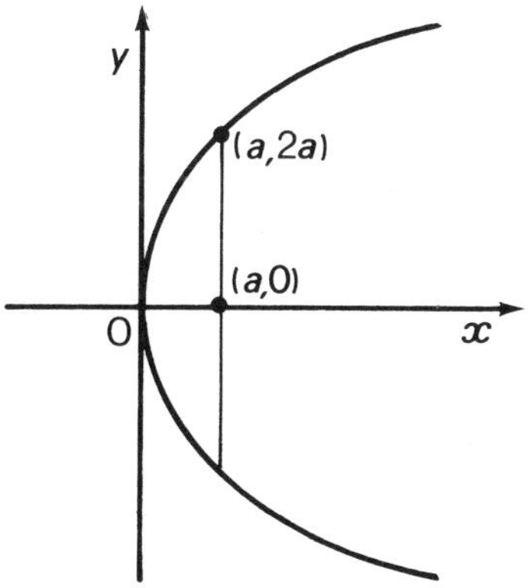

Figure 1.7: The Parabola $y^2 - 4ax = 0$

When considered in general position, all of these conics can be described by the quadratic equation

$$S = ax^2 + 2hxy + by^2 + 2gx + 2fy + c = 0 \tag{1.5}$$

for different values of the coefficients a, b, c, f, g, and h. In particular, the curve is an ellipse if $h^2 < ab$, a parabola if $h^2 = ab$, or an hyperbola if $h^2 > ab$, unless $abc + 2fgh - af^2 - bg^2 - ch^2 = 0$, in which case the conic degenerates to a pair of straight lines, possibly coincident. As with the straight line equation (1.4), the coefficients must be scaled in some conventional way if we require unique coefficients for any given conic.

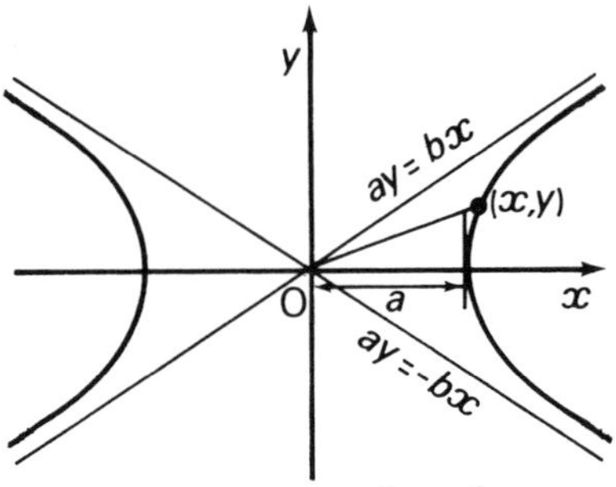

Figure 1.8: The Hyperbola $\frac{x^2}{a^2} - \frac{y^2}{b^2} - 1 = 0$

Apart from their practical importance as the cross-sections of cones and cylinders, the conic sections have comparatively simple analytic properties. Consequently their study forms a large part of most classical texts on two dimensional co-ordinate geometry, such as Robson (1940). Some of their properties are discussed in Section 1.2.

1.1.4 Some important formulae relating points and lines

1.1.4.1 The distance between two points (x_1,y_1) and (x_2,y_2) is calculated as follows.

By Pythagoras, the distance is d, where

$$d = \sqrt{(x_1 - x_2)^2 + (y_1 - y_2)^2}. \tag{1.6}$$

1.1.4.2 The distance between the point (x_1,y_1) and the line $ax + by + c = 0$ is d where

$$d^2 = (ax_1 + by_1 + c)^2/(a^2 + b^2). \tag{1.7}$$

1.1.4.3 The intersection of two lines $a_1x + b_1y + c_1 = 0$ and $a_2x + b_2y + c_2 = 0$ is at (x,y), where

$$x = \frac{b_1c_2 - b_2c_1}{a_1b_2 - a_2b_1} \text{ and } y = \frac{c_1a_2 - c_2a_1}{a_1b_2 - a_2b_1}, \tag{1.8}$$

unless $a_1b_2 = a_2b_1$, in which case the two lines are parallel (or possibly identical).

1.1.4.4 The angle between two lines $a_1x + b_1y + c_1 = 0$ and $a_2x + b_2y + c_2 = 0$ is θ, where

$$\cos\theta = \frac{a_1a_2 + b_1b_2}{\sqrt{(a_1^2 + b_1^2)(a_2^2 + b_2^2)}}. \tag{1.9}$$

1.1.4.5 The two lines above are parallel if

$$a_1b_2 = a_2b_1. \tag{1.10}$$

1.1.4.6 The two lines above are perpendicular if

$$a_1a_2 + b_1b_2 = 0. \tag{1.11}$$

1.1.5 Intersections between lines and curves

To find the intersection between two curves $f(x,y) = 0$ and $g(x,y) = 0$, we must solve these two equations simultaneously. If they are both straight lines, the solution is straightforward, as in equation (1.8), although even here an anomalous case arises when the lines are parallel. To remove this anomaly, it is necessary to use homogeneous co-ordinates. These are discussed later in Chapter 3, which deals with general co-ordinate systems and transformations.

In cases where f and g are non-linear functions of x and y, we usually require an iterative numerical procedure to solve these simultaneous equations. Numerical methods for non-linear equations are discussed in Appendix 4.

1.1.6 Tangents and normals to curves

The tangent line to the curve $y = f(x)$ at the point $P(x_1,y_1)$ can be expressed by the equation

$$y = y_1 + f'(x_1)(x - x_1), \tag{1.12}$$

where $f'(x_1)$ is the value of derivative $\dfrac{df}{dx}$ at $x = x_1$ (see Figure 1.9).

It is apparent from the formula that there will be difficulties when the curve has a vertical or near-vertical tangent at P.

The difficulty can be avoided by using an **implicit** equation $g(x,y) = 0$ to describe the curve. The **implicit** equation of the tangent is then given by

$$g_x(x_1,y_1)(x - x_1) + g_y(x_1,y_1)(y - y_1) = 0, \tag{1.13}$$

where $g_x(x_1,y_1)$ and $g_y(x_1,y_1)$ are the values of $\dfrac{\partial g}{\partial x}$ and $\dfrac{\partial g}{\partial y}$ at P.;

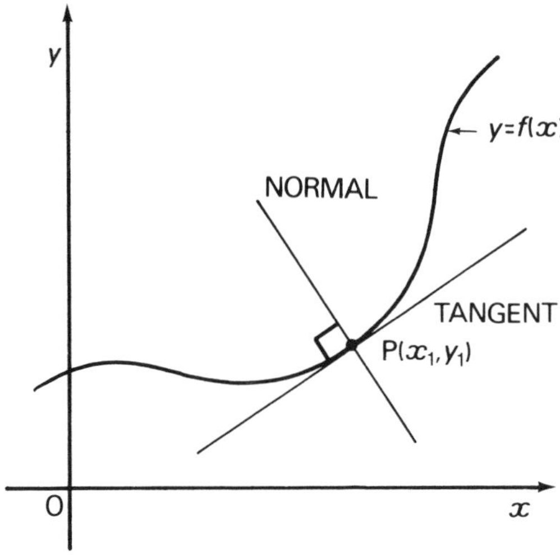

Figure 1.9

Example

The tangent to the circle $x^2 + y^2 - 1 = 0$ at the point (1,0) is calculated as follows.
$g(x,y) = x^2 + y^2 - 1$, so that $g_x = 2x$ and $g_y = 2y$. Hence $g_x(1,0) = 2$ and $g_y(1,0) = 0$. Thus the tangent is given by $2(x-1) + 0(y-0) = 0$, which is the vertical line $x = 1$.

Note that this result cannot be derived when either the circle or the tangent is written in explicit form.

The explicit equation for the normal at P is given by

$$y = y_1 - (x - x_1)/f'(x_1). \tag{1.14}$$

This equation causes problems when the curve is horizontal at P.

The corresponding implicit equation is

$$g_y(x_1,y_1)(x - x_1) - g_x(x_1,y_1)(y - y_1) = 0, \tag{1.15}$$

which again removes the difficulty presented by the explicit form.

1.1.7 Parametric equations of lines and curves

We have seen that the implicit equations for lines and curves, whilst avoiding the difficulties of multiple values and vertical tangents inherent in the explicit form, do not enable us to generate points on the curves directly, and also require numerical procedures to determine intersections. An alternative way of describing lines and curves which treats the co-ordinates x and y symmetrically is the **parametric form.**

The co-ordinates x and y are expressed as functions of an auxiliary parameter t, so that $x = x(t), y = y(t)$. For example, the circle $x^2 + y^2 - 1 = 0$ can be expressed parametrically by the equations

$$x = \cos t$$

and

$$y = \sin t, \tag{1.16}$$

where t takes values in the range $0 \leqslant t < 2\pi$ (see Figure 1.10). Although we normally need to prescribe the range of the parameter t, this can be an advantage if we want to describe a *segment* of a curve. For example, the arc ABC of the circle in Figure 1.10 is completely described by the parametric equations (1.16) and the condition $2\pi/3 \leqslant t \leqslant 7\pi/6$.

The parametric equations enable us to plot points on the curve by evaluating $x(t)$ and $y(t)$ for successive values of t.

If $x(t)$ and $y(t)$ are linear functions of t, the curve is a straight line, and in particular, the equation of the line passing through the points $P_1(x_1,y_1)$ and $P_2(x_2,y_2)$ is given by

$$x = x_1 + t(x_2 - x_1)$$

and

$$y = y_1 + t(y_2 - y_1), \tag{1.17}$$

where the point $P(x,y)$ divides the line joining P_1 and P_2 in the ratio t:1-t as shown in Figure 1.11. This result can be verified by using the similarity of triangles P_1PQ and P_1P_2R.

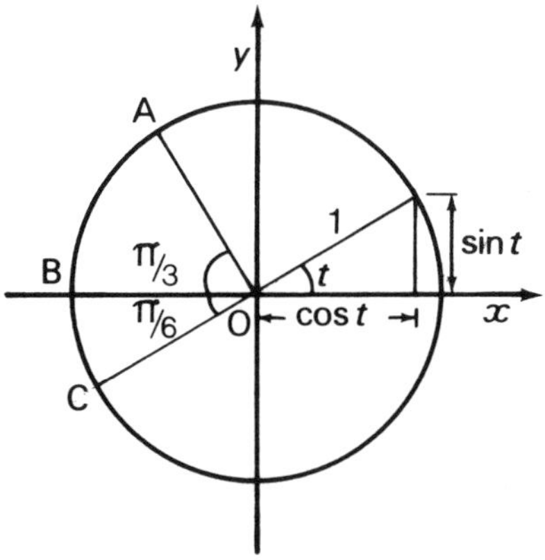

Figure 1.10

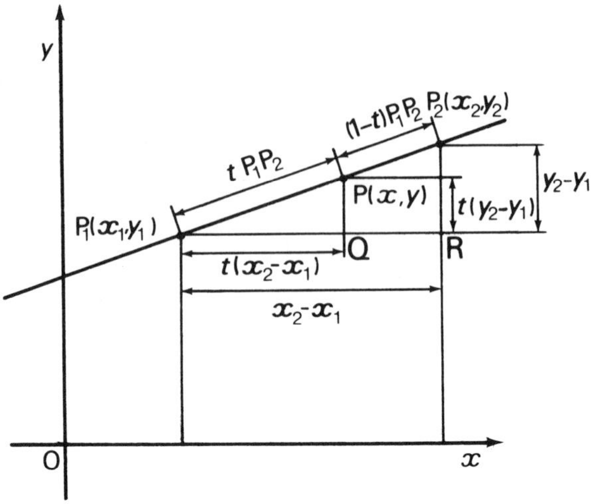

Figure 1.11

The straight line $ax + by + c = 0$ has the parametric equations

$$x = \frac{-ac}{a^2 + b^2} + bt$$

(1.18)

and $$y = \frac{-bc}{a^2 + b^2} - at.$$

Unlike the (scaled) implicit equations, the parametric form is by no means unique, and totally different functions $x(t)$ and $y(t)$ can represent exactly the same curve. The properties of parametric curves will be discussed at greater length in the case of three dimensional curves and lines. However, for completeness we will now give without proof the formulae of the tangents and normals in parametric form.

The **tangent** to the curve $x = x(t)$, $y = y(t)$ at the point P with parameter $t = t_1$, is given by the equations

$$x = x(\tau) = x(t_1) + \tau\dot{x}(t_1)$$

(1.19)

and $$y = y(\tau) = y(t_1) + \tau\dot{y}(t_1),$$

where τ is the parameter on the tangent line, and $\dot{x}(t_1)$, $\dot{y}(t_1)$ are the values of $\frac{dx}{dt}$ and $\frac{dy}{dt}$ at $t = t_1$.

The **normal** to the curve at P is given by the equations

$$x = x(t_1) + \tau\dot{y}(t_1)$$

and

$$y = y(t_1) - \tau\dot{x}(t_1). \tag{1.20}$$

1.1.8 Intersection of two parametric curves

The intersection of two parametric curves $x = x(t)$, $y = y(t)$ and $x = \xi(\tau)$, $y = \eta(\tau)$ requires again the simultaneous solution of two equations in the two unknowns t and τ:

$$x(t) - \xi(\tau) = 0$$

and

$$y(t) - \eta(\tau) = 0, \tag{1.21}$$

so that we are usually no better off in this respect than we were with the implicit equations.

However, if we have the equation of one curve in implicit form and one in parametric form, we can substitute the parametric forms into the implicit equation and obtain a single (usually nonlinear) equation for t.

Example

Intersection between the circle $x^2 + y^2 - 1 = 0$ and the line $x = t, y = 1 - t$.

Now $t^2 + (1 - t)^2 - 1 = 0,$

whence $t^2 + 1 - 2t + t^2 - 1 = 0,$

$2(t^2 - t) = 0,$

and $t = 0 \text{ or } 1.$

Substituting back into the parametric equations, we obtain the two intersections (0,1) and (1,0).

If, conversely, we describe the same circle by $x = \cos t$, $y = \sin t$, and the line by $x + y - 1 = 0$, we obtain the trigonometric equation

$$\cos t + \sin t - 1 = 0.$$

In this case, the solutions $t = 0$ and $t = \pi/2$ are apparent, but in general we require a numerical solution for t when this method is used.

However the method has been included here to indicate that hybrid methods may well be useful in reducing the complexity of the numerical analysis.

1.1.9 Curvature

The radius of curvature ρ of the curve $y = y(x)$ is given by the well-known formula

$$\rho = \frac{(1 + y'^2)^{\frac{3}{2}}}{y''},$$

where the dash denotes differentiation with respect to x. Because the radius of curvature becomes infinite at points of inflection, it is usually better to use the curvature $\kappa = 1/\rho$, which is finite unless there are cusps in the curve.

Thus

$$\kappa = \frac{y''}{(1 + y'^2)^{\frac{3}{2}}}. \tag{1.22}$$

The corresponding formulae for an implicitly defined curve $f(x,y) = 0$ is given by

$$\kappa = -\frac{f_{xx}f_y{}^2 - 2f_{xy}f_xf_y + f_{yy}f_y{}^2}{(f_x{}^2 + f_y{}^2)^{\frac{3}{2}}}, \tag{1.23}$$

where the suffices x and y imply partial differentiation with respect to x and y, for example $f_{xy} = \dfrac{\partial^2 f}{\partial x \partial y}$.

For the parametric curve $x = x(t), y = y(t)$, the expression is

$$\kappa = \frac{\dot{x}\ddot{y} - \dot{y}\ddot{x}}{(\dot{x}^2 + \dot{y}^2)^{\frac{3}{2}}}, \tag{1.24}$$

where the dot implies differentiation with respect to the parameter t.

1.2 SPECIAL TECHNIQUES IN PLANE CO-ORDINATE GEOMETRY

It is impossible in this relatively short book to describe the multitude of special techniques which are used in particular applications. Therefore, the purpose of the following sections is to indicate the wide variety of techniques which can be usefully employed, by describing some which are in current use.

1.2.1 The use of polar co-ordinates for curves with rotational symmetry

When describing shapes possessing some degree of rotational symmetry, it is often convenient to use the polar co-ordinate system shown in Figure 1.12, where the co-ordinates of a point P are given by the radius r(= OP) and the angle θ(= ∠XOP).

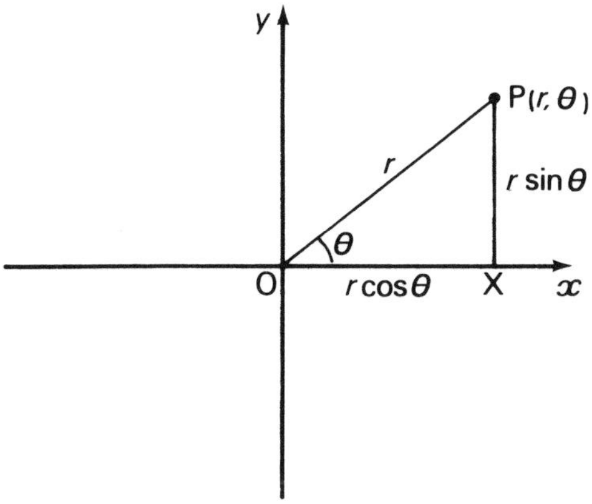

Figure 1.12

The polar co-ordinates are related to the usual Cartesian co-ordinates by the equations

$$x = r\cos\theta$$

and $$y = r\sin\theta. \tag{1.25}$$

The polar equation of a curve is an equation relating r and θ. If the polar equation has the explicit form $r = r(\theta)$, then the equations

$$x = r(\theta)\cos\theta$$

and $$y = r(\theta)\sin\theta \tag{1.26}$$

provide parametric equations for the curve. The tangent and normal to the curve can be determined from equations (1.19) and (1.20).

Many useful profiles can, for example, be described by polar equations of the form

$$r = a + b\cos n\theta, \tag{1.27}$$

which describe a closed curve with n lobes symmetrically disposed on a circular base (see Figure 1.13).

In addition to their application in simple cams, these curves can form a basis of cross-sectional designs in which the coefficients a and b are varied along an axis perpendicular to the Oxy plane.

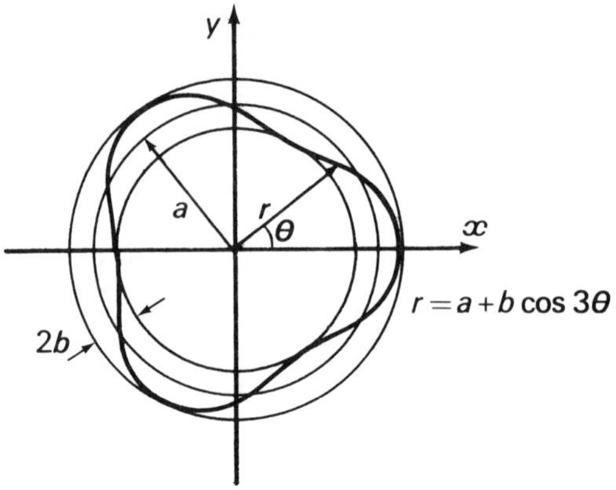

Figure 1.13

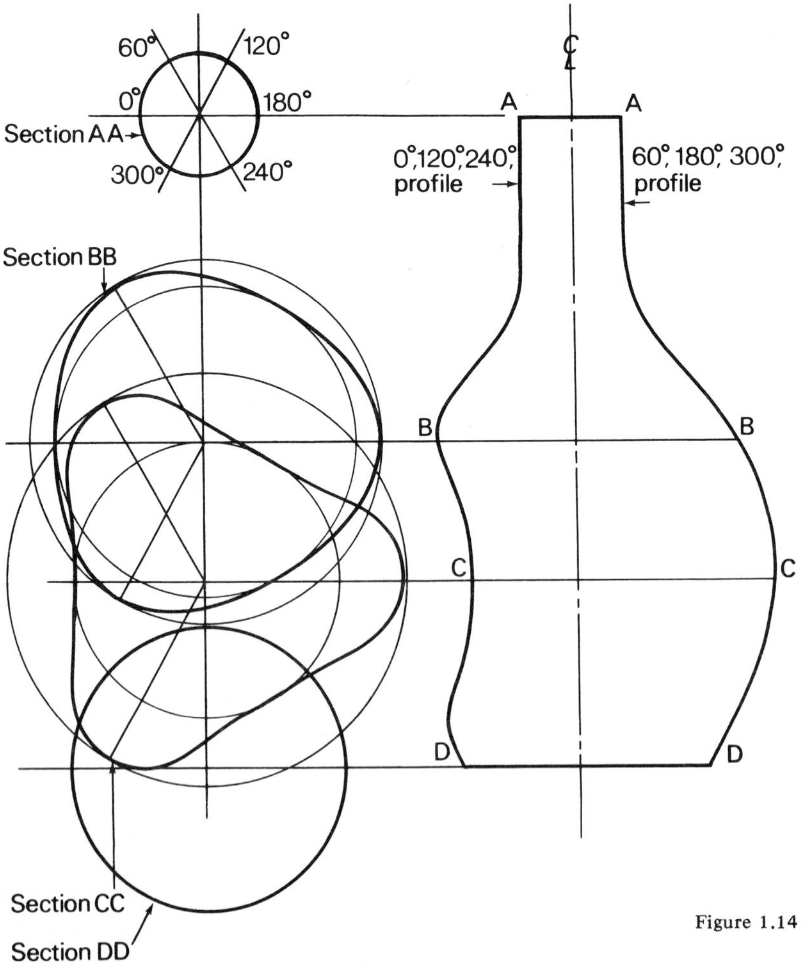

Figure 1.14

For example, all the cross-sections of a well-known whisky bottle take the form of Figure 1.13 for appropriate values of a and b (see Figure 1.14). The subject of cross-sectional designs is taken up in more detail in Chapter 8.

The disadvantages of polar co-ordinates are as follows:
(1) there is no simple way of relating polar co-ordinates referred to different origins.
(2) the equations of tangents and normals cannot be simply described in polar form, so that the parametric Cartesian form must be used.
(3) The polar angle θ of a point $P(x,y)$ can only be found by using inverse trigonometric functions.

1.2.2 Calculation of conic sections satisfying given continuity and tangency requirements

Liming (1944) has described the application of classical methods of conic section theory to the cross-sectional design of aircraft fuselages. His method is to use segments of conics which satisfy conditions of incidence and possibly tangency at several points.

The quadratic equation $ax^2 + 2hxy + by^2 + 2gx + 2fy + c = 0$ describes a general conic section, whose nature may be determined from the coefficients a, b, c, f, g and h. We could, in principle, calculate these when five independent conditions are given, by solving the corresponding equations for the ratios of the coefficients. Thus, if the curve is to pass through (x_1,y_1), the coefficients satisfy the equation

$$ax_1^2 + 2hx_1y_1 + by_1^2 + 2gx_1 + 2fy_1 + c = 0, \qquad (1.28)$$

and if the tangent at (x_1,y_1) makes an angle θ_1 with the Ox axis, then

$$(ax_1 + hy_1 + g)\cos\theta_1 + (hx_1 + by_1 + f)\sin\theta_1 = 0. \qquad (1.29)$$

If five independent conditions of this kind are imposed, we obtain a set of five linear equations for the ratios $a{:}b{:}c{:}f{:}g{:}h$. To remove the necessity of solving these equations, Liming uses the following well-established classical techniques which illustrate some of the advantages of the implicit form.

The first point to note is that if two conics have the equations $S_1(x,y) = 0$ and $S_2(x,y) = 0$ (or for brevity $S_1 = 0$ and $S_2 = 0$), then the equation

$$(1 - \lambda)S_1 + \lambda S_2 = 0 \qquad (1.30)$$

is satisfied for all points lying on both $S_1 = 0$ and $S_2 = 0$. Equation (1.30) then represents another conic (since it is quadratic) which passes through the intersection points of $S_1 = 0$ and $S_2 = 0$, whatever the value of λ (see Figure 1.15).

As λ is varied, a family (or **pencil**) of conics is formed, two of which are $S_1 = 0$ (when $\lambda = 0$) and $S_2 = 0$ (when $\lambda = 1$).

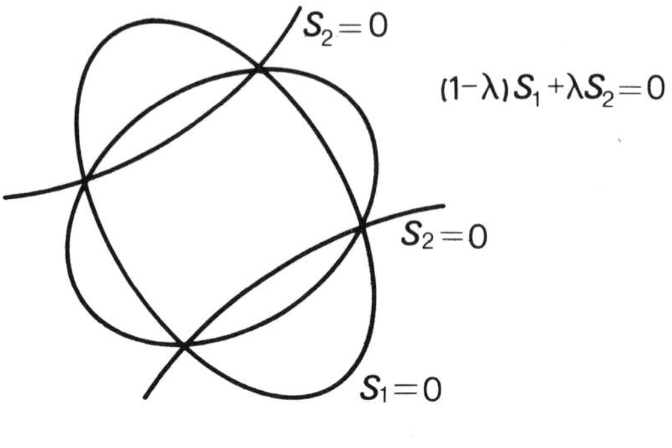

Figure 1.15

The value of the parameter λ can in the general case be determined by specifying another point (not an intersection point) which lies on the curve $(1-\lambda)S_1 + \lambda S_2 = 0$. If the point is $P_1(x_1,y_1)$, then

$$\lambda = \frac{S_1(x_1,y_1)}{S_1(x_1,y_1) - S_2(x_1,y_1)}$$

We next observe that the equation $(a_1x + b_1y + c_1)(a_2x + b_2y + c_2) = 0$, or $\ell_1\ell_2 = 0$, is a quadratic equation satisfied by all points on the **line pair** $\ell_1 = 0$ and $\ell_2 = 0$. It is, in fact, a degenerate conic section obtained by sectioning a cone by a plane through its vertex and parallel to the axis. We may use such line pairs to define non-degenerate conic sections by using equation (1.30).

Thus we see that the equation

$$(1 - \lambda)\ell_1\ell_2 + \lambda\ell_3\ell_4 = 0 \tag{1.31}$$

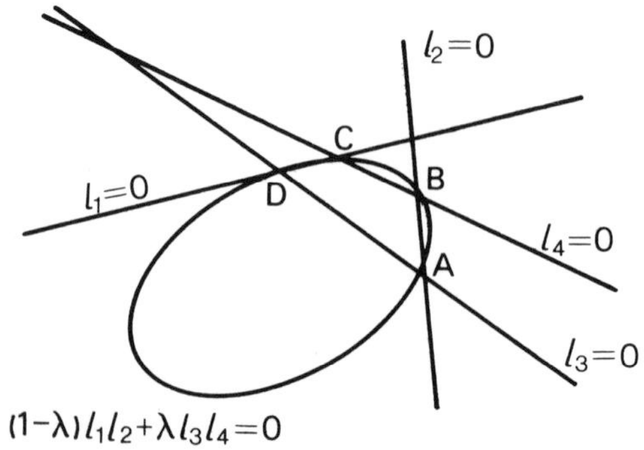

Figure 1.16

represents a family or 'pencil' of conics which pass through the four intersections of the line pairs (ℓ_1,ℓ_2) and (ℓ_3,ℓ_4) (see Figure 1.16). By specifying a fifth point we can fix the value of λ. (See Example 2 at the end of this section).

The method can be adapted to find the conic which passes through two points with given tangents and passing through a given third point. From Figure 1.16 it can be seen that as C moves towards D, the chord CD tends to the tangent at D to the conic shown. Similarly, as B moves towards A, the chord AB tends to the tangent at A. Thus if the lines $\ell_3 = 0$ and $\ell_4 = 0$ are identical, the equation

$$(1 - \lambda)\ell_1\ell_2 + \lambda\ell_3^2 = 0 \tag{1.32}$$

represents a pencil of conics through A and D with ℓ_1 as the tangent at A and ℓ_2 as the tangent at D (see Figure 1.17). The choice of a third point F determines the parameter λ.

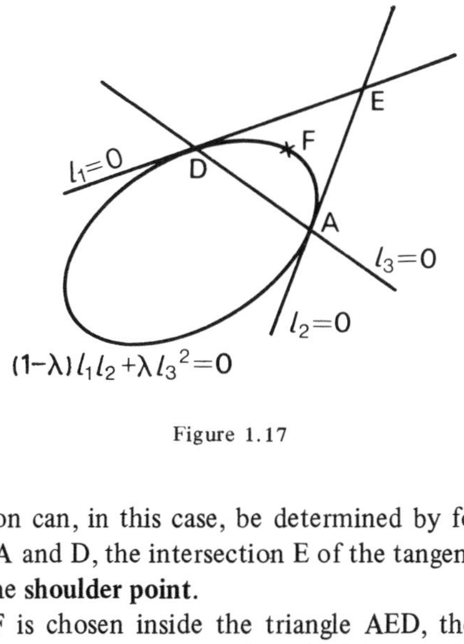

Figure 1.17

The conic section can, in this case, be determined by four points: the two points of tangency A and D, the intersection E of the tangents, and some fourth point F, known as the **shoulder point**.

Provided that F is chosen inside the triangle AED, the conic will always provide a continuous curve between A and D which lies inside the triangle.

If F is the midpoint of the line joining the midpoints of DE and AE, then the conic is a parabola commonly known as the proportional curve. If F lies between the parabola and the line AD, the resulting curve will be an ellipse. If it is outside the parabola, the curve will be hyperbolic.

In Liming's fuselage design method, five points determine the shape of each cross-section (see Figure 1.18). Since the cross-section is symmetric about the vertical axis, only one half needs to be designed, and the tangents at E and A are horizontal. Since C is taken as the point of maximum half-width on the section, the tangent at C is vertical. The cross-section is composed of two conic sections with a common tangent automatically provided at C.

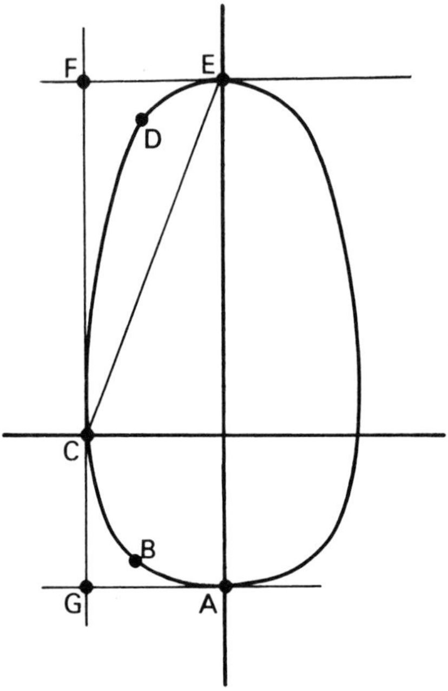

Figure 1.18

The complete fuselage is then described by the five curves traced out by A, B, C, D and E as the plane of cross-section moves along the z-axis.

Example 1

A simple application of Liming's method is the provision of a smooth blend between a circular cross-section in one plane and a square cross-section in another.

Let the circular section be given by $x^2 + y^2 - 1 = 0$ and the square section by the line pair $x = 1, y = 1$ (see Figure 1.19).

Then the conic sections

$$(1 - \lambda)(x^2 + y^2 - 1) + \lambda(x - 1)(y - 1) = 0$$

blend smoothly between the two sections as λ ranges between 0 and 1.

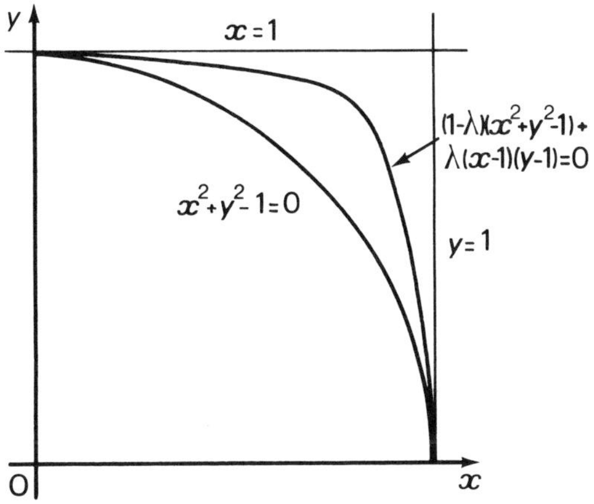

Figure 1.19

Example 2

To determine the conic which passes through the points (0,0), (1,0), (1,1) (0,1) and $(\frac{3}{2}, \frac{1}{2})$.

The first task is to find two line pairs having the first four points as intersections.

The simplest choice is given by the line pairs $\ell_1 = x = 0$, $\ell_2 = x - 1 = 0$ and $\ell_3 = y = 0$, $\ell_4 = y - 1 = 0$. These are shown in Figure 1.20.

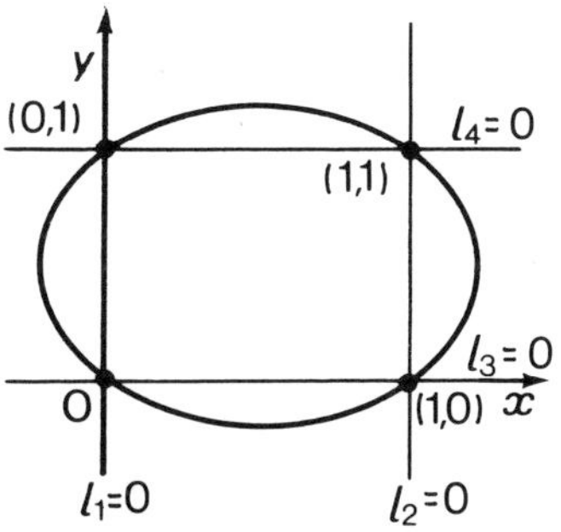

Figure 1.20

Thus every conic section through these points has the equation $(1 - \lambda)x(x - 1) + \lambda y(y - 1) = 0$ for some value of λ.

Since the fifth point is $(\frac{3}{2},\frac{1}{2})$, then

$$(1 - \lambda)\left(\frac{3}{2}\right)\left(\frac{1}{2}\right) + \lambda\left(\frac{1}{2}\right)\left(-\frac{1}{2}\right) = 0,$$

so that $$\lambda = \frac{3}{4}.$$

Hence $$x(x - 1)/4 + 3y(y - 1)/4 = 0.$$

Thus $$x^2 + 3y^2 - x - 3y = 0$$

is the equation of the required conic (in this case an ellipse).

1.2.3 The envelope of a family of curves

In designing parts of moving machinery, or in providing clearance for moving parts, it is often necessary to consider the boundary of the area covered by the superimposed outlines of each part in all its positions during the motion (see Figure 1.21). The required boundary curve is called the **envelope** of the family of outlines, and can be calculated in the following way.

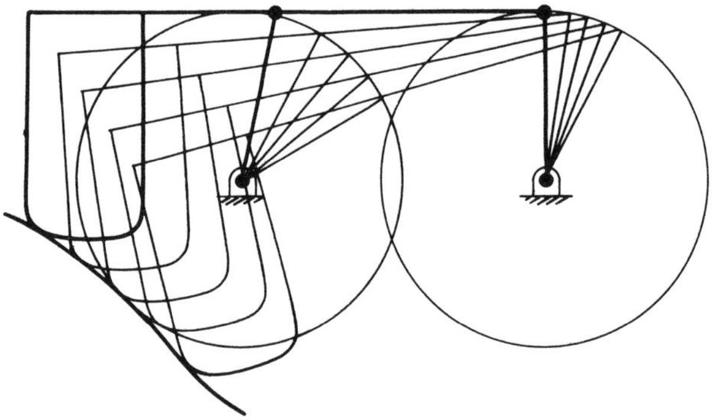

Figure 1.21

If the outline curve has the equation $f(x,y,t) = 0$ at time t, then each of the points of the boundary curve must satisfy this equation for some value of t. The intersection between two curves $f(x,y,t) = 0$ and $f(x,y,t+\delta t) = 0$ will be close to the boundary curve for small δt. At this intersection, since both terms in the numerator are zero there, we can write

$$\frac{f(x,y,t+\delta t) - f(x,y,t)}{\delta t} = 0.$$

It follows that the boundary point satisfies this equation in the limit as $\delta t \to 0$.

Thus the envelope is found by solving the equations

$$f(x,y,t) = 0$$

and

$$\frac{\partial f}{\partial t}(x,y,t) = 0 \qquad (1.33)$$

simultaneously.

Example

Consider the lines (Figure 1.22) given by

$$f(x,y,t) = tx + (1 - t)y + t(t - 1) = 0.$$

Then

$$\frac{\partial f}{\partial t} = x - y + 2t - 1 = 0.$$

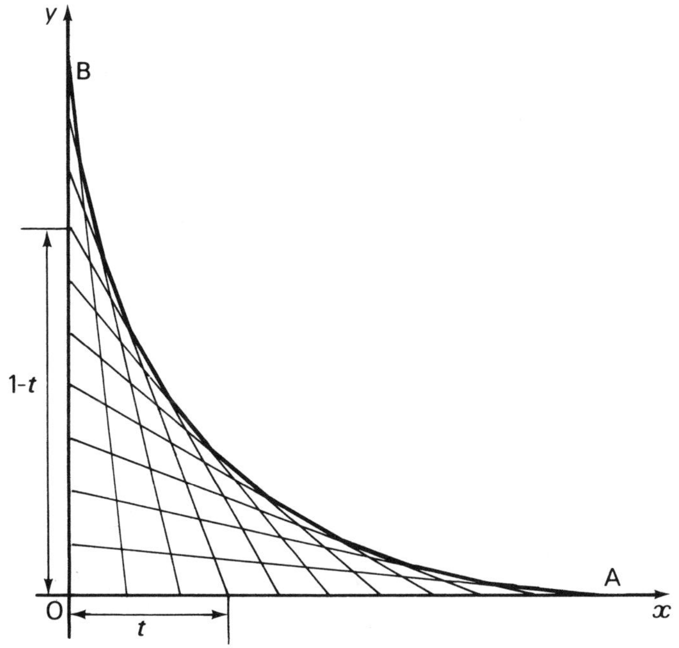

Figure 1.22

We may now eliminate t to obtain the implicit equation of the envelope:

$$x^2 - 2xy + y^2 - 2x - 2y + 1 = 0.$$

Alternatively, we may eliminate x and y in turn to obtain the parametric equations

$$x = (1 - t)^2$$

and

$$y = t^2.$$

(This example describes the classical construction of the so-called **proportional curve** for the special case of the line segments OA and OB.)

If, on the other hand, the curves of the family $f(x,y,v) = 0$ are described parametrically by the equations

$$x = x(u,v)$$

and

$$y = y(u,v), \qquad (1.34)$$

where u is the parameter describing the points on any given curve, and v is the parameter distinguishing the different curves of the family, then $f(x,y,v) = f(x(u,v),v) = F(u,v)$, say, so that $F(u,v) \equiv 0$.

Thus

$$\frac{\partial F}{\partial u} = \frac{\partial f}{\partial x}\frac{\partial x}{\partial u} + \frac{\partial f}{\partial y}\frac{\partial y}{\partial u} = 0$$

and

$$\frac{\partial F}{\partial v} = \frac{\partial f}{\partial x}\frac{\partial x}{\partial v} + \frac{\partial f}{\partial y}\frac{\partial y}{\partial v} + \frac{\partial f}{\partial v} = 0$$

for all points on any member of the family.

For points on the envelope, however, $\dfrac{\partial f}{\partial v} = 0$ also, so that

$$\frac{\partial x}{\partial u}\frac{\partial f}{\partial x} + \frac{\partial y}{\partial u}\frac{\partial f}{\partial y} = 0$$

and

$$\frac{\partial x}{\partial v}\frac{\partial f}{\partial x} + \frac{\partial y}{\partial v}\frac{\partial f}{\partial y} = 0.$$

Considering these as equations for $\dfrac{\partial f}{\partial x}$ and $\dfrac{\partial f}{\partial y}$, we find that, unless $\dfrac{\partial f}{\partial x} = 0$ and $\dfrac{\partial f}{\partial y} = 0$, we must have

$$\frac{\partial x}{\partial u}\frac{\partial y}{\partial v} - \frac{\partial x}{\partial v}\frac{\partial y}{\partial u} = 0. \tag{1.35}$$

If, alternatively, $\frac{\partial f}{\partial x}$ and $\frac{\partial f}{\partial y}$ are both zero, the tangent to the curve $f(x,y,v) = 0$ is not defined and the curve must have a cusp when this occurs.

Thus for a family of smooth curves, we may obtain the parametric equation of the envelope by eliminating either u or v from equation (1.34) with the aid of (1.35).

Example

Consider the family of straight lines, shown in Figure 1.23,

$$x = v + u \cos v$$

and

$$y = u \sin v.$$

Then
$$\frac{\partial x}{\partial u}\frac{\partial y}{\partial v} - \frac{\partial x}{\partial v}\frac{\partial y}{\partial u} = u \cos^2 v - (1 - u \sin v)\sin v = 0.$$

Thus
$$u(\cos^2 v + \sin^2 v) - \sin v = 0,$$

and hence
$$u = \sin v.$$

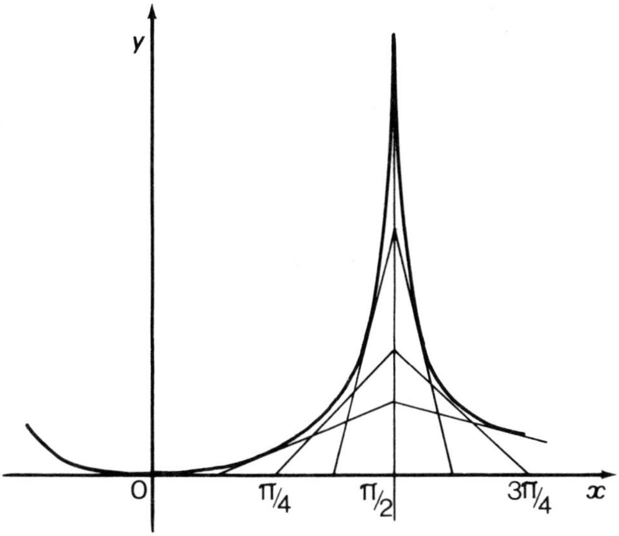

Figure 1.23

Substituting this value of u in the equations of the family, we obtain finally

$$x = v + \sin v \cos v$$

and

$$y = \sin^2 v.$$

These are parametric equations of the envelope (see Figure 1.23).

1.2.4 Intrinsic equations of a curve

Once the initial point of a curve has been defined, the variation with the arc length s of the angle ψ subtended by its tangent on the x-axis is sufficient to define the curve (see Figure 1.24). A relation between s and ψ is called an **intrinsic equation of the curve.**

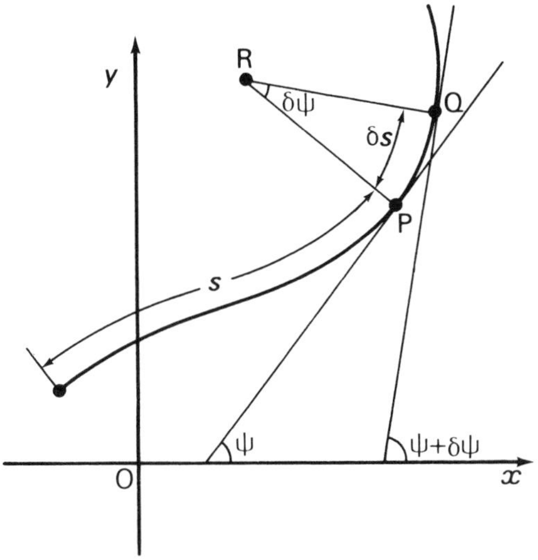

Figure 1.24

Many intrinsic equations have the form $s = s(\psi)$, for example the equation $s = a \tan \psi$ of the catenary.

The **curvature** is obtained from the intrinsic equation by the formula

$$\kappa = \frac{d\psi}{ds}, \tag{1.36}$$

so that in the case of the catenary, where $\psi = \tan^{-1} \dfrac{s}{a}$, we have $\kappa = \dfrac{a}{s^2 + a^2}$.

Alternatively, the curve may be described parametrically in terms of the arc length by the equations $x = x(s)$ and $y = y(s)$. The functions $x(s)$ and $y(s)$ are then related to ψ by the equations

$$\frac{dx}{ds} = \cos\psi \quad \text{and} \quad \frac{dy}{ds} = \sin\psi. \tag{1.37}$$

If we differentiate these equations with respect to s, and substitute κ for $\frac{d\psi}{ds}$, $\frac{dx}{ds}$ for $\cos\psi$ and $\frac{dy}{ds}$ for $\sin\psi$, we obtain the simultaneous differential equations

$$\frac{d^2x}{ds^2} + \kappa(s)\frac{dy}{ds} = 0$$

and

$$\frac{d^2y}{ds^2} - \kappa(s)\frac{dx}{ds} = 0. \tag{1.38}$$

These two second order equations can in principle be solved to determine $x(s)$ and $y(s)$ for any given curvature function $\kappa(s)$. Appropriate numerical procedures have been described by Nutbourne (1972) and Adams (1975). We have freedom to choose the four arbitrary constants, and the curve length L, so that we cannot specify both the end positions and slopes, which impose six conditions. This is to be expected, since the area under the $\kappa(s)$ profile determines the change in slope:

$$\int_0^L \kappa(s)\, ds = \int_0^L \frac{d\psi}{ds}\, ds = \Big[\psi(s)\Big]_0^L$$

Example

As a simple example having an analytic solution, we will calculate the circular arc shown in Figure 1.25 from the constant curve profile. If we measure the arc length from O and describe the circle in an anticlockwise sense, the curvature κ is $+1$.

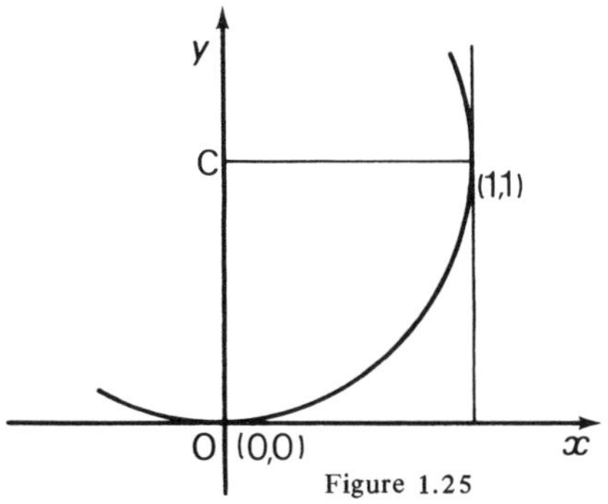

Figure 1.25

Since $\kappa \equiv 1$, the area under the curvature profile is L, and since the change in slope is $\pi/2$, we must put $L = \pi/2$.

We then solve the equations

$$\frac{d^2x}{ds^2} + \frac{dy}{ds} = 0$$

and

$$\frac{d^2y}{ds^2} - \frac{dx}{ds} = 0,$$

subject to the conditions $x(0) = 0, y(0) = 0, x(\pi/2) = 1, y(\pi/2) = 1$.

The result is, eventually,

$$x = \sin s$$

and

$$y = 1 - \cos s,$$

which are the parametric equations of the circle, as expected.

Chapter 2

Three dimensional geometry and vector algebra

2.1 THREE DIMENSIONAL CO-ORDINATES

The Cartesian axis system in 2 dimensions, consisting of two perpendicular lines Ox and Oy, and the corresponding Cartesian co-ordinates (x,y) of a point P in the Oxy plane, will be taken as the starting point for the discussion (Figure 2.1).

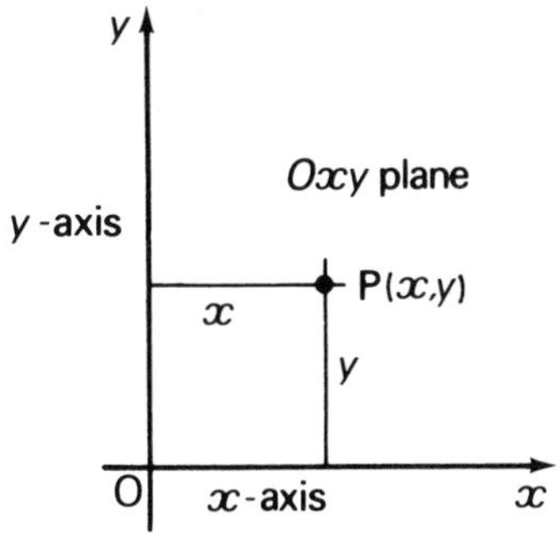

Figure 2.1

If a third dimension is required, then a third axis Oz is added at right angles to the plane Oxy (Figure 2.2). In Figure 2.2, the point O can be regarded

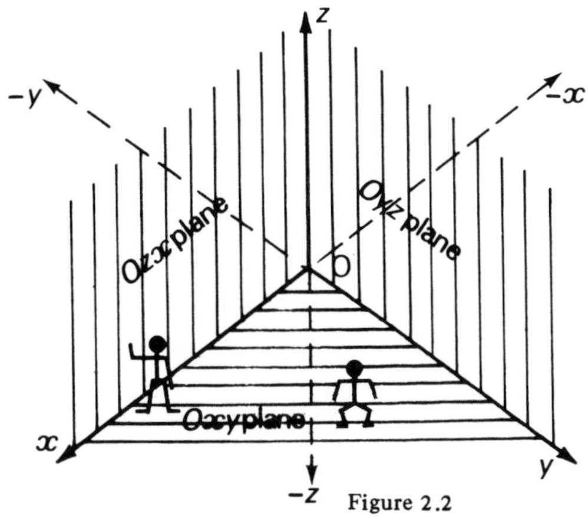

Figure 2.2

as an internal corner of a room, with the planes *Ozx* and *Oyz* as walls and the plane *Oxy* as the floor. The pin men have been added to reduce the visual ambiguity arising from the three dimensional view.

By convention, the positive direction *Oz* is taken as the direction of motion of a right-handed screw or helix which rotates from *Ox* to *Oy* about the *z* axis (Figure 2.3). The axes are then said to be **right-handed.**

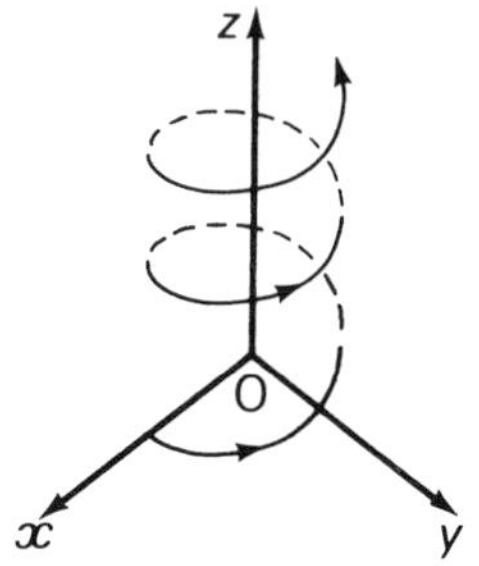

Figure 2.3

A device which avoids the visual ambiguities of non-orthogonal views is illustrated in Figure 2.4, where the direction of a line normal to the plane of the paper is indicated by ⊙ or ⊗, denoting the point or tail feathers of an arrow which points in the direction concerned.

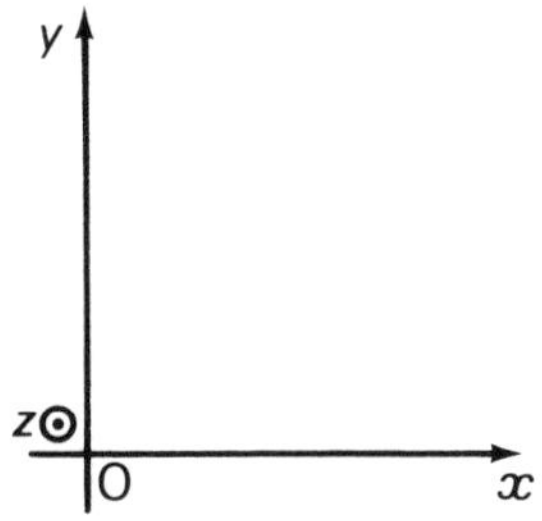

⊙ indicates an arrow out of the paper
⊗ indicates an arrow into the paper

Figure 2.4

The Cartesian co-ordinates of a point P can be found by breaking up the motion from O to P into three mutually orthogonal motions parallel to the axes (Figure 2.5). The order in which these motions take place does not affect the magnitudes of the three displacements, but merely alters the path taken along the sides of the rectangular prism shown in the Figure.

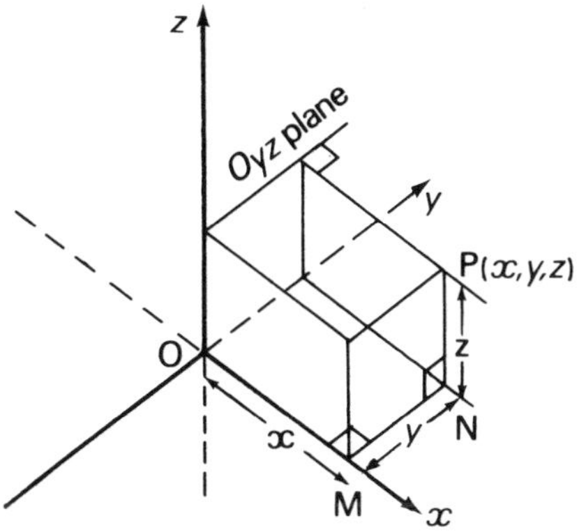

Figure 2.5

The three displacements OM, MN and NP are denoted by (x,y,z) and are the Cartesian co-ordinates of P. If the point P is on the opposite side of the *Oyz* plane to the *Ox* axis, then the displacement OM is reckoned negative, and the co-ordinate x is negative. Similarly, we can have negative y and z co-ordinates. Displacements (x,y,z) measured from O in this way are the **absolute co-ordinates** of P in the axis system *Oxyz*.

For classical treatments of co-ordinate geometry in three dimensions the reader is referred to the texts listed in the bibliography; we shall principally concern ourselves here with methods using vector algebra, and later with differential geometry.

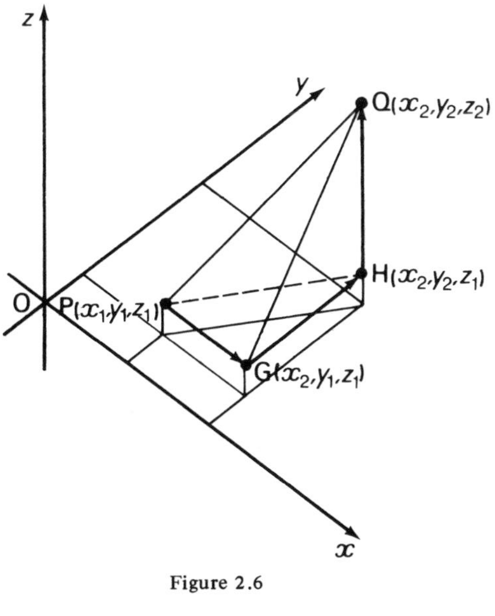

Figure 2.6

2.2 INTRODUCTION TO VECTORS

If we now consider two points P (x_1,y_1,z_1) and Q (x_2,y_2,z_2), we may wish to know the displacement from P to Q or vice-versa. To make the *sense* of the displacement clear we denote the displacement from P to Q by $\overrightarrow{PQ}$ and that from Q to P by $\overrightarrow{QP}$. If the displacement $\overrightarrow{PQ}$ is broken into displacements parallel to the co-ordinate axes as before, the separate displacement PG, GH and HQ determine the **relative co-ordinates** (x_{PQ},y_{PQ},z_{PQ}) at Q with respect to P (see Figure 2.6). We can easily see from Figure 2.6 that

$$x_{PQ} = x_2 - x_1,\ y_{PQ} = y_2 - y_1,\ z_{PQ} = z_2 - z_1\ .$$

We call the *directed* displacement $\overrightarrow{PQ}$ the **relative position vector** of Q with respect to P, and show this on Figure 2.7 as an arrow with its tail at P and its head at Q. Similarly $\overrightarrow{OP}$ and $\overrightarrow{OQ}$ are **absolute position vectors** of P and Q in the axes *Oxyz* and are also denoted by the corresponding arrows in Figure 2.7.

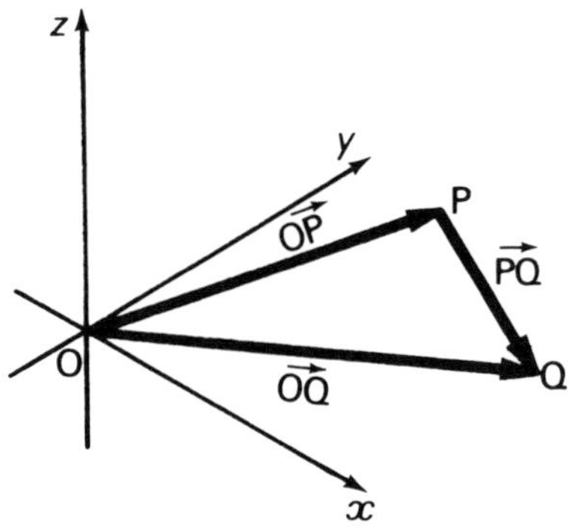

Figure 2.7

We will now write

$$\overrightarrow{OQ} = \overrightarrow{OP} + \overrightarrow{PQ},$$

meaning that the final effect of the displacement $\overrightarrow{OP}$ followed by $\overrightarrow{PQ}$ is the same as that of $\overrightarrow{OQ}$ alone. We have thus informally defined the meaning of equality and of summation in the vector sense. A formal definition will be given later.

It is natural to pursue the analogy with ordinary arithmetic by 'subtracting' $\overrightarrow{OP}$ from both sides and writing

$$\overrightarrow{PQ} = \overrightarrow{OQ} - \overrightarrow{OP}.$$

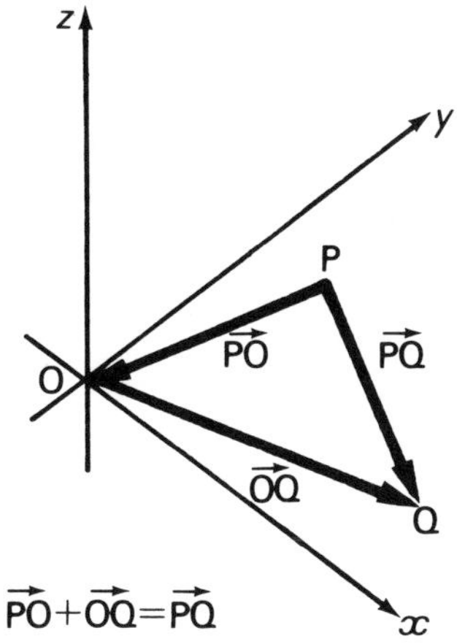

Figure 2.8

In fact, however, we can see from Figure 2.8 that

$$\vec{PQ} = \vec{PO} + \vec{OQ}.$$

It follows that the natural extension of ordinary arithmetic may be made if we define the vector $-\vec{OP}$ to be the same as $\vec{PO}$. This again will be formalised later.

The equation $\vec{PQ} = \vec{OQ} - \vec{OP}$ bears a close relationship to the equations

$$x_{PQ} = x_2 - x_1,$$

$$y_{PQ} = y_2 - y_1,$$

and

$$z_{PQ} = z_2 - z_1,$$

determined earlier. In order to emphasize this, we compress these equations into the single statement

$$\mathbf{r}_{PQ} = \mathbf{r}_2 - \mathbf{r}_1,$$

where $\mathbf{r}_{PQ}$ is a shorthand for (x_{PQ},y_{PQ},z_{PQ}) and similarly $\mathbf{r}_1$, $\mathbf{r}_2$ for (x_1,y_1,z_1) and (x_2,y_2,z_2).

These then correspond directly to $\vec{PQ}$, $\vec{OQ}$ and $\vec{OP}$, as can be seen in Figure 2.9. The notation $\mathbf{r}_{PQ}$ is therefore a compromise between the endpoint notation $\vec{PQ}$ and the co-ordinate notation (x_{PQ},y_{PQ},z_{PQ}).

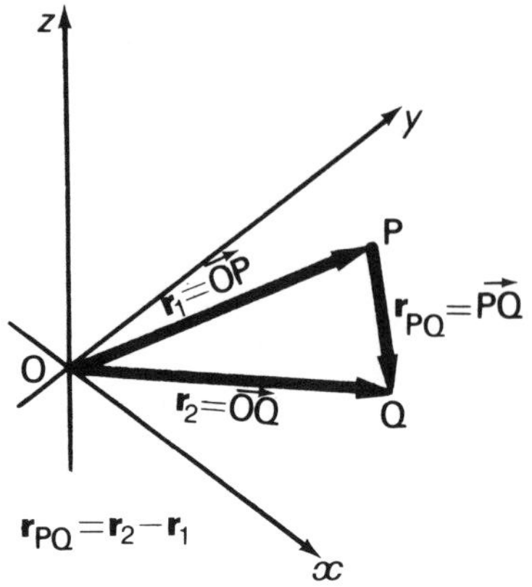

Figure 2.9

Using the endpoint notation, it is useful to write $\frac{1}{2}\overrightarrow{OP}$ to denote a displacement along $\overrightarrow{OP}$ of one half of the magnitude OP, and in general to write $\lambda\overrightarrow{AB}$ for a displacement λ times the magnitude AB in the direction of $\overrightarrow{AB}$. For example, we may wish to consider the midpoint M of PQ (Figure 2.10). Now the displacement of M from O may be written as

$$\overrightarrow{OM} = \overrightarrow{OP} + \overrightarrow{PM}.$$

Hence

$$\begin{aligned}\overrightarrow{OM} &= \overrightarrow{OP} + \tfrac{1}{2}\overrightarrow{PQ}\\ &= \overrightarrow{OP} + \tfrac{1}{2}(\overrightarrow{OQ} - \overrightarrow{OP})\\ &= \tfrac{1}{2}(\overrightarrow{OP} + \overrightarrow{OQ}),\end{aligned}$$

and is thus the average of the two end displacements. Alternatively we may write the absolute position vector $\overrightarrow{OM}$ as $\mathbf{r}_M$, where

$$\mathbf{r}_M = \tfrac{1}{2}(\mathbf{r}_1 + \mathbf{r}_2).$$

In co-ordinate terms, the equations

$$\begin{aligned}x_M &= \tfrac{1}{2}(x_1 + x_2),\\ y_M &= \tfrac{1}{2}(y_1 + y_2),\\ \text{and } z_M &= \tfrac{1}{2}(z_1 + z_2)\end{aligned}$$

may be verified using similar triangles.

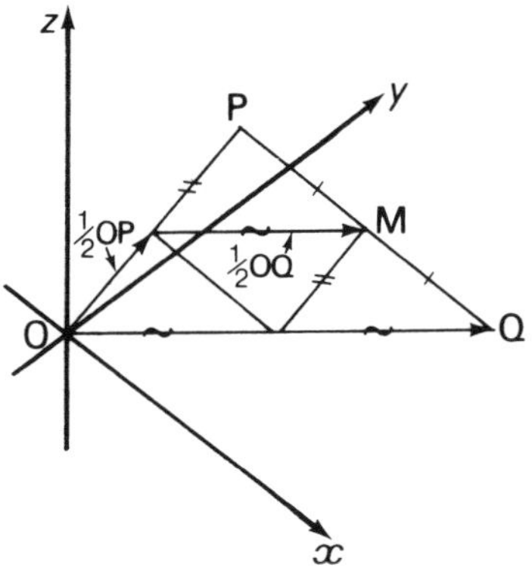

Figure 2.10

Just as these displacements may be represented by arrows indicating their magnitude and direction, and are added by joining the arrows head to tail, certain other physical variables may be represented and combined in the same way. Consider, for example, the motion of a man in a moving train. Figure 2.11 shows the position of the man and the train at times t and $t+\delta t$. The vector $\boldsymbol{\delta}\mathbf{r}$ is a relative position vector describing the movement of the train in this time interval, whereas $\boldsymbol{\delta}\mathbf{r}'$ describes the movement of the man in the same interval *as seen by an observer stationary in the train.* The vector $\boldsymbol{\delta}\mathbf{r}''$ is the total movement observed from the side of the track, and obeys the law of vector addition just described.

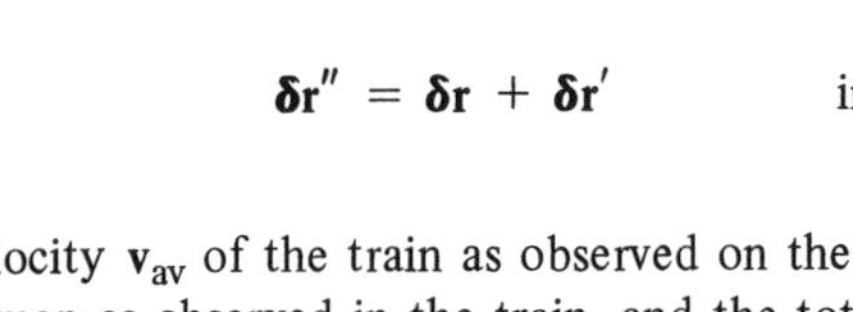

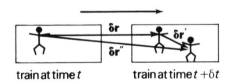

Figure 2.11

Thus $$\boldsymbol{\delta}\mathbf{r}'' = \boldsymbol{\delta}\mathbf{r} + \boldsymbol{\delta}\mathbf{r}'$$ in the vector sense.

The average velocity $\mathbf{v}_{av}$ of the train as observed on the track, the average velocity $\mathbf{v}'_{av}$ of the man as observed in the train, and the total average velocity $\mathbf{v}''_{av}$ of the man as observed from the track are obtained by dividing these displacement vectors by the time interval δt.

Thus $$\mathbf{v}''_{av} = \frac{\delta\mathbf{r}''}{\delta t} = \frac{\delta\mathbf{r}}{\delta t} + \frac{\delta\mathbf{r}'}{\delta t} = \mathbf{v}_{av} + \mathbf{v}'_{av}.$$

Thus average velocities add in the same way as relative position vectors. Letting $\delta t \to 0$, we can see that the instantaneous velocities also do so. By a similar argument we can deduce that accelerations will do the same. Moreover, since force equals mass times acceleration, forces will also add in the same manner.

These are just a few examples of directed physical quantities which obey the same laws of combination as the relative position vector. Any such quantity is called a **vector**. We will now describe the properties of vectors in a more precise and formal manner.

2.3 VECTOR ALGEBRA I – DEFINITIONS AND SIMPLE GEOMETRICAL APPLICATIONS

2.3.1 Definitions

The properties of relative position vectors described in the preceding Section are retained if we define general vectors as follows. **Vectors are quantities having magnitude and direction** which obey certain laws described below. They are usually denoted by lower case letters in **bold print**, for example **a**, **b**, $\boldsymbol{\alpha}$ or $\boldsymbol{\omega}$, but in typewritten or handwritten work the letters are usually underlined, rather than bold.

A vector **a** can be represented by a directed line segment in the direction of **a**, whose length represents the magnitude of **a** to some fixed scale. The **sense** of the vector along the line is indicated by means of an arrowhead, as in Figure 2.12. A vector with zero magnitude is called a **null vector**, and denoted by **0**. No direction is defined for a null vector.

Figure 2.12

As we have just indicated, only quantities *obeying the laws of vector algebra* can be described as vectors; possession of magnitude and direction are not sufficient. These laws are as follows:

1. Equality

Two vectors are equal when they have the same magnitude, direction and sense. Thus they can be represented by parallel lines of equal length when drawn to the same scale. Note that their position in space or on the paper is unimportant for equality.

2. Addition

When two vectors **a** and **b** are given, their sum **a** + **b** is defined by the following construction.

Draw **a** and **b** to the same scale. Join the tail of **b** to the head of **a**. Then the line from the tail of **a** to the head of **b** is the vector **a** + **b** to the same scale as that of **a** and **b**, as shown in Figure 2.13.

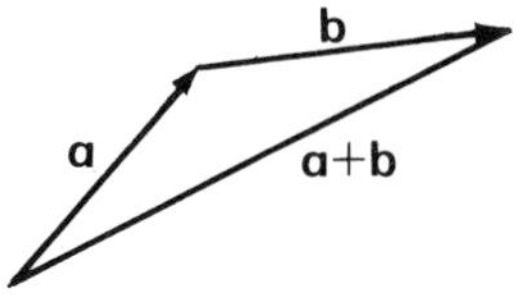

Figure 2.13

3. *Negation*

We define the vector − **a** to be the vector having the same magnitude and direction as **a**, but the opposite sense (see Figure 2.14).

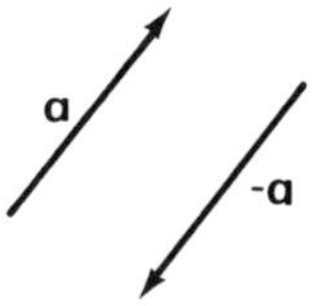

Figure 2.14

4. *Subtraction*

We define **a** − **b** = **a** + (− **b**).

5. *Scalar multiplication*

The vector λ**a** is a vector having the same direction and sense as **a**, but a magnitude λ times that of **a** (see Figure 2.15). Here λ is an ordinary number or **scalar**.

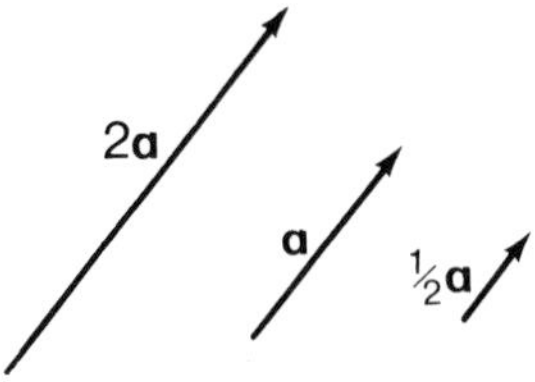

Figure 2.15

2.3.2 Consequences of these definitions

Let **a**, **b** and **c** be vectors, and λ, μ and ν be scalars. Then

i) $$\mathbf{a} + \mathbf{b} = \mathbf{b} + \mathbf{a},$$ (see Figure 2.16)

ii) $$\mathbf{a} + (\mathbf{b} + \mathbf{c}) = (\mathbf{a} + \mathbf{b}) + \mathbf{c},$$

iii) $$\lambda(\mu\mathbf{a}) = \lambda\mu\mathbf{a},$$

iv) $$(\mu + \nu)\mathbf{a} = \mu\mathbf{a} + \nu\mathbf{a},$$

v) $$\lambda(\mathbf{a} + \mathbf{b}) = \lambda\mathbf{a} + \lambda\mathbf{b}$$ (see Figure 2.17).

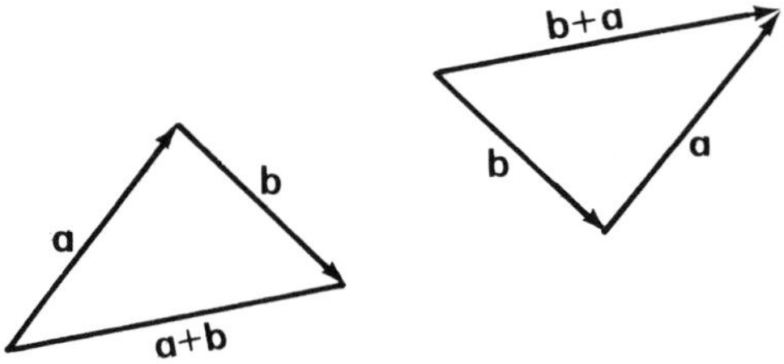

Figure 2.16

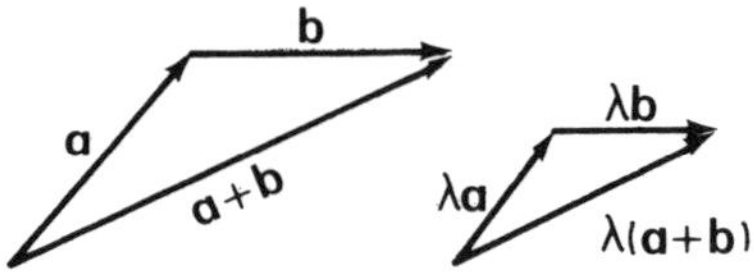

Figure 2.17

2.3.3 Magnitude of a vector. Unit vector

The magnitude (or length) of a vector **a** is denoted by the symbol $|\mathbf{a}|$. Using this notation, we see that the vector $\mathbf{u} = \dfrac{1}{|\mathbf{a}|}\mathbf{a}$ is a vector of unit magnitude or **unit vector**. The vector **u** is then said to be **normalised**; alternatively we can write

$$\mathbf{a} = |\mathbf{a}|\ \mathbf{u}.$$

2.3.4 Cartesian Components of a Vector

Given a rectangular co-ordinate system, we can define **unit vectors** along the positive direction of the x, y and z axes. These are usually denoted by **i, j** and **k** respectively.

These enable us to express any vector in terms of three vectors along the co-ordinate axes (Figure 2.18).

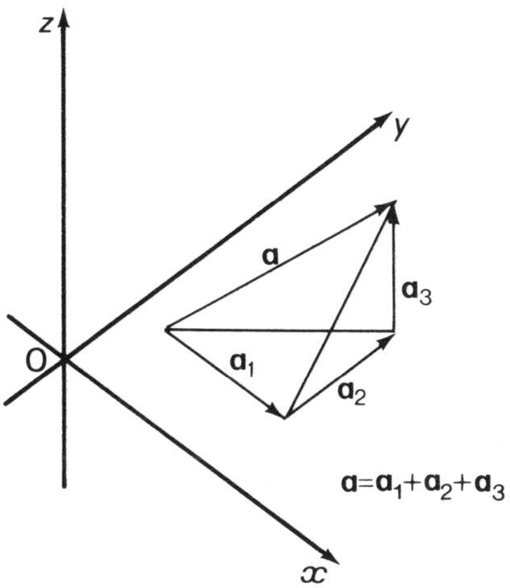

Figure 2.18

These can be visualised as the three edges of a rectangular prism which has **a** as a diagonal (Figure 2.19). In this figure, three of the edges are described by the vectors $\mathbf{a}_1$, $\mathbf{a}_2$ and $\mathbf{a}_3$. If the vectors $\mathbf{a}_1$, $\mathbf{a}_2$ and $\mathbf{a}_3$ are taken in any different order, the path between P and Q comprises different edges of the prism, but the resultant vector **a** is the same. Thus, $\mathbf{a} = \mathbf{a}_1 + \mathbf{a}_2 + \mathbf{a}_3$ uniquely for any given co-ordinate system.

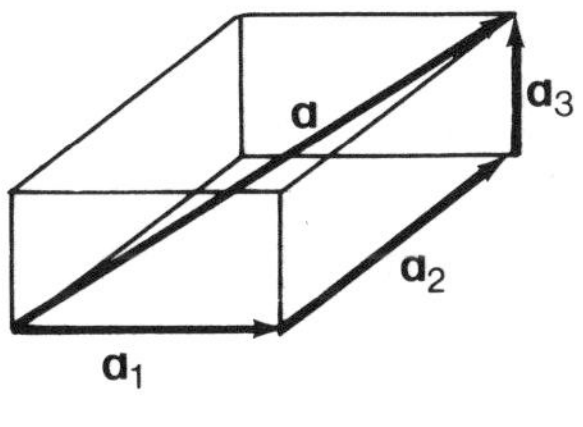

Figure 2.19

Now $\mathbf{a}_1 = a_1\mathbf{i}$, $\mathbf{a}_2 = a_2\mathbf{j}$ and $\mathbf{a}_3 = a_3\mathbf{k}$ where a_1, a_2 and a_3 are the magnitudes of $\mathbf{a}_1$, $\mathbf{a}_2$ and $\mathbf{a}_3$. Thus

$$\mathbf{a} = a_1\mathbf{i} + a_2\mathbf{j} + a_3\mathbf{k}.$$

Then (a_1,a_2,a_3) are the **Cartesian components** of the vector **a** for the co-ordinate axes chosen.

In terms of these Cartesian components, we can deduce the following rules:

i) $\mathbf{a} + \mathbf{b} = (a_1 + b_1)\mathbf{i} + (a_2 + b_2)\mathbf{j} + (a_3 + b_3)\mathbf{k}$ (seeFigure 2.20),

ii) $-\mathbf{a} = -a_1\mathbf{i} - a_2\mathbf{j} - a_3\mathbf{k}$,

iii) $\lambda\mathbf{a} = \lambda a_1\mathbf{i} + \lambda a_2\mathbf{j} + \lambda a_3\mathbf{k}$

iv) $|\mathbf{a}| = \sqrt{a_1^2 + a_2^2 + a_3^2}$ (by Pythagoras' theorem, see Figure 2.21).

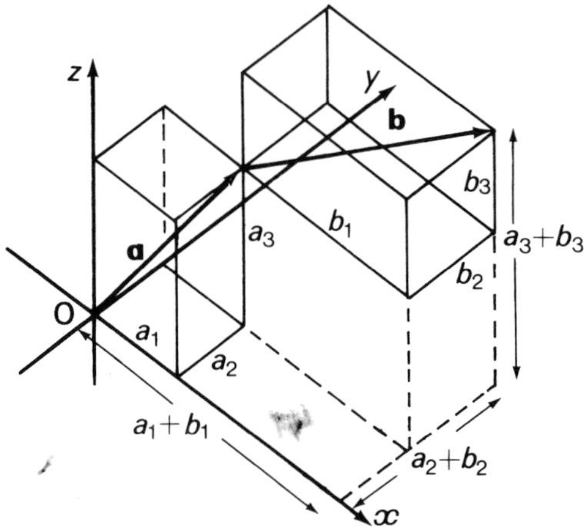

Figure 2.20

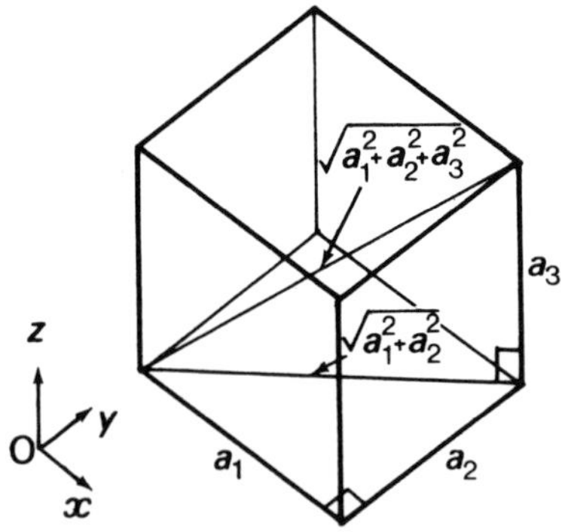

Figure 2.21

The components of a unit vector **u** give the cosines of the angles between the vector direction and the directions of the *Ox*, *Oy* and *Oz* axes, as we shall now demonstrate. Referring to Figure 2.22, we can see that $\mathbf{u} = \overrightarrow{PU}/PU$, which may be resolved as follows:

$$\mathbf{u} = \frac{PW\mathbf{i} + PS\mathbf{j} + PQ\mathbf{k}}{PU}$$

$$= \frac{PW}{PU}\mathbf{i} + \frac{PS}{PU}\mathbf{j} + \frac{PQ}{PU}\mathbf{k}.$$

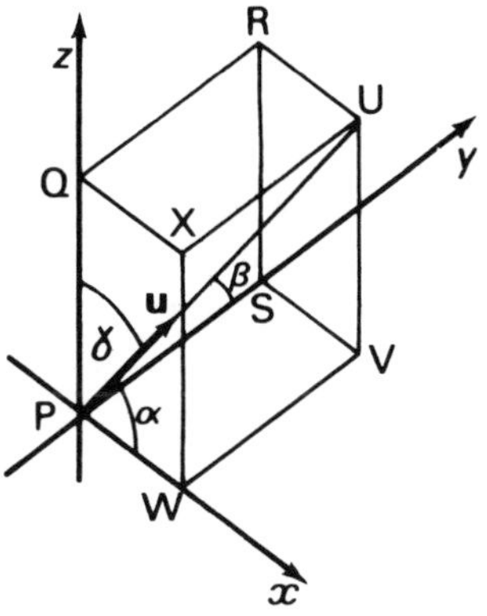

Figure 2.22

Thus $\mathbf{u} = \cos\alpha\mathbf{i} + \cos\beta\mathbf{j} + \gamma\mathbf{k}$, where α, β, γ are the angles between **u** and the positive directions of the Ox, Oy and Oz axes. When associated with the direction of a line, they are called the **direction cosines** of the line, and are often denoted by ℓ, m and n. Then $\mathbf{u} = \ell\mathbf{i} + m\mathbf{j} + n\mathbf{k}$. Any set of numbers $(\lambda\ell, \lambda m, \lambda n)$ proportional to (ℓ, m, n) are called **direction ratios** of the line. Thus the components of any vector parallel to a line may be taken as direction ratios for the line, and by normalising the vector we may obtain the direction cosines.

2.3.5 The vector equations of a straight line

1) The line through a point P_0 with the direction of unit vector **u** (see Figure 2.23) may be described as follows.

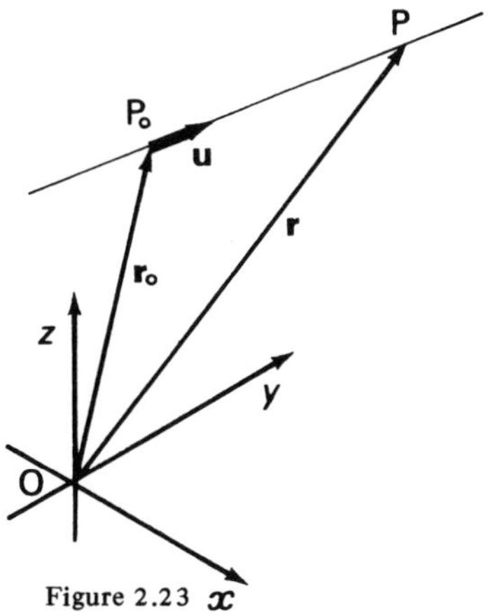

Figure 2.23

Let $\vec{OP}_0$ be denoted by $\mathbf{r}_0$ and the vector $\vec{OP}$ to a general point on the line by $\mathbf{r}$. Then $\vec{P_0P}$ is the vector $\mathbf{r} - \mathbf{r}_0$, and we can see that $\vec{P_0P}$ is a vector in the direction of $\mathbf{u}$, so that $\mathbf{r} - \mathbf{r}_0 = \lambda\mathbf{u}$. Thus the equation of the line is

$$\mathbf{r} = \mathbf{r}_0 + \lambda\mathbf{u}. \tag{2.1}$$

The distance from P_0 to P is $|\mathbf{r} - \mathbf{r}_0| = |\lambda\mathbf{u}| = |\lambda|$.

The different points on the line are distinguished by the value of the **parameter** λ. The direction of the line need not be given by a *unit* vector, but if some general vector $\mathbf{a}$ is used to describe the direction ($\mathbf{r} = \mathbf{r}_0 + \mu\mathbf{a}$), then the parameter μ is now proportional to and not equal to the distance from $\mathbf{r}_0$.

2) In the case of the line through two points P_1 and P_2 (see Figure 2.24), the general point P has the equation

$$\mathbf{r} = \vec{OP} = \vec{OP}_1 + \vec{P_1P} = \vec{OP}_1 + \mu\vec{P_1P_2}.$$

Thus

$$\mathbf{r} = \mathbf{r}_1 + \mu(\mathbf{r}_2 - \mathbf{r}_1),$$

or finally

$$\mathbf{r} = (1 - \mu)\mathbf{r}_1 + \mu\mathbf{r}_2. \tag{2.2}$$

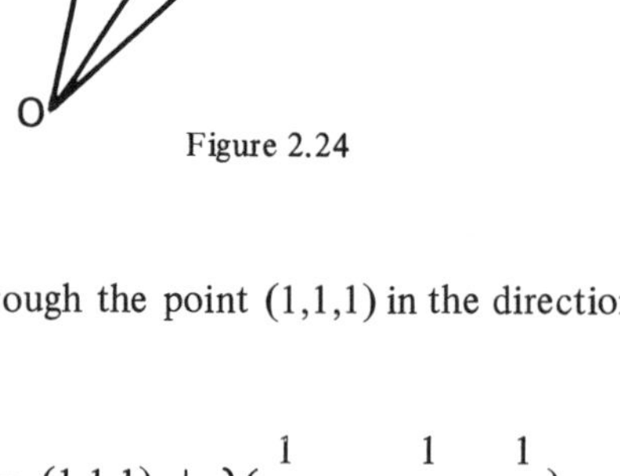

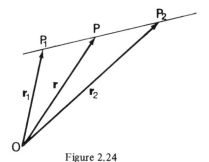

Figure 2.24

2.3.6 Examples

1. The straight line through the point (1,1,1) in the direction $(\frac{1}{\sqrt{3}}, -\frac{1}{\sqrt{3}}, \frac{1}{\sqrt{3}})$ is given by

$$\mathbf{r} = (1,1,1) + \lambda(\frac{1}{\sqrt{3}}, -\frac{1}{\sqrt{3}}, \frac{1}{\sqrt{3}}).$$

In co-ordinate terms

$$x = 1 + \frac{\lambda}{\sqrt{3}}, \quad y = 1 - \frac{\lambda}{\sqrt{3}}, \quad z = 1 + \frac{\lambda}{\sqrt{3}},$$

or $$x = 1 + \mu, \quad y = 1 - \mu, \quad z = 1 + \mu, \quad \text{where } \mu = \frac{\lambda}{\sqrt{3}}.$$

2. The straight line through (1,0,2) and (2,1,1) is given by

$$\mathbf{r} = (1 - \nu)\ (1,0,2) + \nu(2,1,1)\,,$$

or $$x = 1 + \nu, \quad y = \nu, \quad z = 2 - \nu.$$

3. Do the two lines above intersect one another?
If they do, then the three equations

$$1 + \mu = 1 + \nu, \quad \ldots\ldots (1),$$

$$1 - \mu = \nu, \quad \ldots\ldots (2),$$

and $$1 + \mu = 2 - \nu \quad \ldots\ldots (3),$$

are satisfied simultaneously.

Equations (1) and (3) together imply that $1 + \nu = 2 - \nu$,

so that $$\nu = \tfrac{1}{2}.$$

If we substitute $\nu = \frac{1}{2}$ in equation (1), then $\mu = \frac{1}{2}$. These values also satisfy equation (2), so that the two lines do intersect, at $(\frac{3}{2}, \frac{1}{2}, \frac{3}{2})$.

2.4 VECTOR ALGEBRA II – THE SCALAR AND VECTOR PRODUCTS

2.4.1 Introduction

The laws of addition, negation and scalar multiplication of vectors enable us to describe single points and lines in space and to determine such things as midpoints, relative positions and whether lines intersect or not.

When we wish to consider the angles between lines, the shortest distance between skew lines, the equation of a plane, and the projections of points onto given planes, then the vector equivalents of trigonometric relationships are required.

In more general vector situations, we may wish to resolve a vector into components other than the x, y and z components mentioned earlier. We will consider this general situation first, and then apply the results to some of the problems described above.

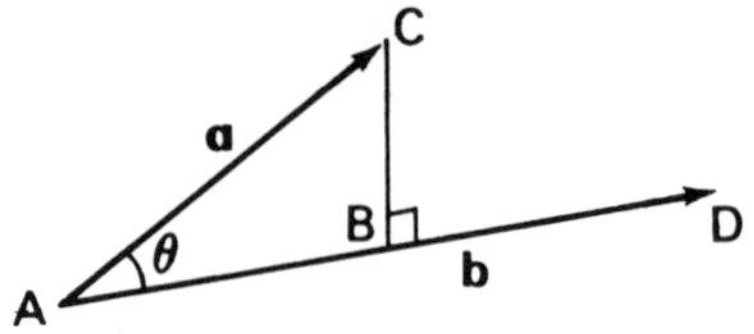

Figure 2.25

2.4.2 Component of a vector along a given direction: the scalar product

A vector **a** can be uniquely resolved into two perpendicular vectors when the direction of one of these is given.

For, if one of the two perpendicular vectors is along the direction of vector **b** ($= \overrightarrow{AB}$), we drop a perpendicular from the head of **a** onto vector **b** (see Figure 2.25), and denote $\overrightarrow{BC}$ by **c**. Then

$$\mathbf{a} = \lambda\mathbf{b} + \mathbf{c},$$

where $\mathbf{b} \perp \mathbf{c}$ (see Figure 2.26). Now if CAB $= \theta$, then

$$AB = AC\cos\theta = |\mathbf{a}|\cos\theta.$$

AB is the **component** of AC resolved along vector **b**.

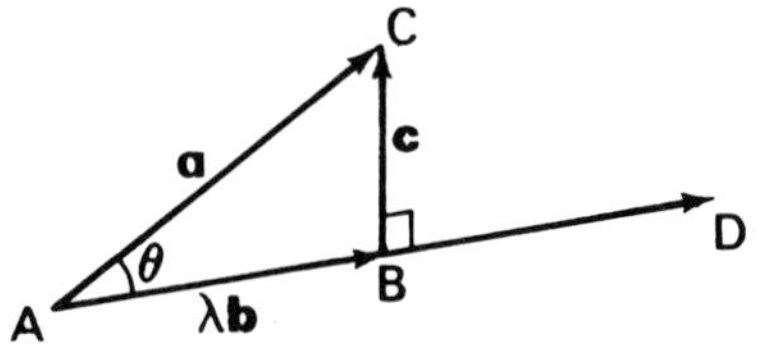

Figure 2.26

To obtain the angle θ, we close the triangle ACD and apply the cosine rule (see Figure 2.27). Thus $CD^2 = AC^2 + AD^2 - 2\,AC.AD\cos\theta$ or

$$|\mathbf{a} - \mathbf{b}|^2 = |\mathbf{a}|^2 + |\mathbf{b}|^2 - 2\,|\mathbf{a}|\,|\mathbf{b}|\cos\theta.$$

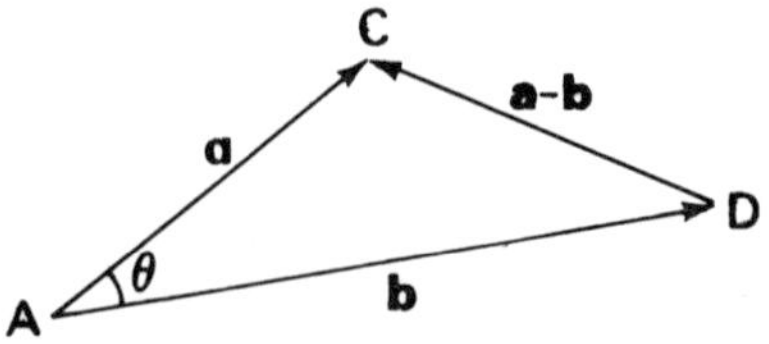

Figure 2.27

Expressing these moduli in terms of the Cartesian components, we obtain the equation

$$(a_1 - b_1)^2 + (a_2 - b_2)^2 + (a_3 - b_3)^2 = (a_1^2 + a_2^2 + a_3^2) + (b_1^2 + b_2^2 + b_3^2) - 2\,|\mathbf{a}|\,|\mathbf{b}|\cos\theta .$$

The angle Θ between two vectors a and b then proves to be

$$\cos\theta = \frac{a_1b_1 + a_2b_2 + a_3b_3}{|\mathbf{a}|\ |\mathbf{b}|}, \tag{2.3}$$

after some cancellation.

The component of a along the direction of b (or the projection of **a** onto **b**) is therefore

$$\mathrm{AB} = |\mathbf{a}|\cos\theta = \frac{a_1b_1 + a_2b_2 + a_3b_3}{|\mathbf{b}|}. \tag{2.4}$$

In particular, the component of a vector **a** along a unit vector **u** is given by $a_1u_1 + a_2u_2 + a_3u_3$.

The scalar quantity $a_1b_1 + a_2b_2 + a_3b_3$ occurs so often in similar applications that we find it useful to define this as the **scalar (or dot) product** of **a** and **b**, denoted by **a.b**.

Thus

$$\mathbf{a.b} = \underset{\textit{component form}}{a_1b_1 + a_2b_2 + a_3b_3} = \underset{\textit{physical significance}}{|\mathbf{a}|\ |\mathbf{b}|\cos\theta}. \tag{2.5}$$

In particular, **a.b** = 0 when **a** is perpendicular to **b** (or if either **a** = **0** or **b** = **0**).

The algebra of scalar products can then be shown to produce the following results for any vectors **a, b, c** and any scalar λ:

i) $\mathbf{a.b} = \mathbf{b.a}$

ii) $\mathbf{a.(b + c)} = \mathbf{a.b} + \mathbf{a.c}$

iii) $(\lambda\mathbf{a}).\mathbf{b} = \mathbf{a}.(\lambda\mathbf{b}) = \lambda(\mathbf{a.b})$

iv) $\mathbf{a.a} = |\mathbf{a}|^2$.

Example

(1). If the directions of two lines are those of the vectors **a** and **b**, then the angle θ between these lines is given by

$$\cos\theta = \frac{\mathbf{a}.\mathbf{b}}{|\mathbf{a}|\ |\mathbf{b}|}.$$

If the two lines concerned are those joining $\mathbf{r}_1 = (0,0,1)$ to $\mathbf{r}_2 = (1,0,-1)$ and $\mathbf{r}_3 = (2,1,-1)$ to $\mathbf{r}_4 = (1,-1,0)$, then we write

$$\mathbf{a} = \mathbf{r}_2 - \mathbf{r}_1 = (1,0,-2) \text{ and } \mathbf{b} = \mathbf{r}_4 - \mathbf{r}_3 = (-1,-2,1).$$

Thus $$\cos\theta = \frac{1 \times (-1) + 0 \times -2 + (-2) \times 1}{\sqrt{5}\ \sqrt{6}} = \frac{-3}{\sqrt{30}}.$$

2.4.3 The vector equation of a plane

The vector equation of general point $P(\mathbf{r})$ on the plane π through a given point $P_0(\mathbf{r}_0)$, and normal to the direction $\mathbf{u}$ can be deduced from the fact that $\mathbf{r} - \mathbf{r}_0$ is a vector in the plane π, so that $\mathbf{r} - \mathbf{r}_0$ is perpendicular to $\mathbf{u}$ (see Figure 2.28).

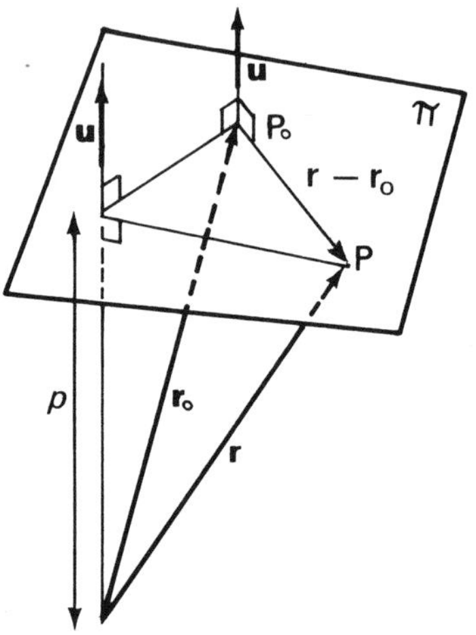

Figure 2.28

Thus $(\mathbf{r} - \mathbf{r}_0).\mathbf{u} = PP_0 \cos\frac{\pi}{2} = 0$, and the equation of the plane is given by

$$\mathbf{r}.\mathbf{u} = \mathbf{r}_0.\mathbf{u} = p, \tag{2.6}$$

where p is the perpendicular distance from the origin to the plane (see Figure 2.28). Note that, since $\mathbf{u}$ appears on both sides of equation (2.6), we may instead use any multiple of $\mathbf{u}$ if this is convenient.

For example, the plane whose normal has the direction of the vector (1,1,0), and which passes through (1,0,0) is given by

$$(x,y,z).(1,1,0) = (1,0,0).(1,1,0),$$

or

$$x + y = 1.$$

(Note that $\mathbf{r.u} = p$ is always a linear equation in x, y and z).

The perpendicular projection of a point $P_1(\mathbf{r}_1)$ onto the plane $\mathbf{r.u} = p$ (see Figure 2.29) is a point $\mathbf{r}$ such that

$$\mathbf{r}_1 - \mathbf{r} = \text{NM}\,\mathbf{u}.$$

Thus

$$\mathbf{r} = \mathbf{r}_1 - (\text{ON} - \text{OM})\mathbf{u} = \mathbf{r}_1 + (\mathbf{r.u} - \mathbf{r}_1\mathbf{.u})\mathbf{u},$$

and hence

$$\mathbf{r} = \mathbf{r}_1 + (p - \mathbf{r}_1\mathbf{.u})\mathbf{u}.$$

This result is useful when making parallel projection drawings of three dimensional objects.

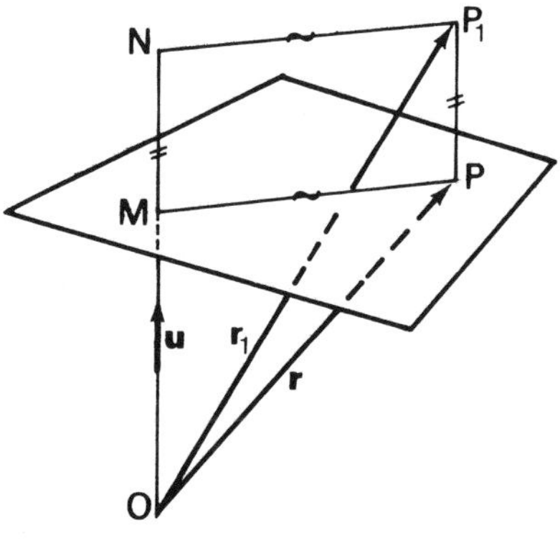

Figure 2.29

2.4.4 Vector perpendicular to two given vectors a and b: the vector product

In many geometrical and mechanical problems we wish to find a vector perpendicular to two given vectors. For instance, we may wish to construct the common perpendicular to two skew lines.

If the vector **v** is perpendicular to **a** and **b**, then

$$\mathbf{v.a} = 0$$

and

$$\mathbf{v.b} = 0,$$

so that $$a_1v_1 + a_2v_2 + a_3v_3 = 0,$$

and $$b_1v_1 + b_2v_2 + b_3v_3 = 0.$$

These can be solved for the ratios $v_1:v_2:v_3$ as follows:

$$\frac{v_1}{a_2b_3-a_3b_2} = \frac{v_2}{a_3b_1-a_1b_3} = \frac{v_3}{a_1b_2-a_2b_1}.$$

Thus the vector $(a_2b_3 - a_3b_2, a_3b_1 - a_1b_3, a_1b_2 - a_2b_1)$ is proportional to **v** and thus is perpendicular to **a** and **b**.

This vector is denoted by **a** X **b** (**a** cross **b**), and is called the **vector product** of **a** and **b**.

Thus $$\mathbf{a} \times \mathbf{b} = (a_2b_3 - a_3b_2)\mathbf{i} + (a_3b_1 - a_1b_3)\mathbf{j} + (a_1b_2 - a_2b_1)\mathbf{k},$$

which may be written conveniently in the form of a determinant (see Appendix 2):

$$\mathbf{a} \times \mathbf{b} = \begin{vmatrix} \mathbf{i} & \mathbf{j} & \mathbf{k} \\ a_1 & a_2 & a_3 \\ b_1 & b_2 & b_3 \end{vmatrix} \tag{2.7}$$

The modulus of **a** X **b** is obtained as follows:

$$|\mathbf{a} \times \mathbf{b}|^2 = (a_2b_3 - a_3b_2)^2 + (a_3b_1 - a_1b_3)^2 + (a_1b_2 - a_2b_1)^2, \quad \text{from (2.7)},$$

$$= a_2^2b_3^2 + a_3^2b_2^2 + a_3^2b_1^2 + a_1^2b_3^2 + a_1^2b_2^2 + a_2^2b_1^2$$

$$- 2a_2a_3b_2b_3 - 2a_3a_1b_3b_1 - 2a_1a_2b_1b_2,$$

$$= (a_1^2 + a_2^2 + a_3^2)(b_1^2 + b_2^2 + b_3^2) - a_1^2b_1^2 - a_2^2b_2^2 - a_3^2b_3^2$$

$$-2a_2a_3b_2b_3 - 2a_3a_1b_3b_1 - 2a_1a_2b_1b_2,$$

$$= |\mathbf{a}|^2|\mathbf{b}|^2 - (a_1b_1 + a_2b_2 + a_3b_3)^2,$$

$$= |\mathbf{a}|^2|\mathbf{b}|^2 - |\mathbf{a}|^2|\mathbf{b}|^2 \cos^2\theta.$$

Hence $|\mathbf{a} \times \mathbf{b}|^2 = |\mathbf{a}|^2|\mathbf{b}|^2 \sin^2\theta$.

Then $|\mathbf{a} \times \mathbf{b}| = |\mathbf{a}|\ |\mathbf{b}|\ |\sin\theta| = |\mathbf{a}|\ |\mathbf{b}| \sin\theta$ if we always choose θ such that $0 < \theta < 180°$. Thus **a** X **b** is a vector perpendicular to **a** and **b** and of magnitude

$|\mathbf{a}|\ |\mathbf{b}|\ \sin\theta$. If we choose a co-ordinate system such that $\mathbf{a} = a\mathbf{i}$ $(a > 0)$ and $\mathbf{b} = b_1\mathbf{i} + b_2\mathbf{j}$, then

$$\mathbf{a} \times \mathbf{b} = \begin{vmatrix} \mathbf{i} & \mathbf{j} & \mathbf{k} \\ a & 0 & 0 \\ b_1 & b_2 & 0 \end{vmatrix} = (0,0,ab_2)\ .$$

Thus $\mathbf{a} \times \mathbf{b}$ is in the positive z-direction if $b_2 > 0$ and in the negative z-direction if $b_2 < 0$ (see Figures 2.30 and 2.31), and the sense of $\mathbf{a} \times \mathbf{b}$ is away from an observer who sees the rotation from $\mathbf{a}$ to $\mathbf{b}$ (for $0 < \theta < 180°$) as clockwise. This is the sense in which the right-handed screw moves when rotated in a clockwise direction.

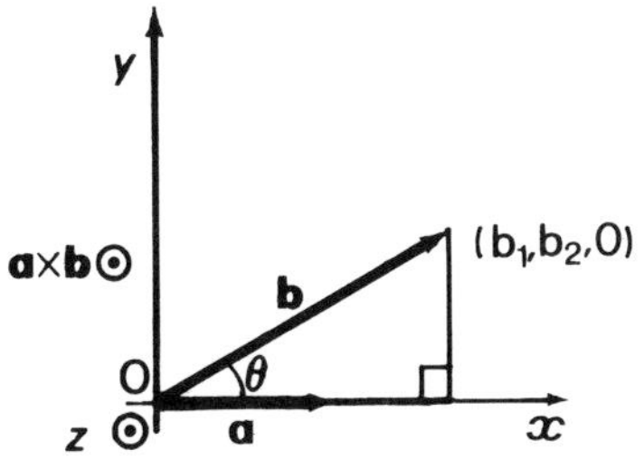

Figure 2.30

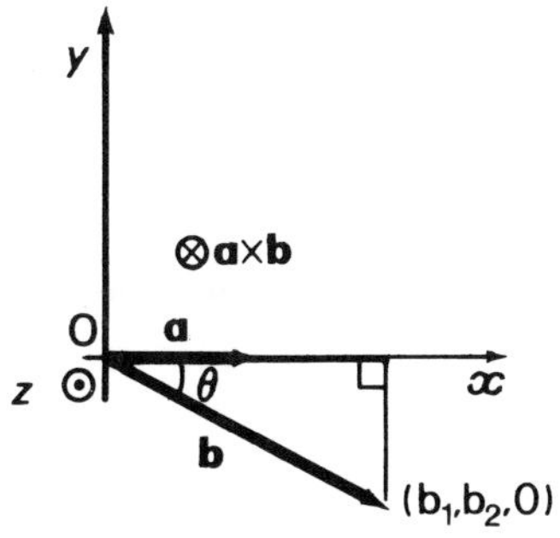

Figure 2.31

We may therefore summarise the properties of the vector product as follows:

Let θ be the smaller angle between the vectors $\mathbf{a}$ and $\mathbf{b}$, so that $0 < \theta < 180°$.

Then
$$\mathbf{a} \times \mathbf{b} = |\mathbf{a}|\ |\mathbf{b}|\ \sin\theta\ \mathbf{u},$$

where $\mathbf{u}$ is a unit vector perpendicular to the plane of $\mathbf{a}$ and $\mathbf{b}$, whose sense is that of the motion of a right-handed screw when turned from **a** to **b** through the angle θ (see Figure 2.32).

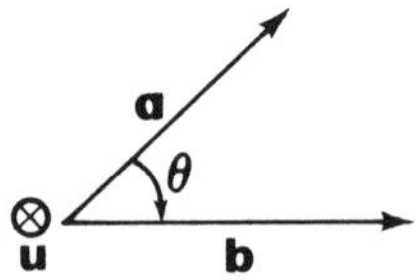

Figure 2.32

We therefore note that $\mathbf{a} \times \mathbf{b} = \mathbf{0}$ when **a** and **b** are parallel vectors, since θ is then zero. In particular $\mathbf{a} \times \mathbf{a} = \mathbf{0}$. However, $\mathbf{a} \times \mathbf{b} = \mathbf{0}$ also if either **a** or **b** are zero, so that we cannot deduce that **a** and **b** are parallel when $\mathbf{a} \times \mathbf{b} = \mathbf{0}$.

The other algebraic properties of the vector product are as follows, where **a,b,c** are any vectors and λ is any scalar:

i) $\mathbf{b} \times \mathbf{a} = -\mathbf{a} \times \mathbf{b}$ (because the sense of rotation is **b** to **a**, not **a** to **b**),

ii) $\mathbf{a} \times (\mathbf{b} + \mathbf{c}) = \mathbf{a} \times \mathbf{b} + \mathbf{a} \times \mathbf{c}$,

iii) $(\lambda\mathbf{a}) \times \mathbf{b} = \mathbf{a} \times (\lambda\mathbf{b}) = \lambda(\mathbf{a} \times \mathbf{b})$,

iv) $\mathbf{i} \times \mathbf{j} = \mathbf{k}, \quad \mathbf{j} \times \mathbf{k} = \mathbf{i}, \quad \mathbf{k} \times \mathbf{i} = \mathbf{j}$.

2.4.5 Applications of vector product

Example 1 **Vector area of a triangle** (Figure 2.33)

The area of triangle $ABC = \frac{1}{2}AC.BN = \frac{1}{2}AC.AB \sin\theta = \frac{1}{2}|\mathbf{a} \times \mathbf{b}|$.

The vector area $\mathbf{A} = \frac{1}{2}\mathbf{a} \times \mathbf{b}$ is often useful, as in Section 4.2.14.

Example 2 **Common perpendicular to two skew lines**

In order to obtain the length of the common perpendicular Q_1Q_2 to the lines P_1R_1 and P_2R_2 (see Figure 2.34), we denote the directions of P_1R_1 and P_2R_2 by the unit vectors $\mathbf{u}_1$ and $\mathbf{u}_2$, the position vectors of P_1 and P_2 by $\mathbf{r}_1$ and $\mathbf{r}_2$, and the direction of $\overrightarrow{Q_1Q_2}$ by the unit vector **u** (unknown).

Then $$\overrightarrow{OQ_1} = \mathbf{r}_1 + P_1Q_1\mathbf{u}_1 = \mathbf{r}_2 + P_2Q_2\mathbf{u}_2 - Q_1Q_2\mathbf{u}.$$

On taking the scalar product of this equation with **u** we obtain

$$\mathbf{r}_1.\mathbf{u} + P_1Q_1\mathbf{u}_1.\mathbf{u} = \mathbf{r}_2.\mathbf{u} + P_2Q_2\mathbf{u}_2.\mathbf{u} - Q_1Q_2.$$

However, since Q_1Q_2 is perpendicular to P_1R_1 and P_2R_2, the products $\mathbf{u}_1.\mathbf{u}$ and $\mathbf{u}_2.\mathbf{u}$ are zero, so that

$$Q_1Q_2 = (\mathbf{r}_2 - \mathbf{r}_1).\mathbf{u}.$$

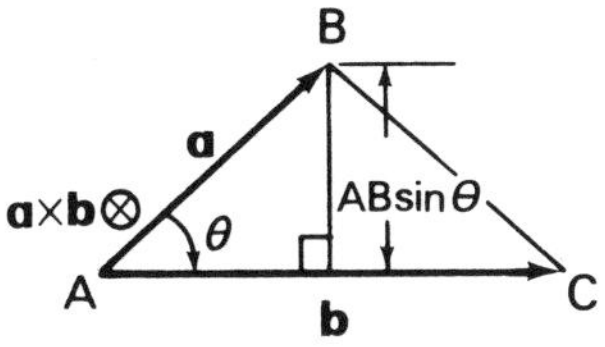

Figure 2.33

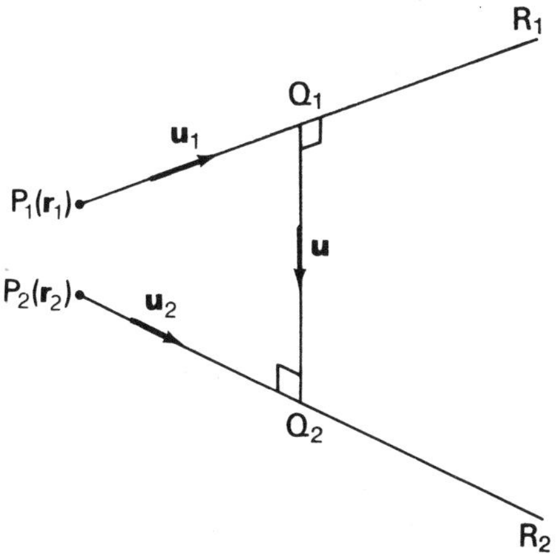

Figure 2.34

Moreover, since $\mathbf{u}$ is perpendicular to $\mathbf{u}_1$ and $\mathbf{u}_2$, it is proportional to $\mathbf{u}_1 \times \mathbf{u}_2$. Since Q_1Q_2 is to be a length, it must be positive, so that finally

$$Q_1Q_2 = \left| \frac{(\mathbf{r}_2 - \mathbf{r}_1) \cdot (\mathbf{u}_1 \times \mathbf{u}_2)}{|\mathbf{u}_1 \times \mathbf{u}_2|} \right| .$$

2.4.6 Triple scalar product

We can combine the scalar and vector products in the triple product

$$\mathbf{a}.(\mathbf{b} \times \mathbf{c}) = a_1(b_2c_3 - b_3c_2) + a_2(b_3c_1 - b_1c_3) + a_3(b_1c_2 - b_2c_1),$$

which may be conveniently written in the determinantel form

$$\mathbf{a}.(\mathbf{b} \times \mathbf{c}) = \begin{vmatrix} a_1 & a_2 & a_3 \\ b_1 & b_2 & b_3 \\ c_1 & c_2 & c_3 \end{vmatrix} . \tag{2.8}$$

From the properties of determinants we can show that

$$\begin{aligned} \mathbf{a}.(\mathbf{b} \times \mathbf{c}) &= \mathbf{b}.(\mathbf{c} \times \mathbf{a}) = \mathbf{c}.(\mathbf{a} \times \mathbf{b}) \\ &= -\,\mathbf{a}.(\mathbf{c} \times \mathbf{b}) = -\,\mathbf{b}.(\mathbf{a} \times \mathbf{c}) = -\,\mathbf{c}.(\mathbf{b} \times \mathbf{a}). \end{aligned} \tag{2.9}$$

Thus cyclic permutations of the symbols produce no change in the value of this triple product.

We may also deduce from either equation (2.8) or (2.9) that a triple scalar product involving two equal or parallel vectors is zero.

Example 1

If $\mathbf{a} = (1,-1,0)$, $\mathbf{b} = (2,0,1)$, and $\mathbf{c} = (3,2,4)$, then

$$\mathbf{a}.(\mathbf{b} \times \mathbf{c}) = \begin{vmatrix} 1 & -1 & 0 \\ 2 & 0 & 1 \\ 3 & 2 & 4 \end{vmatrix}$$

$$= -2 - 2(-4) + 3(-1) = -2 + 8 - 3 = 3.$$

Example 2

The equation of the plane through the points with position vectors $\mathbf{r}_0$, $\mathbf{r}_1$ and $\mathbf{r}_2$ can be obtained by noting that $\mathbf{r}_1 - \mathbf{r}_0$ and $\mathbf{r}_2 - \mathbf{r}_0$ are vectors lying in the plane. Thus the vector $(\mathbf{r}_1 - \mathbf{r}_0) \times (\mathbf{r}_2 - \mathbf{r}_0)$ is normal to the plane, and we may use equation (2.6) by noting that the unit normal vector $\mathbf{u}$ is proportional to $(\mathbf{r}_1 - \mathbf{r}_0) \times (\mathbf{r}_2 - \mathbf{r}_0)$, so that

$$\mathbf{r}.[(\mathbf{r}_1 - \mathbf{r}_0) \times (\mathbf{r}_2 - \mathbf{r}_0)] = \mathbf{r}_0.(\mathbf{r}_1 - \mathbf{r}_0) \times (\mathbf{r}_2 - \mathbf{r}_0).$$

Finally we note that most of the terms in the expansion of the right hand side are zero because they contain the vector $\mathbf{r}_0$ repeated.

Thus a general point $P(\mathbf{r})$ in the plane is given by

$$\mathbf{r}.[(\mathbf{r}_1 - \mathbf{r}_0) \times (\mathbf{r}_2 - \mathbf{r}_0)] = \mathbf{r}_0.(\mathbf{r}_1 \times \mathbf{r}_2).$$

2.4.7 Triple vector product

The product $\mathbf{a} \times (\mathbf{b} \times \mathbf{c})$ can be evaluated by two vector product operations. However, we may evaluate it more simply by use of the identity

$$\mathbf{a} \times (\mathbf{b} \times \mathbf{c}) = (\mathbf{a}.\mathbf{c})\mathbf{b} - (\mathbf{a}.\mathbf{b})\mathbf{c} \tag{2.10}$$

This identity can be proved component by component. For example,

$$[\mathbf{a} \times (\mathbf{b} \times \mathbf{c})]_1 = a_2[(\mathbf{b} \times \mathbf{c})]_3 - a_3[(\mathbf{b} \times \mathbf{c})]_2$$

where the suffices and brackets indicate the components of the vectors concerned.

Thus

$$\begin{aligned}[\mathbf{a} \times (\mathbf{b} \times \mathbf{c})]_1 &= a_2(b_1c_2 - b_2c_1) - a_3(b_3c_1 - b_1c_3) \\ &= (a_2c_2 + a_3c_3)b_1 - (a_2b_2 + a_3b_3)c_1 \\ &= (a_1c_1 + a_2c_2 + a_3c_3)b_1 - (a_1b_1 + a_2b_2 + a_3b_3)c_1 \\ &= (\mathbf{a}.\mathbf{c})b_1 - (\mathbf{a}.\mathbf{b})c_1.\end{aligned}$$

The other components can be obtained by permuting the symbols 1, 2, 3. Since the dot products are unaffected by the permutation, the complete vector takes the form given in equation (2.10).

Example 1 Intersection of three planes.

The three planes $\mathbf{r}.\mathbf{u}_1 = p_1$, $\mathbf{r}.\mathbf{u}_2 = p_2$ and $\mathbf{r}.\mathbf{u}_3 = p_3$ intersect at the point

$$\mathbf{r} = \frac{p_1(\mathbf{u}_2 \times \mathbf{u}_3) + p_2(\mathbf{u}_3 \times \mathbf{u}_1) + p_3(\mathbf{u}_1 \times \mathbf{u}_2)}{\mathbf{u}_1.(\mathbf{u}_2 \times \mathbf{u}_3)},$$

provided that $\mathbf{u}_1.(\mathbf{u}_2 \times \mathbf{u}_3) \neq 0$. This can be verified by showing that $\mathbf{r}$ satisfies the equation of each of the three planes.

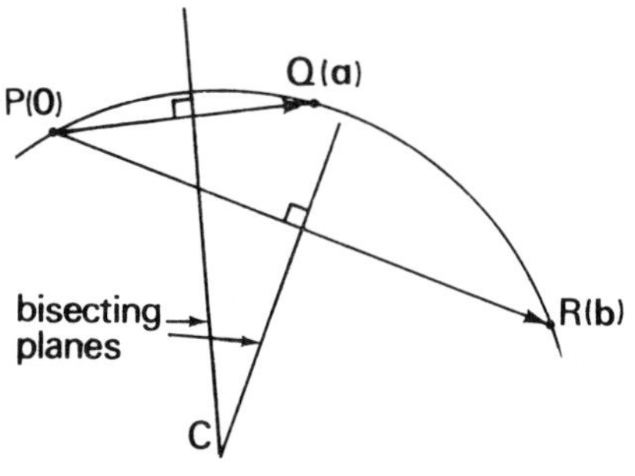

Figure 2.35

Example 2 Circle through 3 given points (see Figure 2.35).

Let the points P, Q and R have position vectors $\mathbf{0}$, $\mathbf{a}$ and $\mathbf{b}$. Then the centre of the circle through P, Q and R lies in the plane $\mathbf{r}.(\mathbf{a} \times \mathbf{b}) = 0$. It also lies on the planes which bisect PQ and PR perpendicularly, that is $\mathbf{r}.\mathbf{a} = \frac{a^2}{2}$ and $\mathbf{r}.\mathbf{b} = \frac{b^2}{2}$. Using the result of the previous example,

$$\mathbf{r} = \frac{a^2(\mathbf{b} \times (\mathbf{a} \times \mathbf{b})) + b^2((\mathbf{a} \times \mathbf{b}) \times \mathbf{a})}{2|\mathbf{a} \times \mathbf{b}|^2}$$

$$= \frac{b^2(a^2 - \mathbf{a}.\mathbf{b})\mathbf{a} + a^2(b^2 - \mathbf{a}.\mathbf{b})\mathbf{b}}{2|\mathbf{a} \times \mathbf{b}|^2} .$$

The radius R is given by

$$R^2 = \mathbf{r}.\mathbf{r} = \frac{b_2(a^2 - \mathbf{a}.\mathbf{b})\,(\mathbf{r}.\mathbf{a}) + a^2(b^2 - \mathbf{a}.\mathbf{b})\,(\mathbf{r}.\mathbf{b})}{2|\mathbf{a} \times \mathbf{b}|^2}$$

$$= \frac{a^2b^2(a^2 - 2\mathbf{a}.\mathbf{b} + b^2)}{4|\mathbf{a} \times \mathbf{b}|^2} .$$

That is
$$R = \frac{|\mathbf{a}|\,|\mathbf{b}|\,|\mathbf{a} - \mathbf{b}|}{2|\mathbf{a} \times \mathbf{b}|} .$$

Chapter 3

Co-ordinate transformations

3.1 INTRODUCTION

Whatever system is used for a computer-aided design, the description of the design must be stored as numerical data in terms of some co-ordinate system or systems. If a number of co-ordinate systems are used for convenience to describe different parts of the design, then the relationships between these systems will have to be stored in order to determine the relative positions and orientations of the parts. These relationships are known as **co-ordinate transformations.**

Moreover, the designer may wish to examine the design at any stage by producing drawings using projections or cross-sections of the object. For illustrative purposes, perspective views may be used. Apart from the cross-sectional views, all these drawings may be obtained by transforming from the **object co-ordinates** into the **picture co-ordinates**.

When the design comes to be manufactured, it is necessary to relate the design co-ordinate system to the machine axes, so that co-ordinate transformations are required here also.

Finally, in designs having some degree of symmetry, it is necessary to store data for *part* of the design only, because the remainder may be generated by symmetry using rotations, translations and reflections. An example of this will be described in Chapter 8. These are **object transformations**, and the underlying co-ordinate system is unchanged.

For further discussion of these transformations, the reader is referred to Roberts (1963), Roberts (1965) and Ahuja and Coons (1968). A comprehensive survey of graphical techniques is given in Newman and Sproull (1973), including a detailed description of the techniques for hidden line suppression, as well as an extensive bibliography on the subject of graphical theory, software and hardware.

We shall describe some of these object transformations, and show how the corresponding co-ordinate transformations are related to them.

3.2 OBJECT TRANSFORMATIONS

3.2.1 Translation

If an object is to be translated without rotation, and the vector displacement of each object point is to be **t**, then the position vector $\mathbf{r}'$ of the displaced point is related to its initial position **r** by the equation

$$\mathbf{r}' = \mathbf{r} + \mathbf{t}. \tag{3.1}$$

On the other hand, if the co-ordinate axes are translated in the same way whilst keeping the object fixed, we see from Figure 3.1 that the co-ordinate vector $\mathbf{x}'$ in the system $O'x'y'z'$ is related to the corresponding vector **x** in $Oxyz$ by the equation

$$\mathbf{x}' = \mathbf{x} - \mathbf{t}. \tag{3.2}$$

We will adopt the convention of describing the co-ordinate transformations by vectors **x** and $\mathbf{x}'$, and object transformations by **r** and $\mathbf{r}'$. Thus $\mathbf{x}'$ and **x** are different representations of the same point P, whereas $\mathbf{r}'$ represents the new point obtained by transforming the object point **r**. The components of both **r** and **x** will be denoted by x, y and z, and those of $\mathbf{r}'$ and $\mathbf{x}'$ by x', y', and z'.

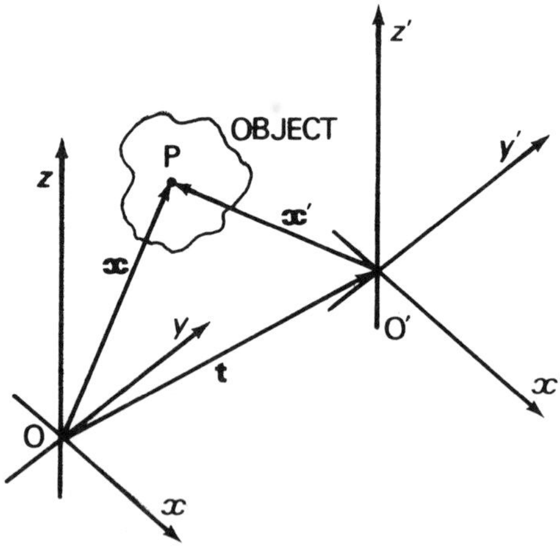

Figure 3.1

Thus co-ordinate axis translation is equivalent to an equal object translation in the opposite sense.

3.2.2 Rotation about the Oz-axis

If all points of an object are rotated through the same angle θ about the

z-axis, we need only consider the relationship between the co-ordinates (x,y) of the original points and (x',y') of the rotated points, since the z co-ordinates are unchanged.

Using polar co-ordinates in the Oxy plane, we note that a point with polar co-ordinates (r,ϕ) rotates to $(r,\phi+\theta)$. Now $x = r \cos \phi$ and $y = r \sin \phi$, so that $x' = r \cos(\phi + \theta) = r \cos \phi \cos \theta - r \sin \phi \sin \theta$. Hence

$$x' = x \cos \theta - y \sin \theta,$$

and similarly

$$y' = x \sin \theta + y \cos \theta. \qquad (3.3)$$

Thus a two-dimensional rotation can be written in the matrix form

$$\mathbf{r}' = \begin{bmatrix} x' \\ y' \end{bmatrix} = \begin{bmatrix} \cos \theta & -\sin \theta \\ \sin \theta & \cos \theta \end{bmatrix} \begin{bmatrix} x \\ y \end{bmatrix} = \mathbf{Ar}. \qquad (3.4)$$

If the co-ordinate axes are rotated in the same way, the effect is the same as if the object had been rotated through angle $-\theta$, so that the co-ordinate transformation is

$$\mathbf{x}' = \begin{bmatrix} \cos \theta & \sin \theta \\ -\sin \theta & \cos \theta \end{bmatrix} \mathbf{x} = \mathbf{Bx}. \qquad (3.5)$$

We note that $\mathbf{A} = \mathbf{B}^T$, so that the object and co-ordinate transformations are again very simply related. The relationship is considered in more detail in the next section where rotation about an arbitrary axis through the origin is considered.

3.2.3 Rotation of a vector about a general axis through the origin

In order to determine the transformation of co-ordinates when the axes are rotated through an angle θ about a line through the origin having direction $\mathbf{u}$, we first consider the rotation of a vector $\mathbf{r}$ about this line. We may then apply this rotation to the axis directions $\mathbf{i}$, $\mathbf{j}$ and $\mathbf{k}$ to find the new axis directions and hence deduce the corresponding co-ordinate transformation.

If the vector $\mathbf{r}$ is rotated as shown in Figure 3.2 we see that the component $\vec{ON}$ along $\mathbf{u}$ is unchanged, but the component $\vec{NX}$ is rotated through θ to $\vec{NX'}$. Since $\vec{ON} = (\mathbf{r.u})$ the vector $\vec{NX}$ is given by $\mathbf{r} - (\mathbf{r.u})\mathbf{u}$. Figure 3.3 shows the view along the direction $\mathbf{u}$, with the vector $\mathbf{u} \times \mathbf{r}$ represented by NP. Now $|\mathbf{r} - (\mathbf{r.u})\mathbf{u}| = |\mathbf{r}| \sin \phi = |\mathbf{u} \times \mathbf{r}|$, so that NX = NX$'$ = NP in Figure 3.3.

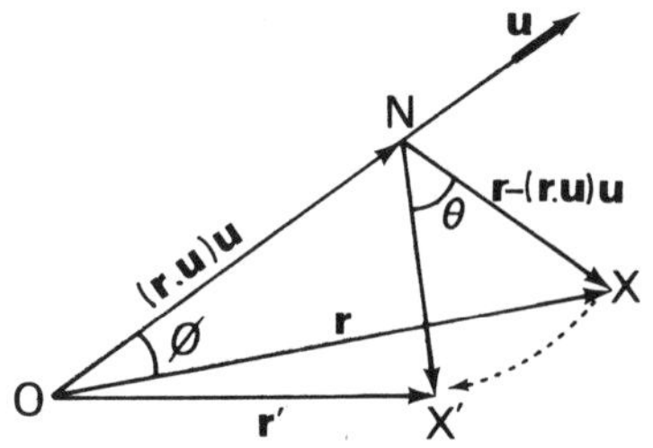

Figure 3.2

We now resolve $\vec{NX}'$ onto $\vec{NX}$ and $\vec{NP}$ and obtain

$$\vec{NX}' = \cos\theta\ [\mathbf{r} - (\mathbf{r.u})\mathbf{u}] + \sin\theta\ (\mathbf{u} \times \mathbf{r}).$$

Denoting the rotated vector $\vec{OX}'$ by $\mathbf{r}'$ we see that $\vec{OX}' = \vec{ON} + \vec{NX}'$, whence

$$\mathbf{r}' = (\mathbf{r.u})\mathbf{u} + \cos\theta\ [\mathbf{r} - (\mathbf{r.u})\mathbf{u}] + \sin\theta(\mathbf{u} \times \mathbf{r}).$$

Using the results of Section A1.8 of Appendix 1, we may translate this result into matrix notation as follows:

$$\mathbf{r}' = \mathbf{u}\,\mathbf{u}^T\mathbf{r} + \cos\theta\,\mathbf{r} - \cos\theta\,\mathbf{u}\,\mathbf{u}^T\mathbf{r} + \sin\theta\,\mathbf{U}\,\mathbf{r},$$

where
$$\mathbf{U} = \begin{bmatrix} 0 & -u_3 & +u_2 \\ +u_3 & 0 & -u_1 \\ -u_2 & +u_1 & 0 \end{bmatrix}.$$

Thus
$$\mathbf{r}' = \mathbf{A}\mathbf{r},$$

where
$$\mathbf{A} = \mathbf{u}\mathbf{u}^T + \cos\theta(\mathbf{I} - \mathbf{u}\mathbf{u}^T) + \sin\theta\,\mathbf{U}. \qquad (3.6)$$

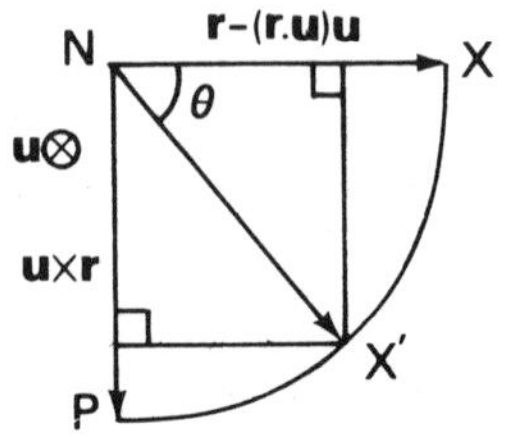

Figure 3.3

The elements of **A** are therefore the following:

$$\begin{bmatrix} u_1^2+\cos\theta(1-u_1^2) & u_1u_2(1-\cos\theta)-u_3\sin\theta & u_3u_1(1-\cos\theta)+u_2\sin\theta \\ u_1u_2(1-\cos\theta)+u_3\sin\theta & u_2^2+\cos\theta(1-u_2^2) & u_2u_3(1-\cos\theta)-u_1\sin\theta \\ u_3u_1(1-\cos\theta)-u_2\sin\theta & u_2u_3(1-\cos\theta)+u_1\sin\theta & u_3^2+\cos\theta(1-u_3^2) \end{bmatrix}.$$

The corresponding matrix for a rotation of $-\theta$ will be found by inspection to be the transpose $\mathbf{A}^T$ of this matrix **A**. It follows that successive rotations $\mathbf{r}' = \mathbf{Ar}$ and $\mathbf{r}'' = \mathbf{A}^T\mathbf{r}'$ should result in returning the vector **r** to its initial position. Thus $\mathbf{r}'' = \mathbf{A}^T\mathbf{r}' = \mathbf{A}^T\mathbf{Ar} = \mathbf{r}$, and this must be true for any vector **r**. It follows that $\mathbf{A}^T\mathbf{A} = \mathbf{I}$, so that the rotation matrix **A** is an *orthogonal* matrix.

We also note that the co-ordinate transformation corresponding to rotation of the axes about the same line is given by

$$\mathbf{x}' = \mathbf{Bx}, \text{ where } \mathbf{B} = \mathbf{A}^T, \tag{3.7}$$

because the co-ordinate rotation is equivalent to the reverse rotation of the object points. The new x-axis direction $\mathbf{i}'$ is given by

$$\mathbf{i}' = \mathbf{Ai} = \begin{bmatrix} a_{11} \\ a_{21} \\ a_{31} \end{bmatrix}.$$

This is the first column of **A**, and hence the first row of **B**. In a similar manner, we can show that all rows of **B** are the direction vectors of the new axes in terms of the old co-ordinate system.

3.2.4 Examples

(1) The matrix corresponding to rotation about the z-axis may be calculated using Equations (3.6) and (3.7).

$$\text{Thus } \mathbf{u} = \begin{bmatrix} 0 \\ 0 \\ 1 \end{bmatrix}, \quad \mathbf{uu}^T = \begin{bmatrix} 0 & 0 & 0 \\ 0 & 0 & 0 \\ 0 & 0 & 1 \end{bmatrix}, \quad \text{and } \mathbf{U} = \begin{bmatrix} 0 & -1 & 0 \\ 1 & 0 & 0 \\ 0 & 0 & 0 \end{bmatrix}.$$

$$\text{Hence } \mathbf{A} = \begin{bmatrix} \cos\theta & -\sin\theta & 0 \\ +\sin\theta & \cos\theta & 0 \\ 0 & 0 & 1 \end{bmatrix}, \text{ in agreement with (3.4).}$$

(2) To rotate the axes through 90° about the **i** direction, note that $\mathbf{i}' = \mathbf{i}$, $\mathbf{j}' = \mathbf{k}, \mathbf{k}' = -\mathbf{j}$.

Then
$$\mathbf{B} = \begin{bmatrix} 1 & 0 & 0 \\ 0 & 0 & 1 \\ 0 & -1 & 0 \end{bmatrix} = \begin{bmatrix} \mathbf{i}^T \\ \hline \mathbf{k}^T \\ \hline -\mathbf{j}^T \end{bmatrix}.$$

3.2.5 Scaling and reflection transformations

If we wish to change the scale of an object in the directions of the individual axes of a co-ordinate system by scaling factors (λ,μ,ν), then

$$\begin{bmatrix} x' \\ y' \\ z' \end{bmatrix} = \begin{bmatrix} \lambda & 0 & 0 \\ 0 & \mu & 0 \\ 0 & 0 & \nu \end{bmatrix} \begin{bmatrix} x \\ y \\ z \end{bmatrix} \qquad (3.8)$$

provides the appropriate transformation. It can be seen that the transformation can be expressed in the same matrix form $\mathbf{r}' = \mathbf{Ar}$ as the rotation transformation.

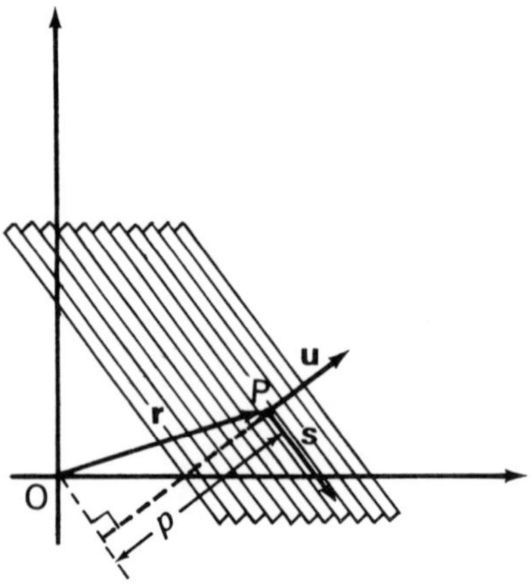

Figure 3.4

If a negative scaling factor is used, the object is then reflected in the co-ordinate plane normal to the corresponding co-ordinate axis. Thus the matrix $\mathbf{A} = \begin{bmatrix} 2 & 0 & 0 \\ 0 & 1 & 0 \\ 0 & 0 & -1 \end{bmatrix}$ represents a magnification of 2 in the x-direction, together with a reflection in the plane Oxy.

3.2.6 Shear transformation

We shall imagine space to be divided into a very large number of thin laminae bounded by parallel slip planes with a common normal direction **u** (see Figure 3.4). On any given slip plane, a point P with position vector **r** satisfies the equation $\mathbf{u}.\mathbf{r} = p$, where P is the perpendicular distance from P to the slip plane through the origin. If the laminae now slip relative to the origin plane by an amount proportional to p and in a uniform dirèction **s**, then the new position $\mathbf{r}'$ of the point P is given by $\mathbf{r}' = \mathbf{r} + \alpha p\mathbf{s} = \mathbf{r} + \alpha(\mathbf{u}.\mathbf{r})\mathbf{s}$, where α is a scalar constant. Using the matrix notation described in Section A1.8 of Appendix 1, we may write

$$\mathbf{r}' = [\mathbf{I} + \alpha\mathbf{s}\mathbf{u}^T]\mathbf{r}. \tag{3.9}$$

Thus the shear transformation can also be represented by a matrix transformation of the form $\mathbf{r}' = \mathbf{A}\mathbf{r}$. For example, if $\mathbf{u} = \mathbf{j}$ and $\mathbf{s} = \mathbf{i}$, then the two dimensional shear transformation is given by

$$\mathbf{A} = \begin{bmatrix} 1 & \alpha \\ 0 & 1 \end{bmatrix}.$$

In this example the shear occurs along planes parallel to the Oxz-plane, and the direction of the shear is parallel to the axis Ox.

3.2.7 The use of homogenous co-ordinates

We have seen that rotation, scaling, reflection and shearing transformations can all be represented by equations of the form $\mathbf{r}' = \mathbf{A}\mathbf{r}$. However, each of these transformations was defined relative to the co-ordinate origin. If we wish to rotate an object about an axis which does not pass through the origin, or reflect it in a plane not containing the origin, these transformations must be combined with translations, which we have so far described by a vector equation of the form $\mathbf{r}' = \mathbf{r} + \mathbf{t}$.

It is often convenient, therefore, to incorporate translations into the general matrix transformation scheme. This is done by representing each position vector $\mathbf{r}$ by the extended vector

$$\mathbf{R} = \begin{bmatrix} x \\ y \\ z \\ 1 \end{bmatrix}.$$

The translation $\mathbf{r}' = \mathbf{r} + \mathbf{t}$ can then be written as

$$\mathbf{R}' = \mathbf{TR}, \text{ where } \mathbf{T} = \left[\begin{array}{ccc|c} 1 & 0 & 0 & t_1 \\ 0 & 1 & 0 & t_2 \\ 0 & 0 & 1 & t_3 \\ \hline 0 & 0 & 0 & 1 \end{array}\right], \qquad (3.10)$$

and $\mathbf{R}'$ is the extended vector corresponding to $\mathbf{r}'$.

The matrix transformations $\mathbf{r}' = \mathbf{Ar}$ can be rewritten in terms of the extended vectors as

$$\mathbf{R}' = \mathbf{MR}, \text{ where } \mathbf{M} = \left[\begin{array}{ccc|c} & & & 0 \\ & \mathbf{A} & & 0 \\ & & & 0 \\ \hline 0 & 0 & 0 & 1 \end{array}\right] \qquad (3.11)$$

Thus $\mathbf{M}$ is simply the matrix $\mathbf{A}$, bordered by six zeros and one 1.

Compound transformations can now be represented as matrix products, as illustrated in the examples in Section 3.2.8. It should be noted that the order of the matrix products is important.

Although the extended vectors $\mathbf{R}$ just described are adequate in most circumstances, they do not enable us to include perspective transformations in the matrix scheme. We shall see in Section 3.4 that the adoption of a vector $\mathbf{P} = w\mathbf{R}$, where w is a scalar which may be varied from point to point, enables us to describe perspective transformations. Moreover, we shall see in Section 5.2 that rational curves and surfaces are most succinctly represented in terms of this vector.

The elements (xw,yw,zw,w) of the vector **P** are **homogenous co-ordinates** of the point (x,y,z). The values of x, y and z must be recovered from **P** by division. Thus

$$x = \frac{P_1}{P_4},\ y = \frac{P_2}{P_4} \text{ and } z = \frac{P_3}{P_4}. \tag{3.12}$$

Consequently the representation of a point is no longer unique, in much the same way that the implicit equation of a line in two dimensions, $ax + by + c = 0$, does not have unique coefficients a, b, c. In each case, only the *ratios* are important.

Any co-ordinate system which represents a point in three dimensions by four co-ordinates (P_1,P_2,P_3,P_4), where any multiple $\lambda\mathbf{P}$ of a given vector **P** represents the same point, is called a **homogenous co-ordinate system.** In general, the homogenous co-ordinates for n-dimensional space consist of $n+1$ numbers.

Because we are concerned with the metrical properties of curves and surfaces, we need to relate the homogeneous co-ordinates **P** to the Cartesian co-ordinates **r**. For our present purposes, the relation (3.12) is a convenient one.

However, many non-metrical properties of geometrical figures can be determined without specifying the relationship between **P** and **r**. These descriptive properties form the subject of **projective geometry**, and the reader is referred to texts such as Semple and Kneebone (1952) for further details of this subject. Apart from unifying the description of co-ordinate transformations, the use of homogeneous co-ordinates removes many anomalous situations which arise in Cartesian co-ordinate geometry, such as the non-intersection of parallel lines and the need for a distinction between central and non-central conics.

In terms of these co-ordinates, translation matrices take the form

$$\mathbf{T} = \left[\begin{array}{ccc|c} t_4 & 0 & 0 & t_1 \\ 0 & t_4 & 0 & t_2 \\ 0 & 0 & t_4 & t_3 \\ \hline 0 & 0 & 0 & t_4 \end{array}\right], \tag{3.13}$$

where $[t_1\ t_2\ t_3\ t_4]^T$ is a homogeneous representation of the translation vector **t**.

The rotation, scaling and reflection transformations can be represented as before by the matrix **M** given in equation (3.9), if we choose to set $w' = w$. However, perspective transformations require a transformation of the w co-ordinate, as will be seen in Section 3.4. In general, therefore, it is sensible to extend equation (3.11) to permit $w' = kw$, and use

$$\mathbf{M} = \left[\begin{array}{ccc|c} & & & 0 \\ & k\mathbf{A} & & 0 \\ & & & 0 \\ \hline 0 & 0 & 0 & k \end{array}\right] \tag{3.14}$$

as the homogeneous co-ordinate equivalent of the matrix **A**.

3.2.8 Examples

The examples are carried out in terms of **R**, and illustrate the use of the special case of homogeneous co-ordinates with $w \equiv 1$.

(1) Obtain the transformation required to rotate an object through 90° about an axis parallel to and in the same sense as the z-axis, but passing through the point (1,2,0).
We first translate the body by a vector (−1,−2,0), so that the desired axis of rotation passes through the origin. The body is then rotated about the z-axis and finally translated by a vector (1,2,0) to bring the axis back to its initial position. Thus

$$\mathbf{R}' = \underbrace{\left[\begin{array}{ccc|c} 1 & 0 & 0 & 1 \\ 0 & 1 & 0 & 2 \\ 0 & 0 & 1 & 0 \\ \hline 0 & 0 & 0 & 1 \end{array}\right]}_{\textit{Translate back}} \underbrace{\left[\begin{array}{ccc|c} 0 & -1 & 0 & 0 \\ 1 & 0 & 0 & 0 \\ 0 & 0 & 1 & 0 \\ \hline 0 & 0 & 0 & 1 \end{array}\right]}_{\textit{Rotate about } z\textit{-axis}} \underbrace{\left[\begin{array}{ccc|c} 1 & 0 & 0 & -1 \\ 0 & 1 & 0 & -2 \\ 0 & 0 & 1 & 0 \\ \hline 0 & 0 & 0 & 1 \end{array}\right]}_{\textit{Translate to origin}} \mathbf{R}$$

Note the order of the matrices: the first transformation is at the right-hand side. The matrices do not in general commute.

(2) Obtain the transformation which reflects an object in the plane $x = 2$.
Here we first translate the reflection plane Π by a vector (−2,0,0) so that it passes through the origin. We now apply a reflection in the plane $x = 0$ and return the reflection plane Π to $x = 2$ by a translation (2,0,0). Then

$$\mathbf{R}' = \underbrace{\begin{bmatrix} 1 & 0 & 0 & 2 \\ 0 & 1 & 0 & 0 \\ 0 & 0 & 1 & 0 \\ 0 & 0 & 0 & 1 \end{bmatrix}}_{\substack{\textit{Translate } \Pi \\ \textit{back to } x = 2}} \underbrace{\begin{bmatrix} -1 & 0 & 0 & 0 \\ 0 & 1 & 0 & 0 \\ 0 & 0 & 1 & 0 \\ 0 & 0 & 0 & 1 \end{bmatrix}}_{\substack{\textit{Reflection} \\ \textit{in } x = 0}} \underbrace{\begin{bmatrix} 1 & 0 & 0 & -2 \\ 0 & 1 & 0 & 0 \\ 0 & 0 & 1 & 0 \\ 0 & 0 & 0 & 1 \end{bmatrix}}_{\substack{\textit{Translate } \Pi \\ \textit{to origin}}} \mathbf{R}.$$

3.3 PLANE PROJECTIONS OF THREE-DIMENSIONAL SPACE

In producing drawings of three-dimensional objects, whether on permanent media or graphical display tubes, two-dimensional projections are required.

In a plane projection, each object point is projected in a defined manner onto the **picture plane**, where it is represented by a **picture point**. If the projection lines joining corresponding object and picture points are all parallel, then we have a **plane parallel projection**, as shown in Figure 3.5. On the other hand, if the projection lines converge on a common point P, the picture obtained is a **central projection** or **perspective view** of the object (see Figure 3.7).

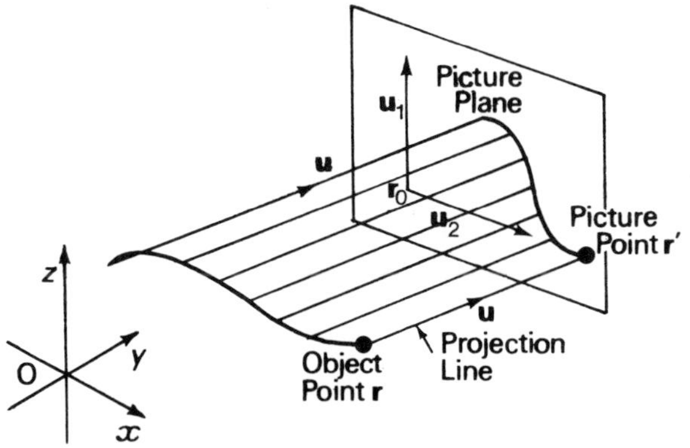

Figure 3.5

3.3.1 Parallel projections

In a plane parallel projection, all object points are projected parallel to some fixed direction $\mathbf{u}$ onto a specified picture plane (see Figure 3.5). We describe a point in the picture plane in terms of some prescribed plane co-ordinate system. The picture plane and its co-ordinate system may be specified by giving its origin $\mathbf{r}_0$ and the directions $\mathbf{u}_1$ and $\mathbf{u}_2$ of its axes in terms of the underlying three dimensional co-ordinate system.

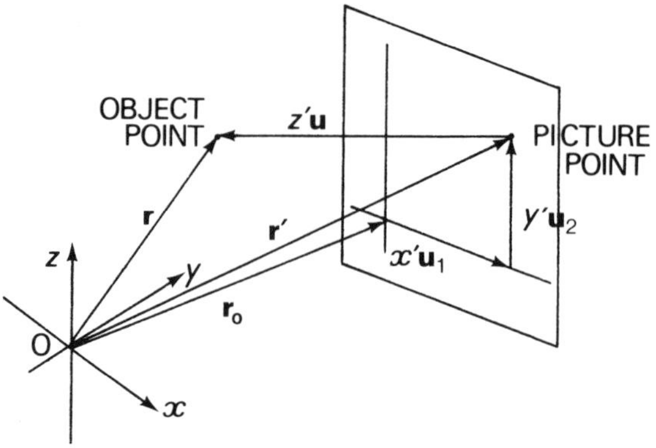

Figure 3.6

Since each object point $\mathbf{r}$ is projected parallel to $\mathbf{u}$ onto the image point $\mathbf{r}'$, we see from Figure 3.5 that

$$\mathbf{r}' = \mathbf{r} - z'\mathbf{u}$$

for some value of z'. If the co-ordinates of the image point in the projection plane are given by (x',y') then Figure 3.6 shows that

$$\mathbf{r}' = \mathbf{r}_0 + x'\mathbf{u}_1 + y'\mathbf{u}_2 ,$$

and hence $$\mathbf{r}' = \mathbf{r} - z'\mathbf{u} = \mathbf{r}_0 + x'\mathbf{u}_1 + y'\mathbf{u}_2 . \qquad (3.15)$$

(Note that $Ox'y'z'$ are not necessarily orthogonal axes.)

We may obtain z' by taking the scalar product of equation (3.15) with $\mathbf{u}_1 \times \mathbf{u}_2$ to eliminate x' and y'. Then

$$z' = \frac{(\mathbf{r} - \mathbf{r}_0) . (\mathbf{u}_1 \times \mathbf{u}_2)}{\mathbf{u}.(\mathbf{u}_1 \times \mathbf{u}_2)}.$$

By taking scalar products of the same equation with $\mathbf{u}_2 \times \mathbf{u}$ and $\mathbf{u}_1 \times \mathbf{u}$ respectively, we can similarly obtain expressions for x' and y'.

Thus $$x' = \frac{(\mathbf{r} - \mathbf{r}_0) . (\mathbf{u}_2 \times \mathbf{u})}{\mathbf{u}.(\mathbf{u}_1 \times \mathbf{u}_2)}$$

and $$y' = \frac{(\mathbf{r} - \mathbf{r}_0) . (\mathbf{u}_1 \times \mathbf{u})}{\mathbf{u}.(\mathbf{u}_2 \times \mathbf{u}_1)}. \qquad (3.16)$$

In most cases the picture plane is chosen perpendicular to the projection lines, so that $\mathbf{u} = \mathbf{u}_1 \times \mathbf{u}_2$, and the equations take the simpler form

$$x' = (\mathbf{r} - \mathbf{r}_0).\mathbf{u}_1,$$

$$y' = (\mathbf{r} - \mathbf{r}_0).\mathbf{u}_2, \quad (3.17)$$

and

$$z' = (\mathbf{r} - \mathbf{r}_0).\mathbf{u}.$$

The information about z' is not needed in order to draw the projection of the object, but it does enable us to reconstruct the object point if we wish to do so, using equation (3.15).

Using matrix notation, we may write the equations (3.17) in the form

$$\mathbf{R}' = \begin{bmatrix} x' \\ y' \\ z' \\ 1 \end{bmatrix} = \left[\begin{array}{c|c} \mathbf{u}_1^T & 0 \\ \hline \mathbf{u}_2^T & 0 \\ \hline \mathbf{u}^T & 0 \\ \hline 0 \;\; 0 \;\; 0 & 1 \end{array}\right] \left[\begin{array}{ccc|c} 1 & 0 & 0 & -x_0 \\ 0 & 1 & 0 & -y_0 \\ 0 & 0 & 1 & -z_0 \\ \hline 0 & 0 & 0 & 1 \end{array}\right] \left[\begin{array}{c} x \\ y \\ z \\ \hline 1 \end{array}\right] = \mathbf{ATR}, \quad (3.18)$$

so that the projection is equivalent to translating the picture plane and object until the picture plane passes through the origin (matrix **T**), and then rotating them (matrix **A**) until the picture axes coincide with the *Ox* and *Oy* axes. The transformation can, of course, be written in terms of homogeneous co-ordinates as discussed in Section 3.2.7.

Thus orthogonal plane parallel projections can be carried out using equivalent object transformations, and the object point can be recovered by the inverse transformations.

3.3.2 Central projections (perspective views)

In a plane perspective view, the picture point $\mathbf{r}'$ corresponding to a given object point $\mathbf{r}$ is the point on the picture plane which is collinear with the object point and some fixed viewpoint $\mathbf{r}_v$, sometimes called the **eyepoint** (see Figure 3.7).

If we again define the picture plane by its origin $\mathbf{r}_0$, and its axes by the direction vectors $\mathbf{u}_1$ and $\mathbf{u}_2$, then

$$\mathbf{r}' = \mathbf{r}_0 + x'\mathbf{u}_1 + y'\mathbf{u}_2$$

as before. Moreover, the picture point lies on the line joining $\mathbf{r}$ to $\mathbf{r}_v$, so that

$$\mathbf{r}' = z'\mathbf{r} + (1 - z')\,\mathbf{r}_v$$

for some value of z. Thus

$$\mathbf{r}' = \mathbf{r}_0 + x'\mathbf{u}_1 + y'\mathbf{u}_2 = z'\mathbf{r} + (1 - z')\mathbf{r}_v. \tag{3.19}$$

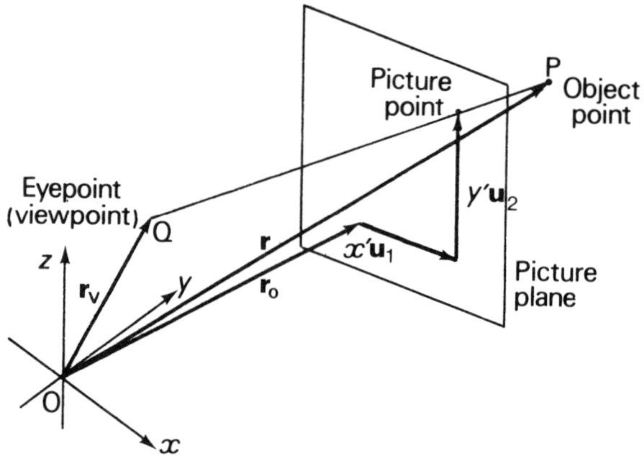

Figure 3.7

We may now obtain x', y' and z' by taking scalar products of this equation with $\mathbf{u}_2 \times (\mathbf{r} - \mathbf{r}_v)$, $\mathbf{u}_1 \times (\mathbf{r} - \mathbf{r}_v)$ and $\mathbf{u}_1 \times \mathbf{u}_2$.

Then

$$x' = \frac{(\mathbf{r} - \mathbf{r}_v).[\mathbf{u}_2 \times (\mathbf{r}_0 - \mathbf{r}_v)]}{(\mathbf{r} - \mathbf{r}_v)\,.\,(\mathbf{u}_1 \times \mathbf{u}_2)},$$

$$y' = \frac{(\mathbf{r} - \mathbf{r}_v)\,.\,[\mathbf{u}_1 \times (\mathbf{r}_0 - \mathbf{r}_v)]}{(\mathbf{r} - \mathbf{r}_v).(\mathbf{u}_2 \times \mathbf{u}_1)}$$

and

$$z' = \frac{(\mathbf{r}_0 - \mathbf{r}_v)\,.\,(\mathbf{u}_1 \times \mathbf{u}_2)}{(\mathbf{r} - \mathbf{r}_v)\,.\,(\mathbf{u}_1 \times \mathbf{u}_2)}. \tag{3.20}$$

The usual arrangement is to choose the origin of the picture co-ordinates such that the line from the viewpoint to the picture origin is perpendicular to the picture plane (see Figure 3.8). If the viewpoint is at distance d from the picture plane, we then have $\mathbf{r}_v = \mathbf{r}_0 + d\mathbf{u}$, where $\mathbf{u} = \mathbf{u}_1 \times \mathbf{u}_2$.

Then

$$x' = \frac{-d(\mathbf{r} - \mathbf{r}_0).\mathbf{u}_1}{(\mathbf{r} - \mathbf{r}_0).\mathbf{u} - d},$$

$$y' = \frac{-d(\mathbf{r} - \mathbf{r}_0).\mathbf{u}_2}{(\mathbf{r} - \mathbf{r}_0).\mathbf{u} - d}$$

and

$$z' = \frac{-d}{(\mathbf{r} - \mathbf{r}_0).\mathbf{u} - d}. \tag{3.21}$$

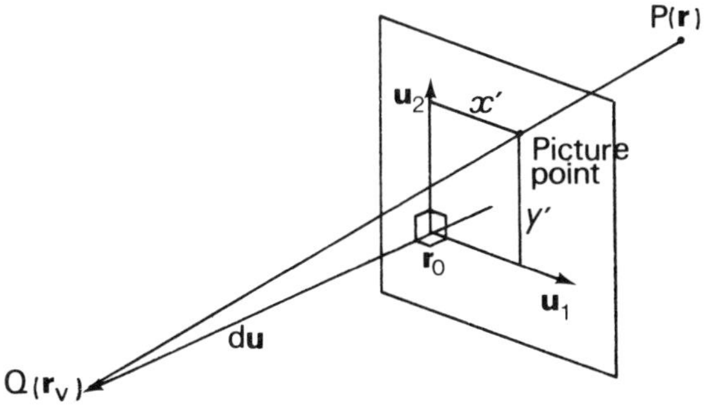

Figure 3.8

Writing $w' = d - (\mathbf{r} - \mathbf{r}_0).\mathbf{u}$, we may express equation (3.21) in terms of the homogeneous co-ordinate vectors $\mathbf{P}$, $\mathbf{P}_0$ and $\mathbf{P}'$ corresponding to the vectors $\mathbf{r}$, $\mathbf{r}_0$ and $\mathbf{r}'$. Thus

$$\mathbf{P}' = \begin{bmatrix} x'w' \\ y'w' \\ z'w' \\ \hline w' \end{bmatrix} = \left[\begin{array}{ccc|c} d & 0 & 0 & 0 \\ 0 & d & 0 & 0 \\ 0 & 0 & 0 & d \\ \hline 0 & 0 & -1 & d \end{array}\right] \left[\begin{array}{c|c} \mathbf{u}_1^T & 0 \\ \hline \mathbf{u}_2^T & 0 \\ \hline \mathbf{u}^T & 0 \\ \hline 0\ 0\ 0 & 1 \end{array}\right] \left[\begin{array}{c|c} & -x_0w_0 \\ \mathbf{I} & -y_0w_0 \\ & -z_0w_0 \\ \hline 0\ 0\ 0 & w_0 \end{array}\right] \begin{bmatrix} xw \\ yw \\ zw \\ \hline w \end{bmatrix}$$

or

$$\mathbf{P}' = \mathbf{VATP}. \tag{3.22}$$

The matrix **V** is the perspective transformation matrix, whereas **T** and **A** simply translate the picture plane (and the object) into the *Oxy* plane and rotate the picture axes to coincide with the *Ox* and *Oy* axes. If the picture plane is already in this standard position, the perspective transformation is simply $\mathbf{P}' = \mathbf{VP}$, or

$$x' = \frac{xd}{d - z}, \quad y' = \frac{yd}{d - z}, \quad z' = \frac{d}{d - z}. \tag{3.23}$$

Since the value of z' is only required in order to enable us to recover the object position **P**, the third row of the matrix **V** can be replaced by any set of numbers for which **V** is non-singular. This accounts for the variety of forms found in the literature.

We shall show in Section 5.2 that the equation of a straight line can be written in terms of homogeneous coordinates as $\mathbf{P} = (1 - u)\mathbf{P}_0 + u\mathbf{P}_1$, where $\mathbf{P}_0$ and $\mathbf{P}_1$ are the homogeneous co-ordinate vectors of two points on the line. By equation (3.22), we see that the projection of the line has the equation $\mathbf{P}' = (1 - u)\mathbf{P}_0' + u\mathbf{P}_1'$, where $\mathbf{P}_0' = \mathbf{VATP}_0$ and $\mathbf{P}_1' = \mathbf{VATP}_1$. If we ignore the z' co-ordinate, we can see that this is the homogeneous equation of the straight line through the picture points corresponding to $\mathbf{P}_0$ and $\mathbf{P}_1$. It follows that we need only transform the vertices of any drawing composed of straight lines.

3.4 OBLIQUE CO-ORDINATES

The use of oblique axes may be an advantage in situations where the basic vectors defining a problem are not orthogonal.

In Figure 3.9, each point in the plane of points A, B and O′ has position vector $\mathbf{r} = \alpha\mathbf{a} + \beta\mathbf{b}$ relative to O′, where $\mathbf{a} = \overrightarrow{O'A}$ and $\mathbf{b} = \overrightarrow{O'B}$, and $\mathbf{a} \times \mathbf{b} \neq 0$. Thus O′ may be regarded as the origin of an oblique axis system with co-ordinates (α,β). The lengths of O′A and O′B determine the scaling along the two axes. Then the midpoint of AB, for example, has co-ordinates $(\frac{1}{2},\frac{1}{2})$.

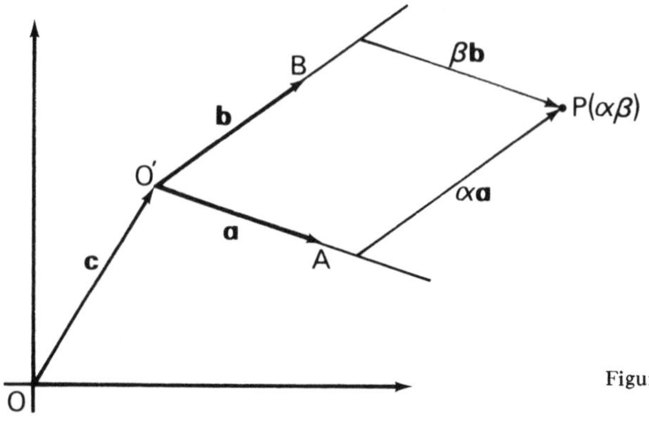

Figure 3.9

The Cartesian equation of the straight line through two given points may be generalised to oblique co-ordinates. Again by similar triangles, we may show that the equation of a point P(α,β) on the line joining $P_1(\alpha_1,\beta_1)$ to $P_2(\alpha_2,\beta_2)$ is given by

$$(\alpha - \alpha_1)(\beta_2 - \beta_1) = (\beta - \beta_1)(\alpha_2 - \alpha_1). \tag{3.24}$$

For example, the line joining A(1,0) and B(0,1) has the equation $\alpha + \beta - 1 = 0$.

The absolute co-ordinates of a point P(α,β) is given by $\mathbf{r} = \alpha\mathbf{a} + \beta\mathbf{b} + \mathbf{c}$, where $\mathbf{c} = \overrightarrow{OO'}$. If O$'$, A and B lie in the Oxy plane, we may express this equation in matrix terms as follows:

$$\mathbf{R} = \begin{bmatrix} x \\ y \\ 1 \end{bmatrix} = \begin{bmatrix} a_1 & b_1 & c_1 \\ a_2 & b_2 & c_2 \\ 0 & 0 & 1 \end{bmatrix} \begin{bmatrix} \alpha \\ \beta \\ 1 \end{bmatrix}, \tag{3.25}$$

where $(a_1,a_2,0)$, $(b_1,b_2,0)$ and $(c_1,c_2,0)$ are the Cartesian components of **a**, **b** and **c** respectively. Since $\mathbf{a} \times \mathbf{b} \neq \mathbf{0}$, the determinant of the transformation matrix is non-singular.

The general conic equation (1.5) may be rewritten in the form

$$S = [x \;\; y \;\; 1] \begin{bmatrix} a & h & g \\ h & b & f \\ g & f & c \end{bmatrix} \begin{bmatrix} x \\ y \\ 1 \end{bmatrix}$$

$$= [\alpha \;\; \beta \;\; 1] \begin{bmatrix} a_1 & a_2 & 0 \\ b_1 & b_2 & 0 \\ c_1 & c_2 & 1 \end{bmatrix} \begin{bmatrix} a & h & g \\ h & b & f \\ g & f & c \end{bmatrix} \begin{bmatrix} a_1 & b_1 & c_1 \\ a_2 & b_2 & c_2 \\ 0 & 0 & 1 \end{bmatrix} \begin{bmatrix} \alpha \\ \beta \\ 1 \end{bmatrix}$$

$$= A\alpha^2 + 2H\alpha\beta + B\beta^2 + 2G\alpha + 2F\beta + C = 0.$$

Thus a general conic is expressed by a quadratic equation in α and β. Moreover, the conditions for degeneracy, and the significance of the discriminant $ab - h^2$ which distinguishes the ellipse, parabola and hyperbola are the same as they are in Cartesian co-ordinates.

However, the metrical properties of distance and angle are lost, so that a circle does *not* necessarily have $A = B$, for example.

The distance between two points $P_1(\alpha_1,\beta_1)$ and $P_2(\alpha_2,\beta_2)$ is given by

$$\begin{aligned} P_1P_2{}^2 &= |\mathbf{r}_2 - \mathbf{r}_1|^2 = |(\alpha_2 - \alpha_1)\mathbf{a} + (\beta_2 - \beta_1)\mathbf{b}|^2 \\ &= (\alpha_2 - \alpha_1)^2\mathbf{a}^2 + 2(\alpha_2 - \alpha_1)(\beta_2 - \beta_1)\mathbf{a}.\mathbf{b} + (\beta_2 - \beta_1)^2\mathbf{b}^2. \end{aligned} \quad (3.26)$$

In three dimensions, we may express the absolute position vector of a point in terms of three non-coplanar vectors **a**, **b** and **c** and a co-ordinate origin with position vector $\mathbf{r}_0$.

Then $\mathbf{r} = \mathbf{r}_0 + \lambda\mathbf{a} + \mu\mathbf{b} + \nu\mathbf{c}$, which may be expressed in matrix terms by

$$\mathbf{R} = \begin{bmatrix} x \\ y \\ z \\ 1 \end{bmatrix} = \left[\begin{array}{ccc|c} \mathbf{a} & \mathbf{b} & \mathbf{c} & \mathbf{r}_0 \\ \hline & \mathbf{0}^T & & 1 \end{array}\right] \left[\begin{array}{c} \lambda \\ \mu \\ \nu \\ \hline 1 \end{array}\right] \quad (3.27)$$

The determinant of the matrix is given by $\mathbf{a}.(\mathbf{b} \times \mathbf{c})$, which is non-zero because **a**, **b** and **c** are not coplanar. The parameters λ, μ and ν are oblique coordinates with respect to axes through $\mathbf{r}_0$, having the directions of **a**, **b** and **c** respectively.

Chapter 4

Three dimensional curve and surface geometry

4.1 PARAMETRIC DESCRIPTION OF CURVES AND SURFACES

4.1.1 Parametric description of curves

The path of a moving point may be described by the values of the position vector **r** at successive instants in time (see Figure 4.1). We denote this relationship between **r** and the time t by the function notation $\mathbf{r} = \mathbf{r}(t)$; in words, **r** is a function of time. This is equivalent to $x = x(t)$, $y = y(t)$, $z = z(t)$ in terms of co-ordinates. The path may be any general twisted space curve.

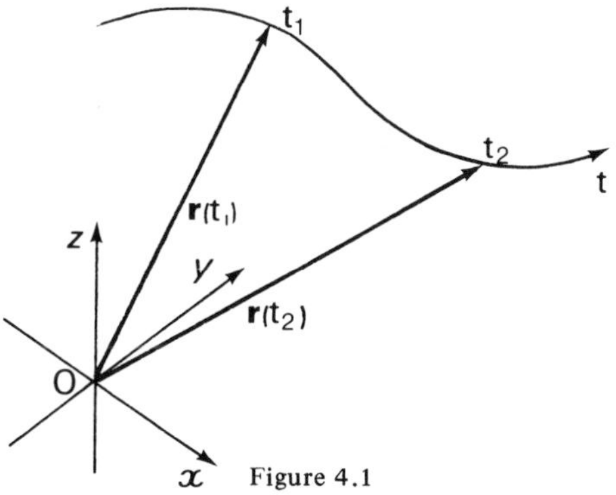

Figure 4.1

Moreover, although a time-dependent function is easiest to visualize, any functional relationship between **r** and a single scalar or **parameter** u may describe a space curve. The parameter simply acts as a co-ordinate or label for points on the curve. It is, of course, desirable that any given value of the parameter should represent only one point on the curve; in other words **r** should be a single-valued function of u. Since the choice of parameter is to a large extent arbitrary, a curve does not have a unique parametric representation.

For example, the expressions

$$\mathbf{r} = a \cos u \, \mathbf{i} + a \sin u \, \mathbf{j} \qquad \text{(see Figure 4.2),}$$

and
$$\mathbf{r} = \frac{a(1 - u^2)}{1 + u^2}\mathbf{i} + \frac{2au}{1 + u^2}\mathbf{j} \qquad \text{(see Figure 4.3)},$$

both represent the circle in the Oxy plane with centre at O and radius a.

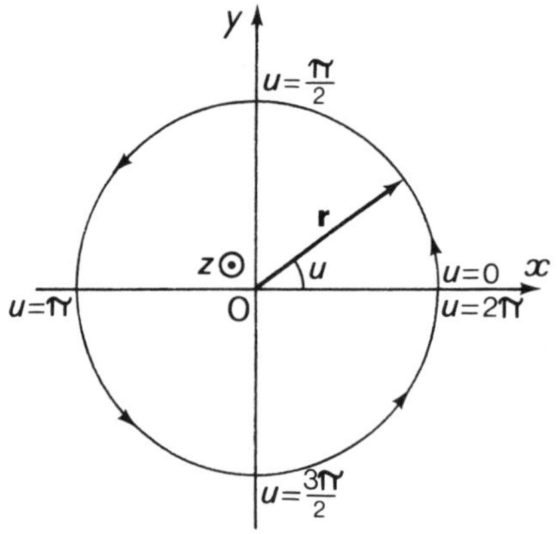

Figure 4.2

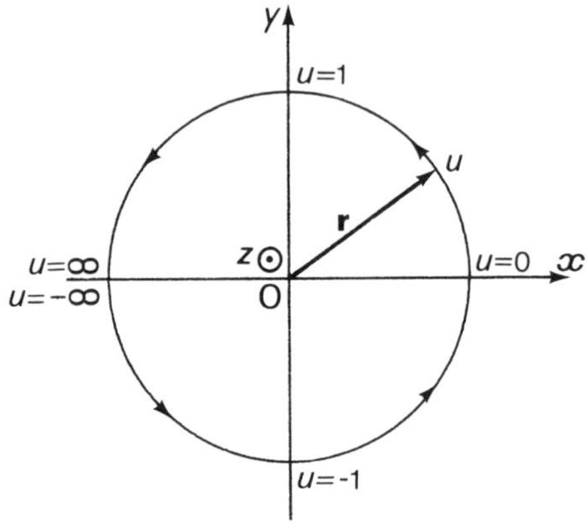

Figure 4.3

The function $\mathbf{r}(u)$ is said to be continuous at a point $u = u_0$ on the curve if the functions $x = x(u)$, $y = y(u)$ and $z = z(u)$ are continuous there, and the corresponding space curve is then also said to be continuous at $u = u_0$.

4.1.2 Examples

(1) The straight line equation $\mathbf{r} = \mathbf{r}_0 + \lambda\mathbf{u}$ derived in Section 3.3.5 is already in parametric form, and since $\mathbf{u}$ is a unit vector then λ is equal to the distance s measured from $\mathbf{r}_0$, so that we can write

$$\mathbf{r} = \mathbf{r}(s) = \mathbf{r}_0 + s\mathbf{u}.$$

(2) The equation

$$\mathbf{r} = a \cos u\, \mathbf{i} + a \sin u\, \mathbf{j} + bu\, \mathbf{k},$$

where a and b are constants, represents a helix. If we project the curve perpendicularly onto the Oxy plane, the projected curve $\mathbf{r} = \mathbf{r}'(u)$ is a circle $\mathbf{r}' = a \cos u\, \mathbf{i} + a \sin u\, \mathbf{j}$, where the projected point moves clockwise round the z-axis direction, following a circular path, as in Figure 4.2. At the same time, the component $bu\, \mathbf{k}$ indicates that the point $\mathbf{r}(u)$ is moving at a constant rate b along the z-axis direction. Hence the circular arc seen in this parallel perpendicular projection on the Oxy plane is in fact stretched out into a helix in three dimensions. Twisted curves of this type are often best visualised with the aid of orthogonal projections similar to that of Figure 4.2.

4.1.3 Parametric description of surfaces

If we now imagine a deformable curve $\mathbf{r} = \mathbf{r}(u)$ which is moving in 3 dimensions (see Figure 4.4), we can see that the successive positions and shapes of the curve generate a surface where each point is distinguished by the time t at which the moving curve passes through it and the parameter u of the point on the moving curve.

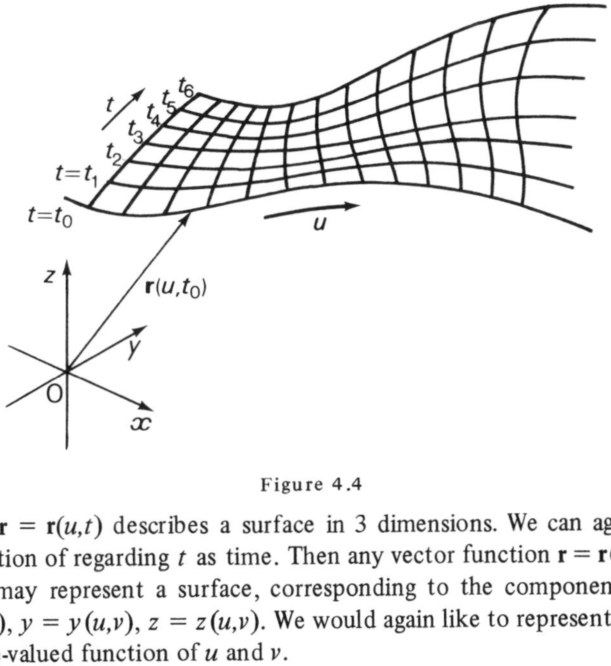

Figure 4.4

Thus $\mathbf{r} = \mathbf{r}(u,t)$ describes a surface in 3 dimensions. We can again remove the restriction of regarding t as time. Then any vector function $\mathbf{r} = \mathbf{r}(u,v)$ of two variables may represent a surface, corresponding to the component functions $x = x(u,v)$, $y = y(u,v)$, $z = z(u,v)$. We would again like to represent the surface by a single-valued function of u and v.

If either u or v is fixed we are left with a single variable parameter, so that the vector equations $\mathbf{r} = \mathbf{r}(u_0,v)$ or $\mathbf{r} = \mathbf{r}(u,v_0)$ represent curves lying on the surface $\mathbf{r} = \mathbf{r}(u,v)$ provided that u_0 and v_0 are constants in the range of u and v which describe the surface. Such curves are called **parametric curves** on the surface; their precise nature will depend upon the particular way in which the parameters are chosen.

4.1.4 Examples

(1) The plane passing through the point $\mathbf{r}_0$ and containing the lines $\mathbf{r} = \mathbf{r}_0 + u\mathbf{n}_1$ and $\mathbf{r} = \mathbf{r}_0 + v\mathbf{n}_2$, where $\mathbf{n}_1$ and $\mathbf{n}_2$ are unit vectors, can be seen in Figure 4.5 to be described by the equation

$$\mathbf{r} = \mathbf{r}_0 + u\mathbf{n}_1 + v\mathbf{n}_2.$$

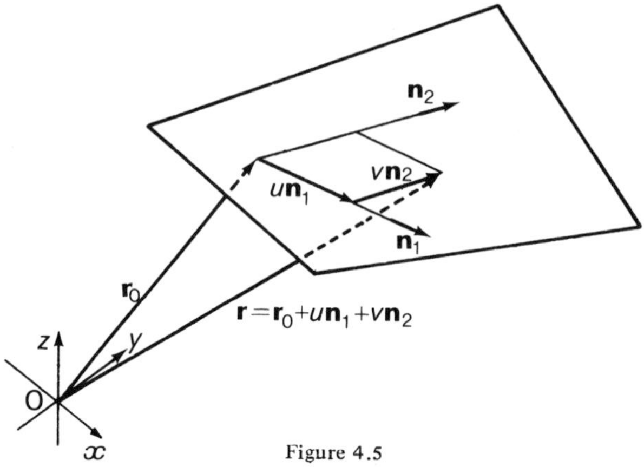

Figure 4.5

The parameters u and v represent the co-ordinates of a general point in a plane co-ordinate system with, in general, oblique axes parallel to $\mathbf{n}_1$ and $\mathbf{n}_2$. If we now write $\mathbf{n} = \dfrac{\mathbf{n}_1 \times \mathbf{n}_2}{|\mathbf{n}_1 \times \mathbf{n}_2|}$, we see that $\mathbf{r}.\mathbf{n} = \mathbf{r}_0.\mathbf{n}$, since $\mathbf{n}.\mathbf{n}_1 = \mathbf{n}.\mathbf{n}_2 = 0$ (see Section 3.4.6). Thus we recover the equation of a plane given in Section 3.4.3.

(2) The equation of a sphere of radius a with its centre at O is given by

$$\mathbf{r} = \overrightarrow{OP} = \mathbf{r}(\theta,\phi) = a \sin\theta \cos\phi\, \mathbf{i} + a \sin\theta \sin\phi\, \mathbf{j} + a\cos\theta\, \mathbf{k},$$

where θ and ϕ are the angles shown in Figure 4.6. In this figure, PN is the perpendicular from P onto the Oxy plane.

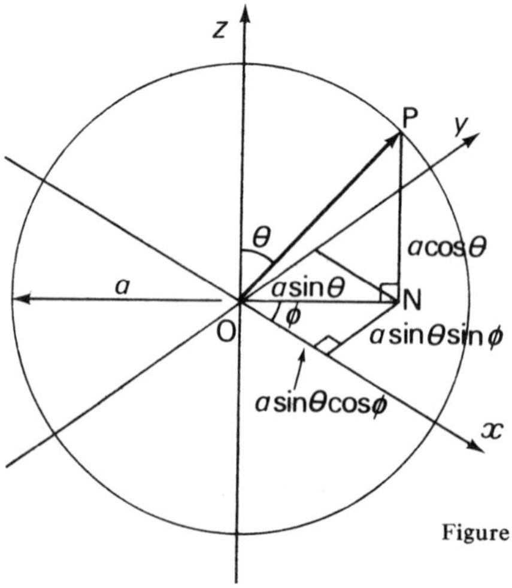

Figure 4.6

(3) The equation of a cone with its vertex at the origin, its axis in direction **n**, and semi-vertical angle α, is given by

$$\mathbf{r} = \overrightarrow{OP} = \mathbf{r}(\ell,\theta) = \ell\mathbf{n} + \ell \tan\alpha \cos\theta\, \mathbf{n}' + \ell \tan\alpha \sin\theta\, (\mathbf{n} \times \mathbf{n}'),$$

where ℓ, θ and $\mathbf{n}'$ are shown in Figure 4.7, $\mathbf{n}'$ being an arbitrary unit vector perpendicular to **n**.

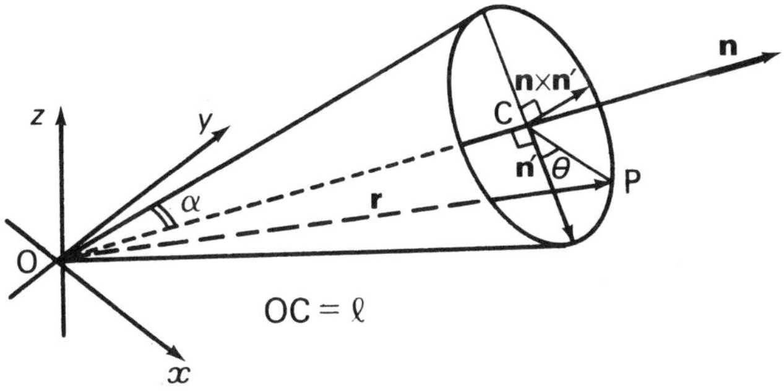

Figure 4.7

4.1.5 Advantages of the parametric description of curves and surfaces

Although there are other ways of describing curves and surfaces (for example, the sphere in Section 1.4.4 may be described implicitly by the equation $x^2 + y^2 + z^2 = a^2$), the parametric method possesses definite advantages over other methods where graphical display is required or where numerical control tapes are to be produced.

For example, points on the curve or surface may be readily computed sequentially along the curve or along parametric lines in the surface for display purposes. On the other hand, implicitly defined curves usually require the solution of a non-linear equation for each point (although this process can be accelerated by using the previously computed points to make a good initial approximation at each step). Moreover, the computation of cutter offsets and similar related curves for numerical control purposes can also be much simpler when parametric methods are used. Another significant advantage is that translation or rotation of the axes or the object can usually be carried out by translating or rotating the vectors defining the curve, without modifying the functions of the parameters used. These co-ordinate transformations have been discussed in Chapter 3.

Finally, we shall see later that the parametric method lends itself to the piecewise description of curves and surfaces, as used for example in Bézier's UNISURF design system.

However, the advantages of implicit equations which were noted in Chapter 2 should not be overlooked, and some aspects of these equations are discussed in Section 4.3.

4.1.6 Surfaces of revolution

A surface of revolution may be described by the rotation of a plane curve about the axis of symmetry. If we take this axis to be the z-axis, and let the surface intersect the Oxz plane in the plane curve $\mathbf{r} = \mathbf{r}_c(u) = p(u)\mathbf{i} + z(u)\mathbf{k}$, then Figure 4.8 shows that the surface of revolution has the equation

$$\mathbf{r} = p(u)\cos\theta\,\mathbf{i} + p(u)\sin\theta\,\mathbf{j} + z(u)\,\mathbf{k}. \tag{4.1}$$

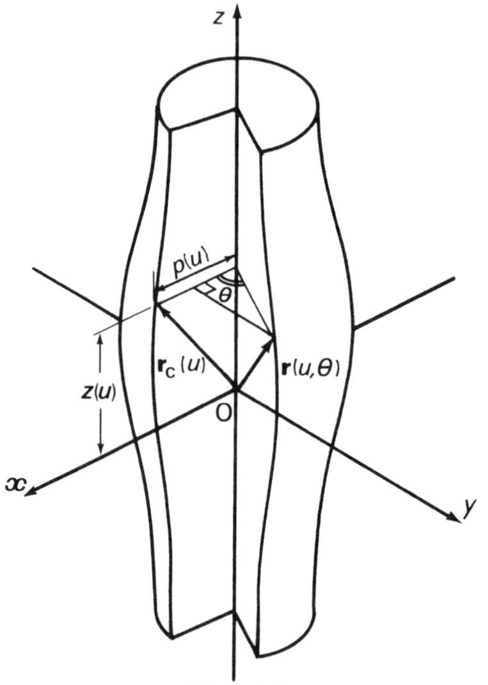

Figure 4.8

4.1.7 Examples

(1) *Ellipsoid of revolution* (Figure 4.9)

The elliptic cross-section curve $\dfrac{x^2}{a^2} + \dfrac{z^2}{b^2} = 1$ is given by

$$\mathbf{r}_c = \mathbf{r}_c(u) = a\sin u\,\mathbf{i} + b\cos u\,\mathbf{k},$$

so that $\quad \mathbf{r} = \mathbf{r}(u,\theta) = a\sin u\cos\theta\,\mathbf{i} + a\sin u\sin\theta\,\mathbf{j} + b\cos u\,\mathbf{k}.$

When $a = b$ we have the sphere again.

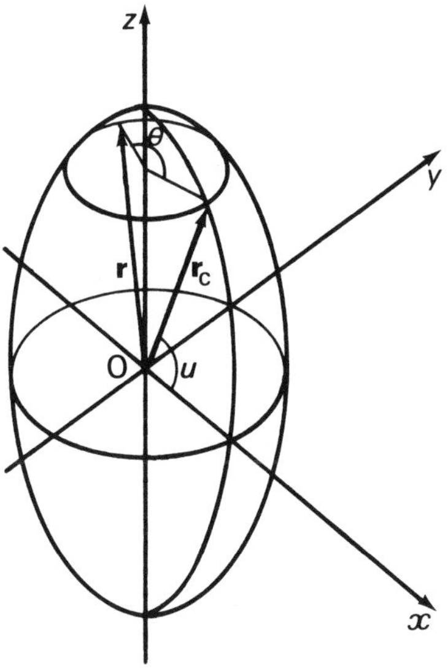

Figure 4.9

(2) *Torus* (Figure 4.10)
In this case, the cross-section is the circle $(x - R)^2 + z^2 = a^2$, with radius a and centre at $R\mathbf{i}$. In parametric terms

$$\mathbf{r}_c = \mathbf{r}_c(u) = (R + a\cos u)\mathbf{i} + a\sin u\,\mathbf{k}.$$

Thus the equation of the toroidal surface is

$$\begin{aligned}\mathbf{r} &\doteq \mathbf{r}(u,\theta) \\ &= (R + a\cos u)\cos\theta\,\mathbf{i} + (R + a\cos u)\sin\theta\,\mathbf{j} + a\sin u\,\mathbf{k}\end{aligned}$$

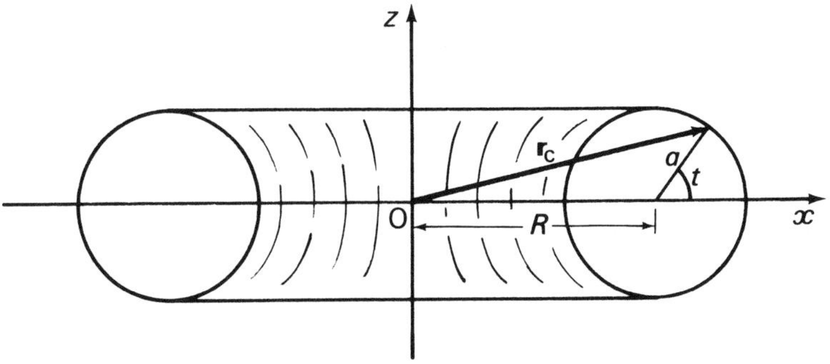

Figure 4.10

4.2 ELEMENTARY DIFFERENTIAL GEOMETRY

4.2.1 Differentiation of a vector

If a vector **a** is a function of some variable t (that is, if its length and direction are functions of t), then its components are functions of t:

$$\mathbf{a}(t) = a_1(t)\mathbf{i} + a_2(t)\mathbf{j} + a_3(t)\mathbf{k}.$$

Then, if the derivatives of $a_1(t), a_2(t)$ and $a_3(t)$ exist, we define the derivative of **a** as follows:

$$\frac{d\mathbf{a}}{dt} = \frac{da_1}{dt}\mathbf{i} + \frac{da_2}{dt}\mathbf{j} + \frac{da_3}{dt}\mathbf{k}. \tag{4.2}$$

Higher derivatives can be defined in a similar manner; for example

$$\frac{d^3\mathbf{a}}{dt^3} = \frac{d^3a_1}{dt^3}\mathbf{i} + \frac{d^3a_2}{dt^3}\mathbf{j} + \frac{d^3a_3}{dt^3}\mathbf{k}.$$

The consequences of this definition are that, for any vector functions $\mathbf{a}(t)$ and $\mathbf{b}(t)$ and any scalar function $k(t)$, we have the following properties:

$$\frac{d}{dt}(\mathbf{a} + \mathbf{b}) = \frac{d\mathbf{a}}{dt} + \frac{d\mathbf{b}}{dt} \tag{4.3}$$

$$\frac{d}{dt}(k\mathbf{a}) = \frac{dk}{dt}\mathbf{a} + k\frac{d\mathbf{a}}{dt} \tag{4.4}$$

$$\frac{d}{dt}(\mathbf{a}.\mathbf{b}) = \mathbf{a}.\frac{d\mathbf{b}}{dt} + \frac{d\mathbf{a}}{dt}.\mathbf{b} \tag{4.5}$$

$$\frac{d}{dt}(\mathbf{a} \times \mathbf{b}) = \mathbf{a} \times \frac{d\mathbf{b}}{dt} + \frac{d\mathbf{a}}{dt} \times \mathbf{b}. \tag{4.6}$$

Similar definitions can be made for the partial derivatives of vector functions of two or more variables. Let $\mathbf{b} = \mathbf{b}(u,v)$, so that

$$\mathbf{b} = b_1(u,v)\mathbf{i} + b_2(u,v)\mathbf{j} + b_3(u,v)\mathbf{k}.$$

Then

$$\frac{\partial \mathbf{b}}{\partial u} = \frac{\partial b_1}{\partial u}\mathbf{i} + \frac{\partial b_2}{\partial u}\mathbf{j} + \frac{\partial b_3}{\partial u}\mathbf{k},$$

and
$$\frac{\partial^2\mathbf{b}}{\partial u\partial v} = \frac{\partial^2 b_1}{\partial u\partial v}\mathbf{i} + \frac{\partial^2 b_2}{\partial u\partial v}\mathbf{j} + \frac{\partial^2 b_3}{\partial u\partial v}\mathbf{k},$$

for example. Note that, if $\mathbf{u}(t)$ is a unit vector, then $\mathbf{u}.\mathbf{u} = 1$. Thus

$$\frac{\mathrm{d}(\mathbf{u}.\mathbf{u})}{\mathrm{d}t} = 0 = 2\mathbf{u}.\frac{\mathrm{d}\mathbf{u}}{\mathrm{d}t},$$

so that for any unit vector $\mathbf{u}$

$$\mathbf{u}.\frac{\mathrm{d}\mathbf{u}}{\mathrm{d}t} = 0. \tag{4.7}$$

Thus $\mathbf{u}$ and $\frac{\mathrm{d}\mathbf{u}}{\mathrm{d}t}$ are mutually perpendicular vectors when $\mathbf{u}$ is a unit vector.

Example

Let
$$\mathbf{a} = (t^2 + 1)\mathbf{i} + (3t - 5)\mathbf{j} + t^3\mathbf{k}.$$

Then
$$\frac{\mathrm{d}\mathbf{a}}{\mathrm{d}t} = 2t\mathbf{i} + 3\mathbf{j} + 3t^2\mathbf{k}$$

and
$$\frac{\mathrm{d}^2\mathbf{a}}{\mathrm{d}t^2} = 2\mathbf{i} + 6t\mathbf{k}.$$

4.2.2 Taylor's theorem for vector functions

The value of a function $f(t)$ in the vicinity of the given value of t at which the value of f and its derivatives are known can be approximated by using Taylor's theorem, as is shown in Goult *et al* (1973), for example.

If the function and its first n derivatives exist for all points t' in the range $t \leqslant t' \leqslant t + \delta t$, and the nth derivative is continuous in the same range and differentiable in the range $t < t' < t + \delta t$, we may show that

$$f(t + \delta t) = f(t) + \delta t f^{(1)}(t) + \frac{\delta t^2}{2} f^{(2)}(t) + \ldots + \frac{\delta t^n}{n!} f^{(n)}(t)$$
$$+ \frac{\delta t^{n+1}}{(n + 1)!} f^{(n+1)}(t + \theta\delta t),$$

where $0 < \theta < 1$. The symbol $f^{(r)}(t)$ denotes the value of the rth derivative of f at t. Then the value of f at a neighbouring point $t + \delta t$ may be approximated in terms of values of f and its derivatives at t.

Provided that the $(n + 1)$st derivative is bounded, we may say that the last term is 'of order $n + 1$ in δt' and we denote this by writing, for example in the case where $n = 2$,

$$f(t + \delta t) = f(t) + \delta t f^{(1)}(t) + \frac{\delta t^2}{2} f^{(2)}(t) + O(\delta t^3).$$

Since a vector-valued function is differentiated by calculating the derivatives of its scalar components, Taylor's theorem can be applied to such a function whenever each of the component functions satisfy the requirements of Taylor's theorem. Then

$$\mathbf{a}(t + \delta t) = \mathbf{a}(t) + \delta t \mathbf{a}^{(1)}(t) + \frac{\delta t^2}{2} \mathbf{a}^{(2)}(t) + \ldots$$

$$\ldots + \frac{\delta t^n}{n!} \mathbf{a}^{(n)}(t) + O(\delta t^{n+1}). \tag{4.8}$$

4.2.3 Tangent to a curve

The curve in Figure 4.11 is a general (possibly twisted) space curve. The vector $\delta \mathbf{r} = \mathbf{r}(u + \delta u) - \mathbf{r}(u)$ represents the chord PQ joining two points P and Q with parameters u and $u + \delta u$ on the curve $\mathbf{r} = \mathbf{r}(u)$. As $\delta u \to 0$, the vector $\delta \mathbf{r}/\delta u$ has a direction which approaches the direction of the tangent at P, if the curve has a well-defined tangent there. If we choose the arc length s as our parameter, then the chord length $|\delta \mathbf{r}|$ and the arc length δs become equal in the limit. Then $\dfrac{d\mathbf{r}}{ds} = \lim\limits_{\delta s \to 0} \dfrac{\delta \mathbf{r}}{\delta s} = \mathbf{T}$ is a vector of unit length in the direction of the tangent to the curve, known as the **unit tangent vector**. For simplicity, we will assume that this and all derivatives of the function $\mathbf{r}(s)$ (or $\mathbf{r}(u)$) required in this chapter exist.

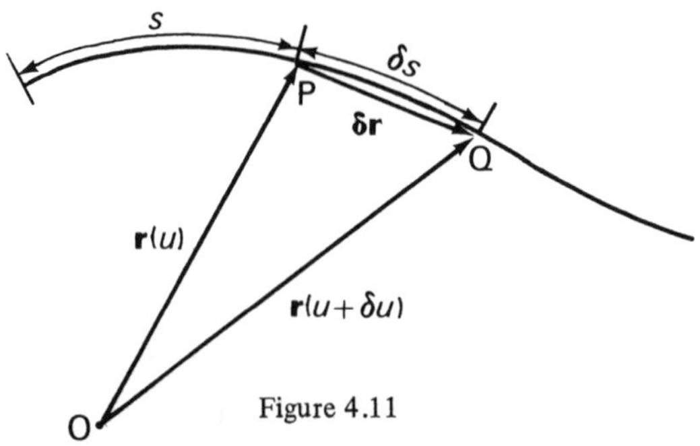

Figure 4.11

For a general parameter u, we note that $\frac{d\mathbf{r}}{du}$ is proportional to $\mathbf{T}$, so that

$$\mathbf{T} = \frac{d\mathbf{r}}{du} \Big/ \left| \frac{d\mathbf{r}}{du} \right|. \tag{4.9}$$

provided that $\frac{d\mathbf{r}}{du} \neq 0$.

Moreover, since $\mathbf{T} = \frac{d\mathbf{r}}{ds} = \frac{d\mathbf{r}}{du} \Big/ \frac{ds}{du}$, then $\frac{ds}{du} = \left|\frac{d\mathbf{r}}{du}\right|$. If we denote differentiation with respect to u by a dot, then $\dot{\mathbf{r}} = \dot{s}\mathbf{T}$.

The **tangent line** to the curve $\mathbf{r} = \mathbf{r}_1(u)$ at the point $u = u_0$ is the line through $\mathbf{r}_1(u_0)$ with direction $\mathbf{T}_1(u_0)$. Its equation is therefore

$$\mathbf{r} = \mathbf{r}_1(u_0) + \lambda \mathbf{T}_1(u_0), \tag{4.10}$$

where λ is the distance of the point $\mathbf{r}$ from the point of tangency.

The parametrisation of a curve should if possible be chosen such that $\frac{d\mathbf{r}}{du} \neq 0$ at any point. This means that s should increase strictly monotonically with u so that $u' > u$ implies $s(u') > s(u)$. A simple example where this is not the case is provided by the segment

$$\mathbf{r} = [1 - f(u)]\,\mathbf{r}_0 + f(u)\mathbf{r}_1, \; 0 < u < 1,$$

in which $f(0) = 0, f(1) = 1$.

We see that

$$\mathbf{r} - \mathbf{r}_0 = f(u)\,(\mathbf{r}_1 - \mathbf{r}_0),$$

so that the point $P(\mathbf{r})$ lies on the line joining $P_0(\mathbf{r}_0)$ and $P_1(\mathbf{r}_1)$ for any value of u, whatever function $f(u)$ is taken.

Then $\frac{d\mathbf{r}}{du} = \frac{df}{du}(\mathbf{r}_1 - \mathbf{r}_0)$ may be zero if $f(u)$ has any stationary points in the range $0 \leqslant u \leqslant 1$ of the segment. For example, if $f(u)$ takes the form shown in Figure 4.12, the parametrisation of points in the range u_1 to u_4 is not unique.

In this example, the curve has a well-defined tangent, and the fault lies with the parametrisation chosen. In other cases, where the curve has a cusp at some point $u = u_0$, the derivative $\frac{d\mathbf{r}}{du}$ is either discontinuous at $u = u_0$, or is zero (see Section 5.1.2).

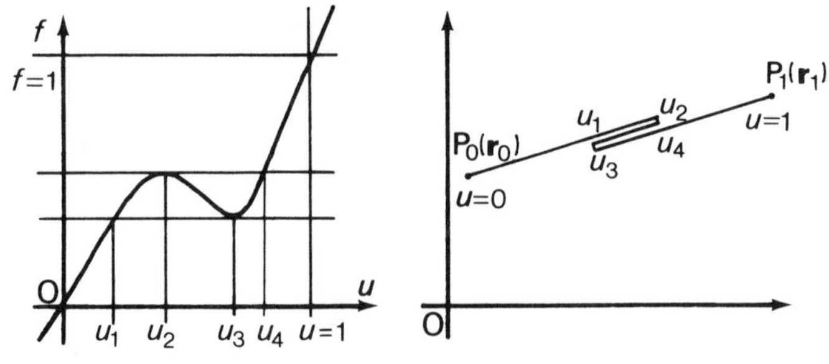

Figure 4.12

Example 1

Consider the circle in Figure 4.13, where

$$\mathbf{r} = a \cos \theta \mathbf{i} + a \sin \theta \mathbf{j}.$$

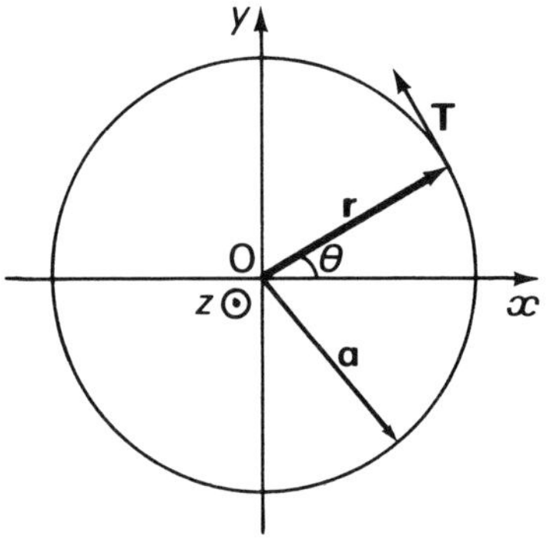

Figure 4.13

Then the unit tangent **T** is given by

$$\mathbf{T} = \frac{\mathrm{d}\mathbf{r}}{\mathrm{d}\theta} \Bigg/ \left|\frac{\mathrm{d}\mathbf{r}}{\mathrm{d}\theta}\right| = \frac{- a \sin \theta \mathbf{i} + a \cos \theta \mathbf{j}}{\sqrt{a^2 \sin^2\theta + a^2 \cos^2\theta}},$$

or
$$\mathbf{T} = - \sin \theta \mathbf{i} + \cos \theta \mathbf{j}.$$

Example 2

For the curve $\mathbf{r} = \mathbf{r}_1(u) = u^3\mathbf{a} + u^2\mathbf{b} + u\mathbf{c} + \mathbf{d}$, where **a**, **b**, **c** and **d** are constant vectors, we compute the equation of the tangent line at $u = 0$ as follows:

$$\frac{\mathrm{d}\mathbf{r}_1}{\mathrm{d}u} = 3u^2\mathbf{a} + 2u\mathbf{b} + \mathbf{c},$$

so that $\dfrac{d\mathbf{r}_1}{du} = \mathbf{c}$ at $u = 0$, so that the unit vector $\mathbf{T}_1$ to the curve is

$$\mathbf{T}_1(0) = \frac{\mathbf{c}}{|\mathbf{c}|} .$$

Also $\mathbf{r}_1(0) = \mathbf{d}$, and hence the tangent line at $u = 0$ is given by

$$\mathbf{r} = \mathbf{d} + \frac{\lambda\mathbf{c}}{|\mathbf{c}|}.$$

4.2.4 The principal normal and binormal to a curve

For a two-dimensional curve, a well-defined normal exists. However, for a three-dimensional curve any vector perpendicular to the tangent vector $\mathbf{T}$ is a normal vector.

In particular, because $\mathbf{T}$ is a unit vector we see from equation (4.7) that the vector $\dot{\mathbf{T}} = (d\mathbf{T}/du)$ is normal to $\mathbf{T}$. The unit vector $\mathbf{N}$ in the direction of $\dot{\mathbf{T}}$ is known as the **principal normal** vector.

When the parameter is the arc length s, we write

$$\frac{d\mathbf{T}}{ds} = \kappa\mathbf{N},$$

where κ is known as the **curvature** of the curve. By convention $\kappa > 0$, the sense of $\dot{\mathbf{T}}$ being defined by that of the vector $\mathbf{N}$. Thus

$$\dot{\mathbf{T}} = \frac{d\mathbf{T}}{du} = \dot{s}\kappa\mathbf{N}.$$

The vector product $\mathbf{T} \times \mathbf{N}$ defines a third unit vector perpendicular to $\mathbf{T}$ and $\mathbf{N}$, known as the **binormal vector B**.

The three vectors $\mathbf{T}$, $\mathbf{N}$, and $\mathbf{B}$ form a right-handed set of mutually orthogonal unit vectors, so that $\mathbf{B} = \mathbf{T} \times \mathbf{N}$, $\mathbf{T} = \mathbf{N} \times \mathbf{B}$ and $\mathbf{N} = \mathbf{B} \times \mathbf{T}$. The planes through a given point on the curve which contain the vectors $\mathbf{T}$ and $\mathbf{N}$, $\mathbf{N}$ and $\mathbf{B}$, and $\mathbf{B}$ and $\mathbf{T}$ respectively are known as the **osculating plane**, the **normal plane** and the **rectifying plane** (see Figure 4.14).

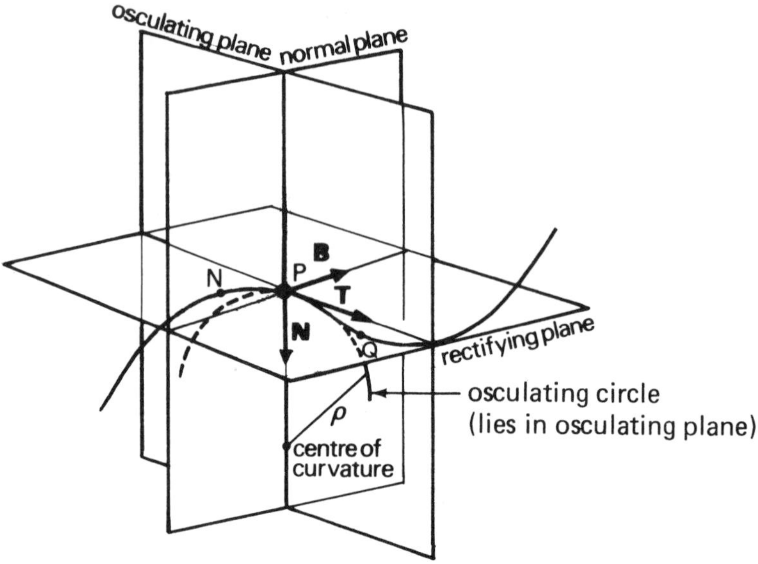

Figure 4.14

The **osculating circle** of the curve at P is the limit as $\delta u \to 0$ of the circle drawn through the points N, P and Q on the curve which have parameters $u - \delta u$, u and $u + \delta u$. We may show that this circle lies on the osculating plane by noting that the plane of the circle contains both $\vec{PQ}$ and $\vec{PN}$. Thus the normal to the plane of the circle has the direction of the vector product of $\vec{PQ}$ and $\vec{PN}$ (see Figure 4.15). Now

$$\begin{aligned}\vec{PQ} \times \vec{PN} &= [\mathbf{r}(u + \delta u) - \mathbf{r}(u)] \times [\mathbf{r}(u - \delta u) - \mathbf{r}(u)] \\ &= [\delta u\, \dot{\mathbf{r}}(u) + \frac{\delta u^2}{2}\ddot{\mathbf{r}}(u)] \times [-\delta u\, \dot{\mathbf{r}}(u) + \frac{\delta u^2}{2}\ddot{\mathbf{r}}(u)] + O(\delta u^4) \\ &= \delta u^3 [\dot{\mathbf{r}}(u) \times \ddot{\mathbf{r}}(u)] + O(\delta u^4),\end{aligned}$$

by the use of Taylor's Theorem. Thus the normal to the plane of the osculating circle has the direction of $\dot{\mathbf{r}} \times \ddot{\mathbf{r}}$. But $\dot{\mathbf{r}} = \dot{s}\mathbf{T}$, so that $\ddot{\mathbf{r}} = \ddot{s}\mathbf{T} + \dot{s}\dot{\mathbf{T}}$. Hence

$$\dot{\mathbf{r}} \times \ddot{\mathbf{r}} = \dot{s}^2\mathbf{T} \times \dot{\mathbf{T}} = \dot{s}^3\kappa\mathbf{T} \times \mathbf{N} = \dot{s}^3\kappa\mathbf{B},$$

and the vector **B** is the normal to the osculating plane.

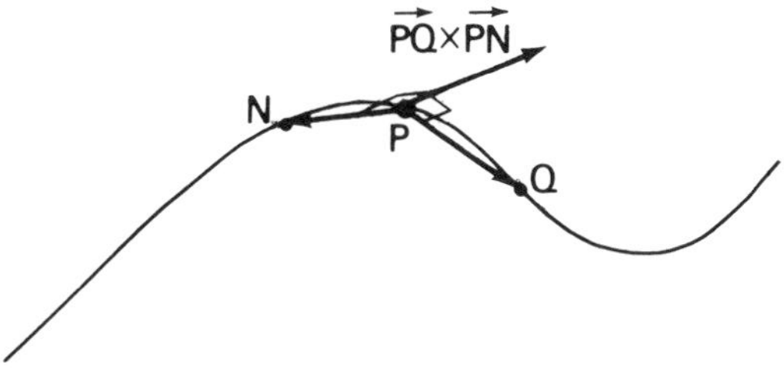

Figure 4.15

The radius of the osculating circle may be derived from the equation $\rho = |\mathbf{a}|\,|\mathbf{b}|\,|\mathbf{a}-\mathbf{b}|\,/2|\mathbf{a}\times\mathbf{b}|$ given in Example 2 of Section 3.4.7. Putting in the appropriate expressions for $\overrightarrow{PQ}$ and $\overrightarrow{PN}$ in place of **a** and **b** and taking the limit as $\delta u \to 0$, we obtain

$$\rho = \frac{\dot{s}^3}{|\dot{\mathbf{r}} \times \ddot{\mathbf{r}}|} = \frac{\dot{s}^3}{\dot{s}^3\kappa} = \frac{1}{\kappa}.$$

Then the curvature κ is the reciprocal of the radius ρ of the osculating circle; ρ is referred to as the **radius of curvature** of the curve at P.

Example 1

For the circle in Example 1 of Section 4.2.3 we may calculate the curvature and the unit normal as follows. As previously,

$$\mathbf{r} = a\cos\theta\mathbf{i} + a\sin\theta\,\mathbf{j},$$

and

$$\mathbf{T} = -\sin\theta\mathbf{i} + \cos\theta\mathbf{j}.$$

Thus

$$\frac{d\mathbf{T}}{d\theta} = -\cos\theta\mathbf{i} - \sin\theta\mathbf{j},$$

and

$$\kappa\mathbf{N} = \frac{d\mathbf{T}}{ds} = \frac{d\mathbf{T}}{d\theta}\frac{d\theta}{ds} = \frac{1}{a}\frac{d\mathbf{T}}{d\theta} = -\frac{\cos\theta}{a}\mathbf{i} - \frac{\sin\theta}{a}\mathbf{j}.$$

Now

$$\kappa = \left|\frac{d\mathbf{T}}{ds}\right| = \frac{1}{a} = \frac{1}{\text{radius of curvature}},$$

and

$$\mathbf{N} = -\cos\theta\mathbf{i} - \sin\theta\mathbf{j}.$$

Example 2

In the case of the helix

$$\mathbf{r} = \mathbf{r}_1(\theta) = a\cos\theta\,\mathbf{i} + a\sin\theta\,\mathbf{j} + b\theta\,\mathbf{k},$$

the unit tangent vector is given by

$$\mathbf{T}_1 = \frac{d\mathbf{r}_1}{d\theta}\Big/\left|\frac{d\mathbf{r}_1}{d\theta}\right| = \frac{-a\sin\theta\,\mathbf{i} + a\cos\theta\,\mathbf{j} + b\mathbf{k}}{\sqrt{a^2 + b^2}}.$$

The equation of the normal plane at the point with parameter θ is given by

$$\mathbf{r}.\mathbf{T}_1(\theta) = \mathbf{r}_1(\theta).\mathbf{T}_1(\theta) \tag{4.11}$$

or $-xa\sin\theta + ya\cos\theta + zb =$

$$-a^2\sin\theta\cos\theta + a^2\sin\theta\cos\theta + b^2\theta\,.$$

Hence $$xa\sin\theta - ya\cos\theta - zb + b^2\theta = 0$$

is the equation of the normal plane at parameter value θ.

4.2.5 The torsion of a curve; Frenet-Serret formulae

In the case of a plane curve, the osculating plane is the plane of the curve, and the binormal vector **B** is fixed. Thus $\dfrac{d\mathbf{B}}{ds} = \mathbf{0}$ for a plane curve.

On the other hand, when the curve is not plane the vector **B** is no longer constant, and an understanding of the twisted nature of the curve requires the evaluation of $\dfrac{d\mathbf{B}}{ds}$.

We now note that $\mathbf{B}.\mathbf{T} = 0$, so that

$$\frac{d\mathbf{B}}{ds}\,.\,\mathbf{T} + \mathbf{B}\,.\frac{d\mathbf{T}}{ds} = 0,$$

and hence $\dfrac{d\mathbf{B}}{ds}.\mathbf{T} = 0$ since $\mathbf{B}.\dfrac{d\mathbf{T}}{ds} = \kappa\mathbf{B}.\mathbf{N} = 0$. Moreover, because **B** is a unit vector, $\dfrac{d\mathbf{B}}{ds}\,.\,\mathbf{B} = 0$. Thus $\dfrac{d\mathbf{B}}{ds}$ is perpendicular to **T** and **B** and is therefore in the direction of **N**. By convention, we write

$$\frac{d\mathbf{B}}{ds} = -\tau\mathbf{N},$$

where τ is called the **torsion** of the curve. Now

$$\frac{d\mathbf{N}}{ds} = \frac{d}{ds}(\mathbf{B} \times \mathbf{T}) = \frac{d\mathbf{B}}{ds} \times \mathbf{T} + \mathbf{B} \times \frac{d\mathbf{T}}{ds}$$

$$= \frac{d\mathbf{B}}{ds} \times \mathbf{T} - \kappa\mathbf{T},$$

since $\frac{d\mathbf{T}}{ds} = \kappa\mathbf{N}$. Thus

$$\frac{d\mathbf{N}}{ds} = \tau\mathbf{B} - \kappa\mathbf{T},$$

so that the torsion of a curve is positive when the curve normal turns out of the osculating plane into the positive direction of **B**.

We have now derived all the principal equations of the differential geometry of space curves. These are collectively known as the **Frenet-Serret** formulae, and we restate them below:

$$\frac{d\mathbf{r}}{ds} = \mathbf{T}, \qquad (4.12)$$

$$\frac{d\mathbf{T}}{ds} = \kappa\mathbf{N}, \qquad (4.13)$$

$$\frac{d\mathbf{N}}{ds} = \tau\mathbf{B} - \kappa\mathbf{T}, \qquad (4.14)$$

$$\frac{d\mathbf{B}}{ds} = -\tau\mathbf{N}. \qquad (4.15)$$

When dealing with a curve $\mathbf{r}(u)$ it is usually more convenient to work in terms of derivatives with respect to u rather than s, however, in which case the equivalent equations are

$$\mathbf{T} = \dot{\mathbf{r}}/\dot{s}, \qquad (4.16)$$

$$\kappa\mathbf{B} = (\dot{\mathbf{r}} \times \ddot{\mathbf{r}})/\dot{s}^3, \qquad (4.17)$$

$$\mathbf{N} = \mathbf{B} \times \mathbf{T}, \qquad (4.18)$$

$$\tau = \dot{\mathbf{r}}.(\ddot{\mathbf{r}} \times \dddot{\mathbf{r}})/\dot{s}^6\kappa^2, \tag{4.19}$$

where
$$\dot{s} = |\dot{\mathbf{r}}|. \tag{4.20}$$

All these have been derived earlier with the exception of equation (4.19), whose derivation follows that given in Section 4.2.4 of equation (4.17).

4.2.6 Examples

Example 1

For the circle $\mathbf{r} = a\cos\theta\,\mathbf{i} + a\sin\theta\,\mathbf{j}$,

$$\mathbf{B} = \mathbf{T} \times \mathbf{N} = \begin{vmatrix} \mathbf{i} & \mathbf{j} & \mathbf{k} \\ -\sin\theta & \cos\theta & 0 \\ -\cos\theta & -\sin\theta & 0 \end{vmatrix} = \mathbf{k}\ ,$$

using the results of the examples in Section 4.2.3 and 4.2.4. Then $\dfrac{d\mathbf{B}}{ds} = \mathbf{0}$, which is to be expected since there is no torsion in a plane curve.

Example 2

For the helix $\mathbf{r} = a\cos\theta\mathbf{i} + a\sin\theta\mathbf{j} + b\theta\mathbf{k}$,

$$\frac{d\mathbf{r}}{d\theta} = -a\sin\theta\mathbf{i} + a\cos\theta\mathbf{j} + b\mathbf{k},$$

so that
$$\frac{ds}{d\theta} = \left|\frac{d\mathbf{r}}{d\theta}\right| = \sqrt{a^2 + b^2}$$

and
$$\mathbf{T} = (-a\sin\theta\mathbf{i} + a\cos\theta\mathbf{j} + b\mathbf{k})/\sqrt{a^2 + b^2}.$$

Thus
$$\frac{d\mathbf{T}}{ds} = \frac{1}{\sqrt{a^2 + b^2}}\frac{d\mathbf{T}}{d\theta} = \frac{-a\cos\theta\mathbf{i} - a\sin\theta\mathbf{j}}{(a^2 + b^2)} = \kappa\mathbf{N},$$

where
$$\kappa = \left|\frac{d\mathbf{T}}{ds}\right| = \frac{a}{a^2 + b^2},$$

and
$$\mathbf{N} = -\cos\theta\mathbf{i} - \sin\theta\mathbf{j}.$$

Then $$\mathbf{B} = \mathbf{T} \times \mathbf{N} = \frac{1}{\sqrt{a^2 + b^2}} \begin{vmatrix} \mathbf{i} & \mathbf{j} & \mathbf{k} \\ -a \sin\theta & a\cos\theta & b \\ -\cos\theta & -\sin\theta & 0 \end{vmatrix}$$

$$= \frac{b \sin\theta\mathbf{i} - b\cos\theta\mathbf{j} + a\mathbf{k}}{\sqrt{a^2 + b^2}}.$$

Finally, $$\frac{d\mathbf{B}}{ds} = \frac{1}{\sqrt{a^2 + b^2}} \frac{d\mathbf{B}}{d\theta} = \frac{b\cos\theta\mathbf{i} + b\sin\theta\mathbf{j}}{(a^2 + b^2)}.$$

Since $\dfrac{d\mathbf{B}}{ds} = -\tau\mathbf{N}$, we see that the torsion of the helix, which is a measure of its departure from a plane curve, is given by $\dfrac{b}{a^2 + b^2}$.

4.2.7 Differential Geometry of Surfaces

In order to calculate cutter offsets for 3D numerical control programming, for example, we need to determine the normal to the part surface being machined. Because the machining of a surface $\mathbf{r} = \mathbf{r}(u,v)$ is often performed by following the parametric curves u = constant and v = constant, the tangents to these curves are also of interest in some applications.

The tangent vector to a parametric curve $\mathbf{r} = \mathbf{r}(u,v_0)$, where v_0 is a constant, is a multiple of the vector $\dfrac{\partial\mathbf{r}}{\partial u}$. Similarly, the tangent vector to the curve $\mathbf{r} = \mathbf{r}(u_0,v)$ is a multiple of $\dfrac{\partial\mathbf{r}}{\partial v}$. The tangent plane at the intersection of these curves at $\mathbf{r}(u_0,v_0)$ contains these two tangent vectors, so that the normal to the surface is a multiple of their vector product (see Figure 4.16). The unit normal vector $\mathbf{n}$ is then given by

$$\mathbf{n} = \pm\left(\frac{\partial\mathbf{r}}{\partial u} \times \frac{\partial\mathbf{r}}{\partial v}\right) \Bigg/ \left|\frac{\partial\mathbf{r}}{\partial u} \times \frac{\partial\mathbf{r}}{\partial v}\right|, \tag{4.21}$$

where the derivatives are evaluated at $u = u_0$, $v = v_0$. The sense of $\mathbf{n}$ must be chosen to suit the application.

The exceptional points where the partial derivatives do not exist, or where $\dfrac{\partial\mathbf{r}}{\partial u} \times \dfrac{\partial\mathbf{r}}{\partial v} = \mathbf{0}$, correspond either to singularities of the parametrisation or to ridges or cusps in the surface.

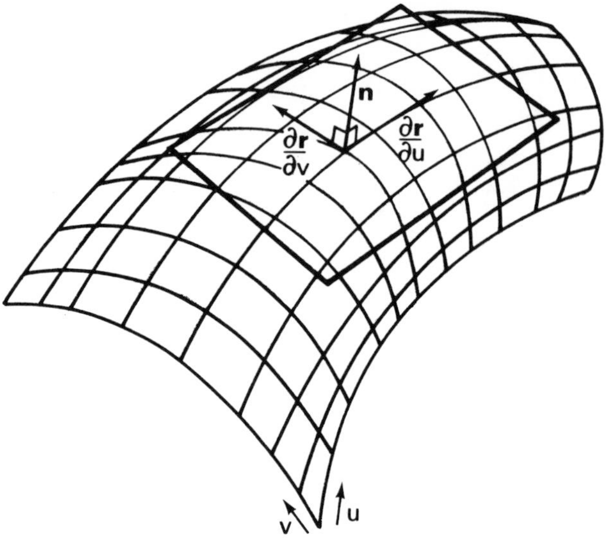

Figure 4.16

Example 1

On the paraboloid of revolution given by

$$\mathbf{r} = \mathbf{r}(t,\phi) = 2at\cos\phi\,\mathbf{i} + 2at\sin\phi\,\mathbf{j} + at^2\,\mathbf{k}$$

we have
$$\frac{\partial \mathbf{r}}{\partial t} \times \frac{\partial \mathbf{r}}{\partial \phi} = \begin{vmatrix} \mathbf{i} & \mathbf{j} & \mathbf{k} \\ 2a\cos\phi & 2a\sin\phi & 2at \\ -2at\sin\phi & 2at\cos\phi & 0 \end{vmatrix},$$

$$= -4a^2t\cos\phi\mathbf{i} - 4a^2t\sin\phi\mathbf{j} + 4a^2t\mathbf{k}.$$

Then
$$\left|\frac{\partial \mathbf{r}}{\partial t} \times \frac{\partial \mathbf{r}}{\partial \phi}\right|^2 = 16a^4t^2(1 + t^2),$$

and hence
$$\mathbf{n} = \frac{1}{\sqrt{1 + t^2}}(-t\cos\phi, -t\sin\phi, 1).$$

Example 2

Application to the analytic derivation of the cutter path for a spherical end mill.

Let us assume that we are cutting along the curve $u = u_0$ on the surface $\mathbf{r} = \mathbf{r}(u,v)$ using a spherical end mill of radius R (see Figure 4.17). Then the tip of the cutter follows the path

$$\mathbf{r} = \mathbf{r}_c(u_0,v) = \underset{\text{contact point}}{\mathbf{r}(u_0,v)} + \underset{\text{cutter offset}}{R(\mathbf{n} - \mathbf{u})},$$

where $\mathbf{n}$ is the unit normal at $u = u_0$ and $\mathbf{u}$ is a unit vector along the cutter axis. Note that the sense of the surface normal must be chosen so that $\mathbf{n}$ points *out* of the part being machined.

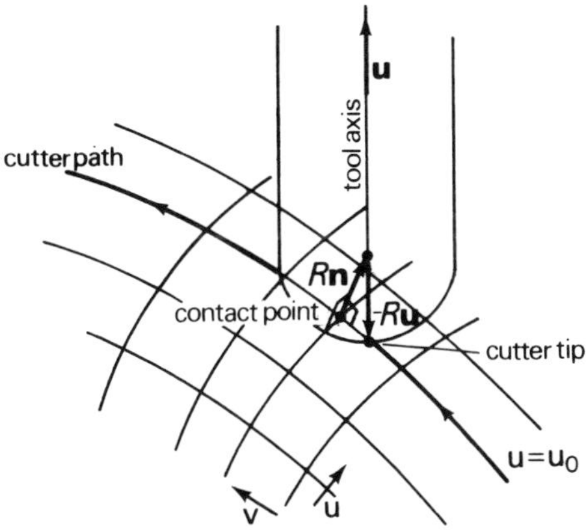

Figure 4.17

4.2.8 Metrical properties of surfaces

We now consider the metrical properties of curves lying on a parametric surface, and the area of a region of the surface.

A curve on the parametric surface $\mathbf{r} = \mathbf{r}(u,v)$ may be represented parametrically by the equations $u = u(t)$, $v = v(t)$, which may be summarised as

$$\mathbf{u} = \mathbf{u}(t),$$

where $\mathbf{u} = [u(t), v(t)]^T$. Whereas $\mathbf{r}(u,v)$ denotes a general point on the surface, we will use $\mathbf{r}(t)$ to denote a point on the curve.

A tangent vector to this curve is given by $\dot{\mathbf{r}}$, which we may expand by the chain rule to give

$$\dot{\mathbf{r}} = \frac{\partial \mathbf{r}}{\partial u}\dot{u} + \frac{\partial \mathbf{r}}{\partial v}\dot{v} = \mathbf{A}\dot{\mathbf{u}},$$

where

$$\mathbf{A} = \begin{bmatrix} \dfrac{\partial x}{\partial u} & \dfrac{\partial x}{\partial v} \\[2ex] \dfrac{\partial y}{\partial u} & \dfrac{\partial y}{\partial v} \\[2ex] \dfrac{\partial z}{\partial u} & \dfrac{\partial z}{\partial v} \end{bmatrix} = \left[\frac{\partial \mathbf{r}}{\partial u} \,\middle|\, \frac{\partial \mathbf{r}}{\partial v} \right].$$

The length of this tangent vector is then given by

$$\dot{s}^2 = |\dot{\mathbf{r}}|^2 = \dot{\mathbf{r}}^T \dot{\mathbf{r}} = \dot{\mathbf{u}}^T \mathbf{A}^T \mathbf{A} \dot{\mathbf{u}} = \dot{\mathbf{u}}^T \mathbf{G} \dot{\mathbf{u}},$$

where

$$\mathbf{G} = \mathbf{A}^T \mathbf{A} = \begin{bmatrix} \dfrac{\partial \mathbf{r}}{\partial u} \cdot \dfrac{\partial \mathbf{r}}{\partial u} & \dfrac{\partial \mathbf{r}}{\partial u} \cdot \dfrac{\partial \mathbf{r}}{\partial v} \\[2ex] \dfrac{\partial \mathbf{r}}{\partial v} \cdot \dfrac{\partial \mathbf{r}}{\partial u} & \dfrac{\partial \mathbf{r}}{\partial v} \cdot \dfrac{\partial \mathbf{r}}{\partial v} \end{bmatrix}. \tag{4.22}$$

The matrix **G** is the **first fundamental matrix** of the surface. Its importance is seen in the following metrical formulae.

The *unit tangent vector* along the curve $\mathbf{u} = \mathbf{u}(t)$ is given by

$$\mathbf{T} = \dot{\mathbf{r}} / |\dot{\mathbf{r}}| = \mathbf{A}\dot{\mathbf{u}} / (\dot{\mathbf{u}}^T \mathbf{G} \dot{\mathbf{u}})^{1/2}. \tag{4.23}$$

The *length* of the curve segment $\mathbf{u} = \mathbf{u}(t)$, $t_0 < t < t_1$ is given by

$$s = \int_{t_0}^{t_1} \frac{ds}{dt} dt = \int_{t_0}^{t_1} |\dot{\mathbf{r}}| \, dt = \int_{t_0}^{t_1} (\dot{\mathbf{u}}^T \mathbf{G} \dot{\mathbf{u}})^{1/2} \, dt. \tag{4.24}$$

If two curves $\mathbf{u} = \mathbf{u}_1(t)$ and $\mathbf{u} = \mathbf{u}_2(t)$ lie on the surface and intersect at angle θ, then

$$\mathbf{T}_1 \cdot \mathbf{T}_2 = \frac{\dot{\mathbf{u}}_1^T \mathbf{A}^T \mathbf{A} \dot{\mathbf{u}}_2}{(\dot{\mathbf{u}}_1^T \mathbf{G} \dot{\mathbf{u}}_1)^{1/2} (\dot{\mathbf{u}}_2^T \mathbf{G} \dot{\mathbf{u}}_2)^{1/2}},$$

so that the *angle of intersection* θ between the two curves is given by

so that

$$\cos\theta = \frac{\dot{\mathbf{u}}_1^T \mathbf{G} \dot{\mathbf{u}}_2}{(\dot{\mathbf{u}}_1^T \mathbf{G} \dot{\mathbf{u}}_1)^{1/2} (\dot{\mathbf{u}}_2^T \mathbf{G} \dot{\mathbf{u}}_2)^{1/2}}. \tag{4.25}$$

Finally, we consider the area enclosed between the parametric curves $u = u_0, u = u_0 + \delta u, v = v_0$ and $v = v_0 + \delta v$ (see Figure 4.18). Approximating this curved surface by the plane parallelogram shown, we have the approximate area

$$\delta S \simeq \left| \frac{\partial \mathbf{r}}{\partial u} \times \frac{\partial \mathbf{r}}{\partial v} \right| \delta u \, \delta v.$$

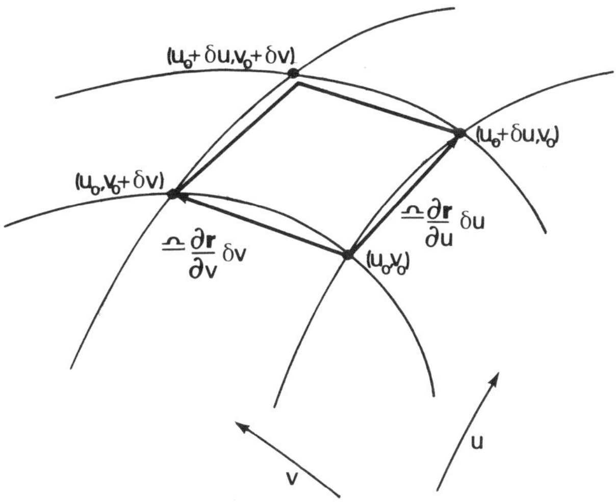

Figure 4.18

We now note that

$$\left| \frac{\partial \mathbf{r}}{\partial u} \times \frac{\partial \mathbf{r}}{\partial v} \right|^2 = \left| \frac{\partial \mathbf{r}}{\partial u} \right|^2 \left| \frac{\partial \mathbf{r}}{\partial v} \right|^2 - \left(\frac{\partial \mathbf{r}}{\partial u} \cdot \frac{\partial \mathbf{r}}{\partial v} \right)^2 = g_{11}g_{22} - g_{12}g_{21} = |\mathbf{G}|, \tag{4.26}$$

where the g_{ij} are elements of the first fundamental matrix **G**.
Thus the area of a bounded region of the surface corresponding to a region R in the u-v plane may be obtained from the integral

$$S = \iint_R (|\mathbf{G}|)^{1/2} \, du \, dv. \tag{4.27}$$

We shall now derive a condition for the tangent vector **T** to be well-defined on all curves $\mathbf{u} = \mathbf{u}(t)$, in terms of tangents to the curves of constant u and constant v. From equation 4.23 we have the requirement that $\dot{\mathbf{u}}^T \mathbf{G} \, \dot{\mathbf{u}} > \mathbf{0}$ for every point on the surface. It can be shown that this will be so provided $\dot{\mathbf{u}} \neq \mathbf{0}$, $g_{11} > 0$ and $|\mathbf{G}| > 0$. Assuming that $\dfrac{\partial \mathbf{r}}{\partial u}$ is well-defined everywhere, we see from equation (4.22) that $g_{11} > 0$ automatically, and since we have seen that $|\mathbf{G}| = \left|\dfrac{\partial \mathbf{r}}{\partial u} \times \dfrac{\partial \mathbf{r}}{\partial v}\right|^2$ it follows that the tangent to a general curve $\mathbf{u} = \mathbf{u}(t)$ is well-defined at all points for which

$$\dot{\mathbf{u}} \neq 0$$

and

$$\frac{\partial \mathbf{r}}{\partial u} \times \frac{\partial \mathbf{r}}{\partial v} \neq 0\,. \tag{4.28}$$

At points where the second condition holds but not the first, the tangent plane and normal to the surface are well-defined (see Section 4.2.7) but the curve $\mathbf{u} = \mathbf{u}(t)$ contains a kink so that $\mathbf{T}$ does not exist there. Where the first condition holds but not the second the surface has no well-defined tangent plane, either because it contains a sharp ridge or because the lines of constant u and constant v are parallel there. Then the failure of the condition (4.28) may stem either from an inherent geometrical property of the surface or from an unfortunate choice of parametrisation.

4.2.9 Curvature of a surface

For a general space curve $\mathbf{r} = \mathbf{r}(t)$, the curvature κ is obtained by differentiating $\mathbf{r}(t)$ twice. Thus $\dot{\mathbf{r}} = \dot{s}\mathbf{T}$ and $\ddot{\mathbf{r}} = \ddot{s}\mathbf{T} + \dot{s}^2\kappa\mathbf{N}$, as shown in Sections 4.2.3 and 4.2.4.

For a curve $\mathbf{u} = \mathbf{u}(t)$ on the surface $\mathbf{r} = \mathbf{r}(u,v)$,

$$\ddot{\mathbf{r}} = \ddot{s}\mathbf{T} + \dot{s}^2\kappa\mathbf{N} = \frac{\partial^2 \mathbf{r}}{\partial u^2}\dot{u}^2 + 2\frac{\partial^2 \mathbf{r}}{\partial u \partial v}\dot{u}\dot{v} + \frac{\partial^2 \mathbf{r}}{\partial v^2}\dot{v}^2 + \frac{\partial \mathbf{r}}{\partial u}\ddot{u} + \frac{\partial \mathbf{r}}{\partial v}\ddot{v}\,.$$

This result is obtained by a further application of the chain rule to the expression for $\dot{\mathbf{r}}$ given in Section 4.2.8. The component of this vector in the direction of the surface normal $\mathbf{n}$ is given by

$$\ddot{\mathbf{r}}.\mathbf{n} = \dot{s}^2\kappa\mathbf{N}.\mathbf{n} = \mathbf{n}.\frac{\partial^2 \mathbf{r}}{\partial u^2}\dot{u}^2 + 2\mathbf{n}.\frac{\partial^2 \mathbf{r}}{\partial u \partial v}\dot{u}\dot{v} + \mathbf{n}.\frac{\partial^2 \mathbf{r}}{\partial v^2}\dot{v}^2,$$

since $\mathbf{n}$ is perpendicular to $\mathbf{T}$, $\dfrac{\partial \mathbf{r}}{\partial u}$ and $\dfrac{\partial \mathbf{r}}{\partial v}$. In matrix notation the curvature is given by

$$\dot{s}^2\kappa\mathbf{N}.\mathbf{n} = \dot{\mathbf{u}}^T\mathbf{D}\dot{\mathbf{u}}, \tag{4.29}$$

where

$$\mathbf{D} = \begin{bmatrix} \mathbf{n} \cdot \dfrac{\partial^2 \mathbf{r}}{\partial u^2} & \mathbf{n} \cdot \dfrac{\partial^2 \mathbf{r}}{\partial u \partial v} \\ \mathbf{n} \cdot \dfrac{\partial^2 \mathbf{r}}{\partial v \partial u} & \mathbf{n} \cdot \dfrac{\partial^2 \mathbf{r}}{\partial v^2} \end{bmatrix}.$$

The matrix **D** is the **second fundamental matrix** of the surface. Since $\dfrac{\partial^2 \mathbf{r}}{\partial u \partial v} = \dfrac{\partial^2 \mathbf{r}}{\partial v \partial u}$ for the functions we shall be concerned with, the matrix **D** will be symmetric.

The **normal curvature** κ_n of the surface in the direction $\mathbf{A}\dot{\mathbf{u}}$ is the curvature of the intersection curve between the surface and the plane containing the surface normal **n** and the tangent vector $\dot{\mathbf{r}} = \mathbf{A}\dot{\mathbf{u}}$ (see Figure 4.19). For such a curve **n** is parallel to **N**, so that we may define the normal curvature κ_n by

$$\kappa_n = \frac{\dot{\mathbf{u}}^T \mathbf{D} \dot{\mathbf{u}}}{\dot{s}^2} = \frac{\dot{\mathbf{u}}^T \mathbf{D} \dot{\mathbf{u}}}{\dot{\mathbf{u}}^T \mathbf{G} \dot{\mathbf{u}}}. \tag{4.30}$$

With this definition, the curvature κ_n is positive when the curve is turning towards the positive direction of the surface normal.

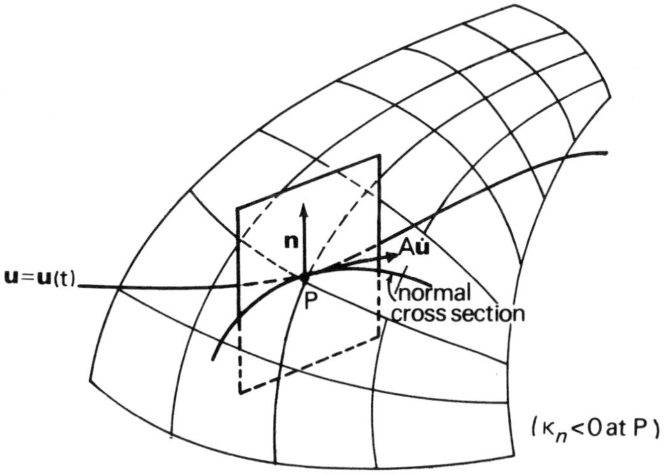

Figure 4.19

The direction in which κ_n takes its maximum and minimum values with respect to $\dot{\mathbf{u}}$ are called the **principal directions** of normal curvature, and we may deduce from equation (4.30) that these occur when

$$(\mathbf{D} - \kappa_n \mathbf{G})\dot{\mathbf{u}} = \mathbf{0},$$

or

$$(d_{11} - \kappa_n g_{11})\dot{u} + (d_{12} - \kappa_n g_{12})\dot{v} = 0$$

and

$$(d_{21} - \kappa_n g_{21})\dot{u} + (d_{22} - \kappa_n g_{22})\dot{v} = 0. \tag{4.31}$$

Eliminating $\dot{u}$ and $\dot{v}$, we obtain

$$|\mathbf{G}|\,\kappa_n^2 - (g_{11}d_{22} + d_{11}g_{22} - 2g_{12}d_{12})\kappa_n + |\mathbf{D}| = 0, \tag{4.32}$$

from which the maximum and minimum (or principal) curvatures may be obtained and substituted into (4.31) to obtain the ratio $\dot{u}:\dot{v}$ for the principal directions.

The product of the principal curvatures is known as the **Gaussian curvature** K of the surface. By the properties of the roots of quadratic equations, we see from (4.32) that

$$K = \frac{|\mathbf{D}|}{|\mathbf{G}|}.$$

The principal directions of curvature may alternatively be obtained by eliminating κ_n in (4.31). The result is the quadratic equation

$$(d_{11}g_{12} - d_{12}g_{11})\dot{u}^2 + (d_{11}g_{22} - d_{22}g_{11})\dot{u}\dot{v} + (d_{12}g_{22} - d_{22}g_{12})\dot{v}^2 = 0, \tag{4.33}$$

from which the ratios $\dot{u}:\dot{v}$ for the two principal directions may be calculated. Note that every term in (4.33) involves an element of $\mathbf{D}$, so that we may use $\dfrac{\partial \mathbf{r}}{\partial u} \times \dfrac{\partial \mathbf{r}}{\partial v}$ instead of the unit normal to simplify the working. The principal curvatures may then be obtained from either of the equations (4.31).

If the principal curvatures are not equal, it can be shown that the principal directions are necessarily orthogonal. Let the curvatures and directions be $\kappa_{n1}, \kappa_{n2}, \dot{\mathbf{u}}_1$ and $\dot{\mathbf{u}}_2$. Then, since $\mathbf{D}\dot{\mathbf{u}}_2 = \kappa_{n2}\mathbf{G}\dot{\mathbf{u}}_2$, we have

$$\dot{\mathbf{u}}_1^T\mathbf{D}\dot{\mathbf{u}}_2 = \kappa_{n2}\dot{\mathbf{u}}_1^T\mathbf{G}\dot{\mathbf{u}}_2.$$

Similarly,

$$\dot{\mathbf{u}}_2^T\mathbf{D}\dot{\mathbf{u}}_1 = \kappa_{n1}\,\dot{\mathbf{u}}_2^T\,\mathbf{G}\,\dot{\mathbf{u}}_1.$$

On taking transposes of both sides of this last relation and using the symmetry of $\mathbf{D}$ and $\mathbf{G}$ we obtain

$$\dot{\mathbf{u}}_1^T\,\mathbf{D}\,\dot{\mathbf{u}}_2 = \kappa_{n1}\,\dot{\mathbf{u}}_1^T\,\mathbf{G}\,\dot{\mathbf{u}}_2.$$

Comparison of the first and third equations now shows that

$$\kappa_{n2}\dot{\mathbf{u}}_1{}^T\mathbf{G}\dot{\mathbf{u}}_2 = \kappa_{n1}\dot{\mathbf{u}}_1{}^T\mathbf{G}\dot{\mathbf{u}}_2 .$$

Since $\kappa_{n1} \neq \kappa_{n2}$, we conclude that

$$\dot{\mathbf{u}}_1{}^T\mathbf{G}\dot{\mathbf{u}}_2 = 0 .$$

By equation (4.25) it follows that the directions are orthogonal.

Points where $\kappa_{n1} = \kappa_{n2}$ are called **umbilics.** A curve whose tangent is always in a principal direction is called a **line of curvature**.

4.2.10 Parameter transformations

If the parametrisation of a surface is transformed by the equations $u' = u'(u,v)$ and $v' = v'(u,v)$, we obtain the new derivatives

$$\frac{\partial \mathbf{r}}{\partial u'} = \frac{\partial \mathbf{r}}{\partial u}\frac{\partial u}{\partial u'} + \frac{\partial \mathbf{r}}{\partial v}\frac{\partial v}{\partial u'}$$

$$\qquad (4.34)$$

and

$$\frac{\partial \mathbf{r}}{\partial v'} = \frac{\partial \mathbf{r}}{\partial u}\frac{\partial u}{\partial v'} + \frac{\partial \mathbf{r}}{\partial v}\frac{\partial v}{\partial v'} ,$$

so that

$$\mathbf{A}' = \left[\frac{\partial \mathbf{r}}{\partial u'} \,\middle|\, \frac{\partial \mathbf{r}}{\partial v'} \right] = \mathbf{AP} ,$$

where

$$\mathbf{P} = \begin{bmatrix} \dfrac{\partial u}{\partial u'} & \dfrac{\partial u}{\partial v'} \\[2ex] \dfrac{\partial v}{\partial u'} & \dfrac{\partial v}{\partial v'} \end{bmatrix}$$

is called the **Jacobian matrix** of the transformation.

It follows from (4.22) that the new fundamental matrix $\mathbf{G}'$ is given by

$$\mathbf{G}' = \mathbf{A}'^T\mathbf{A}' = \mathbf{P}^T\mathbf{A}^T\mathbf{AP} = \mathbf{P}^T\mathbf{GP}. \qquad (4.35)$$

From equation (4.35), we see by the properties of determinants that $|\mathbf{G}'| = |\mathbf{P}|^2\,|\mathbf{G}|$. Using this result, and equations (4.34), we can show that the unit surface normal $\mathbf{n}$ is invariant under the transformation, as we could expect.

The transformation of the second fundamental matrix can similarly be shown to be given by

$$\mathbf{D}' = \mathbf{P}^T\mathbf{D}\mathbf{P}, \tag{4.36}$$

by differentiating equations (4.34) and using the invariance of $\mathbf{n}$. From equations (4.35) and (4.36), it can be shown that the principal curvatures and directions are invariant under the transformation.

We conclude that the unit normal vector $\mathbf{n}$ and the principal directions and curvatures are independent of the parameters used, and are therefore geometric properties of the surface itself. They should be continuous if the surface is to be tangent and curvature continuous, as we shall see in Chapter 7, where the construction of composite surfaces is considered.

4.2.11 Envelope of a family of space curves

If the equation $\mathbf{r} = \mathbf{r}(u;\alpha)$ represents a family of space curves with curve parameter u and family parameter α, there *may* be a curve $\mathbf{r} = \mathbf{r}_e(\alpha)$ which is tangential to each member of the family. In this case, the curve $\mathbf{r} = \mathbf{r}_e(\alpha)$ is known as the **envelope** of the family, and we have parametrised it here by describing each point by the family parameter of the member which is tangential to the envelope at that point (see Figure 4.20).

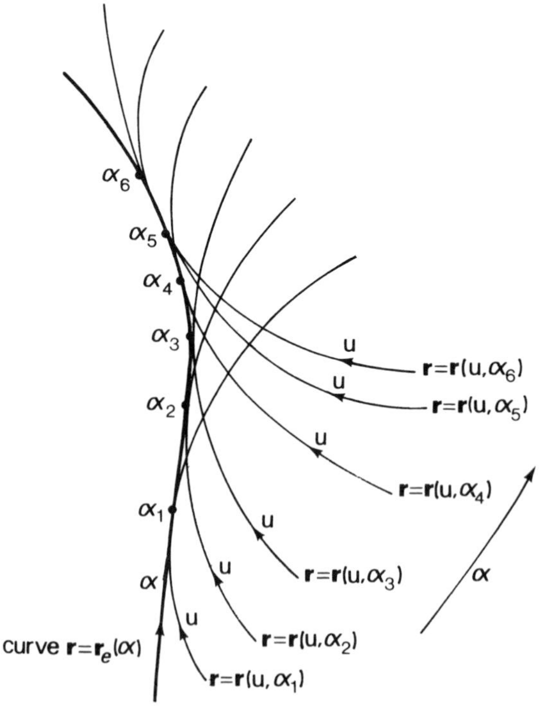

Figure 4.20

Now let the curve $\mathbf{r} = \mathbf{r}(u,\alpha)$ contact the envelope at $u = u_1(\alpha)$, so that $\mathbf{r} = \mathbf{r}(u_1(\alpha),\alpha) = \mathbf{r}_e(\alpha)$ there. The tangent to the envelope is proportional to

$$\frac{d\mathbf{r}_e}{d\alpha} = \frac{\partial \mathbf{r}}{\partial u}\frac{du_1}{d\alpha} + \frac{\partial \mathbf{r}}{\partial \alpha}.$$

However, the tangent to the curve $\mathbf{r}(u,\alpha)$ is proportional to $\frac{\partial \mathbf{r}}{\partial u}$, and this tangent must be parallel to that of the envelope at their point of contact, so that $\frac{\partial \mathbf{r}}{\partial u} \times \frac{d\mathbf{r}_e}{d\alpha} = \mathbf{0}$ there and finally

$$\frac{\partial \mathbf{r}}{\partial u} \times \frac{\partial \mathbf{r}}{\partial \alpha} = \mathbf{0}. \tag{4.37}$$

If we can solve this equation for α in terms of u or vice-versa, we may substitute for u or α in $\mathbf{r}(u,\alpha)$ to obtain the equation of the envelope.

It should be noted that not all families of curves possess an envelope. In particular, the families of parametric curves on a surface $\mathbf{r} = \mathbf{r}(u,v)$ normally satisfy the condition

$$\frac{\partial \mathbf{r}}{\partial u} \times \frac{\partial \mathbf{r}}{\partial v} \neq 0,$$

as we saw in Sections 4.2.7 and 4.2.8.

In fact a consequence of equation (4.37) is that, if we regard the family $\mathbf{r} = \mathbf{r}(u,\alpha)$ as generating a surface, then if an envelope exists, there must be no well-defined normal vector at any point on the envelope curve. This may be a feature of the surface parametrisation, as in the case of a family of plane curves, but is often the symptom of a (curved) ridge in the surface.

4.2.12 Ruled surfaces

A **ruled surface** is a surface generated by a family of straight lines, and can therefore be expressed in general by the equation

$$\mathbf{r} = \mathbf{r}(u,v) = \mathbf{r}_0(u) + v\mathbf{n}(u), \tag{4.38}$$

where $\mathbf{r}_0(u)$ is a given point on the line whose parameter value is u, and $\mathbf{n}(u)$ is its direction vector. The parameter v gives the distance of the point $\mathbf{r}(u,v)$ from $\mathbf{r}_0(u)$.

An alternative expression based on rulings joining corresponding points on two space curves $\mathbf{r} = \mathbf{r}_0(u)$ and $\mathbf{r} = \mathbf{r}_1(u)$ is given by

$$\mathbf{r} = \mathbf{r}(u,v) = (1 - v)\mathbf{r}_0(u) + v\mathbf{r}_1(u). \tag{4.39}$$

In both cases, the curves $\mathbf{r} = \mathbf{r}_0(u)$ and $\mathbf{r} = \mathbf{r}_1(u)$ are known as **directrices**, and the rulings are called **generators**.

4.2.13 Developable surfaces

A **developable surface** is one which may be unrolled onto a plane without distortion.

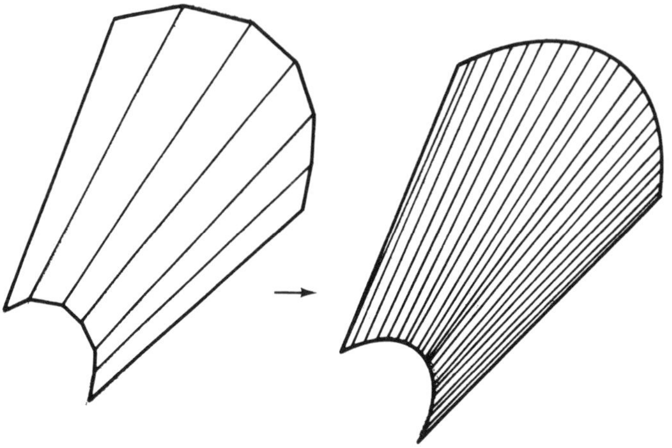

Figure 4.21

If we imagine the surface to be approximated by a plane sheet with a sequence of folds upon it (see Figure 4.21), we observe that adjacent fold lines are either parallel or intersect at some point which does not lie on the sheet. If we now increase the number of folds indefinitely we obtain a smooth developable surface. The fold lines now form a continuous family, so that the surface is ruled. Consider first the case where adjacent fold lines intersect. Taking these lines to be $\mathbf{r} = \mathbf{r}_0(u) + v\mathbf{n}(u)$ and $\mathbf{r} = \mathbf{r}_0(u + \delta u) + w\mathbf{n}(u + \delta u)$, and by using the result obtained in Section 3.4.5 Example 2, we may deduce that the lines intersect when

$$[\mathbf{r}_0(u + \delta u) - \mathbf{r}_0(u)] \cdot [\mathbf{n}(u) \times \mathbf{n}(u + \delta u)] = 0.$$

Taking the limit as $\delta u \to 0$, we obtain the condition

$$\dot{\mathbf{r}}_0 . (\mathbf{n} \times \dot{\mathbf{n}}) = 0 \tag{4.40}$$

for the ruled surface $\mathbf{r} = \mathbf{r}_0(u) + v\mathbf{n}(u)$ to be developable.

We may show by substitution in equation (4.29) that (4.40) is also the condition which ensures that $|\mathbf{D}|$, and hence the Gaussian curvature K, is zero.

Conversely, it may be shown that $K = 0$ is a sufficient condition for a surface to be developable.

By equation (4.37), we may obtain the equation of the envelope of the rulings by solving

$$[\dot{\mathbf{r}}_0(u) + v\dot{\mathbf{n}}(u)] \times \mathbf{n}(u) = \mathbf{0}. \qquad (4.41)$$

A necessary condition for the solution of this equation to exist is that $\dot{\mathbf{r}}_0$, $\mathbf{n}$ and $\dot{\mathbf{n}}$ be coplanar; equation (4.40) shows that this condition is satisfied.

On setting $\mathbf{N} = \mathbf{n} \times \dot{\mathbf{n}}$ we may solve (4.41) for v as follows, provided $\mathbf{N} \neq \mathbf{O}$:

$$v = \mathbf{N}.(\dot{\mathbf{r}}_0 \times \mathbf{n})/N^2.$$

The surface is now seen to be generated by a family of lines whose envelope is the curve

$$\mathbf{r} = \mathbf{r}_e(u) = \mathbf{r}_0(u) + \mathbf{N}.(\dot{\mathbf{r}}_0 \times \mathbf{n})\mathbf{n}(u)/N^2. \qquad (4.42)$$

Since we may satisfy (4.40) by choosing $\mathbf{n}(u)$ as the unit tangent $\mathbf{T}_0$ to the curve $\mathbf{r} = \mathbf{r}_0(u)$, it follows that the surface generated by the tangents to any space curve (known as the **convolute** of the curve) is developable. Hence the general developable may be expressed in the form

$$\mathbf{r} = \mathbf{r}_0(u) + \lambda\mathbf{T}_0(u). \qquad (4.43)$$

Equation (4.40) is also satisfied when $\dot{\mathbf{r}}_0 = \mathbf{0}$, and when $\mathbf{N} = \mathbf{0}$. In the first case, $\mathbf{r}_0$ is constant. Thus all the rulings pass through the point $\mathbf{r}_0$, so that the surface is a **cone** with its vertex at $\mathbf{r}_0$.

In the second case $\dot{\mathbf{n}} = \lambda\mathbf{n}$, but since $\mathbf{n}$ is a unit vector, $\mathbf{n}.\dot{\mathbf{n}} = 0$. It follows that $\lambda = 0$, so that $\mathbf{n}$ is a constant vector. Equation (4.40) is again satisfied for all $\mathbf{r}_0(u)$, and the surface is a **cylinder**.

Applying equation (4.40) to the ruled surface in (4.39), we obtain the condition

$$(\mathbf{r}_0 - \mathbf{r}_1) \cdot (\dot{\mathbf{r}}_0 \times \dot{\mathbf{r}}_1) = 0 \qquad (4.44)$$

for this surface to be developable. Regarding the parametrisation of $\mathbf{r} = \mathbf{r}_0(u)$ to be fixed, we may choose the parametrisation of the second curve in such a way that the condition is satisfied. The parameter u is assigned to the point on $\mathbf{r}_1$ at which a plane tangent to $\mathbf{r}_0$ at parameter value u is tangent to $\mathbf{r}_1$ (see Figure 4.22). This is the basis of the tangent plane generation of a developable surface containing two given curves.

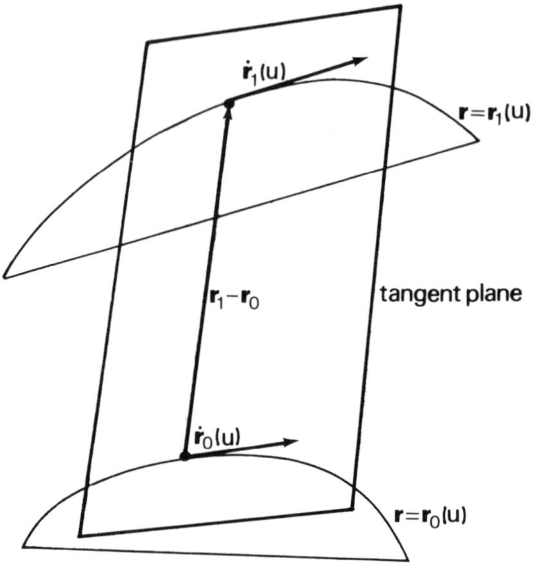

Figure 4.22

Algorithms for calculating the plane developments of these surfaces are considered in Chapter 9.

4.2.14 Arc lengths, areas and volumes for parametric curves

In order to calculate integrated quantities such as arc lengths, areas and volumes, we first approximate the curves by a sequence of chords and the surfaces by a patchwork of plane facets. Provided that these elements are sufficiently small, summation over the elements provides an adequate approximation to the integrals we are seeking.

In the case of the curve $\mathbf{r} = \mathbf{r}(u)$, the line element used is the chord $\delta\mathbf{r} = \mathbf{r}(u + \delta u) - \mathbf{r}(u)$, which we in turn approximate by $\dot{\mathbf{r}}\delta u$, where δu is small. The length of this element is approximately $\dot{s}\delta u$, so that the **arc length** L of the segment $u_0 < u < u_1$ is given by

$$L = \int_{u_0}^{u_1} \dot{s}\, \mathrm{d}u = \int_{u_0}^{u_1} |\dot{\mathbf{r}}(u)|\, \mathrm{d}u \tag{4.45}$$

where the summation over the elements has been replaced by an integration.

In the case of a plane curve whose plane contains the origin, we may calculate the area subtended at the origin by summing the triangular areas subtended by the elements $\delta\mathbf{r}$. The vector area of a triangle has been defined in Section 4.3.5. In Figure 4.23 we see that the vector area of the triangle OPQ subtended at the origin by the element PQ is approximately $\frac{1}{2}\mathbf{r}(u) \times \dot{\mathbf{r}}(u)\,\delta u$, so that the total **vector area A** subtended by the segment $u_0 < u < u_1$ is given by

$$\mathbf{A} = \tfrac{1}{2}\int_{u_0}^{u_1} \mathbf{r}(u) \times \dot{\mathbf{r}}(u)\,\mathrm{d}u. \tag{4.46}$$

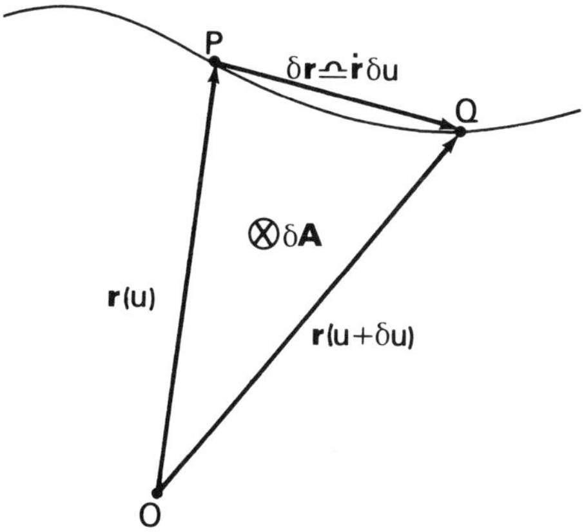

Figure 4.23

If the curve lies in the *Oxy* plane, this equation reduces to the following equation for the scalar area *A*:

$$A = \pm\tfrac{1}{2}\int_{u_0}^{u_1} (x\dot{y} - \dot{x}y)\,\mathrm{d}u. \tag{4.47}$$

The sign of this scalar area is positive if the curve is parametrised in an anti-clockwise direction in the *Oxy* plane.

Other integrals for laminae bounded by parametric curves can be calculated in a similar manner. For example, the centroid of the elementary triangle OPQ is approximately at $\frac{2}{3}\,\mathbf{r}(u)$, so that the **centroid** $\bar{\mathbf{r}}$ of the total area subtended by the segment is given by

$$\bar{\mathbf{r}} = \frac{1}{3A}\int_{u_0}^{u_1} \mathbf{r}|\mathbf{r} \times \dot{\mathbf{r}}|\,\mathrm{d}u.$$

The area integral is far simpler than the other two because no square roots are involved; even for cubic polynomial functions, the evaluation of arc length involves a numerical integration.

We now consider the integrals associated with the surface $\mathbf{r} = \mathbf{r}(u,v)$. The appropriate surface element is formed by the two triangles ABC and BCD shown in Figure 4.24. For large δu and δv, these triangles are not coplanar. However, for a small element these each have vector area $\frac{1}{2}\left(\frac{\partial \mathbf{r}}{\partial u} \times \frac{\partial \mathbf{r}}{\partial v}\right) \delta u\, \delta v$. Thus the total vector area of the element ABCD is approximately $\left(\frac{\partial \mathbf{r}}{\partial u} \times \frac{\partial \mathbf{r}}{\partial v}\right) \delta u\, \delta v$.

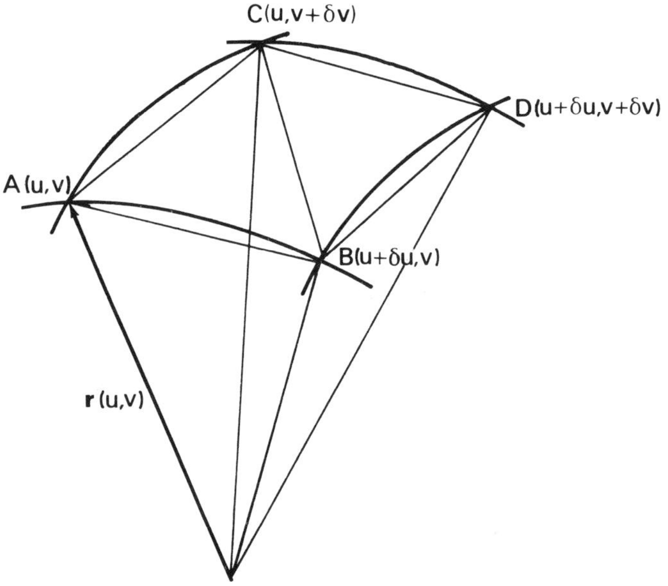

Figure 4.24

This leads us to the equation (4.27) in Section 4.2.8 for the surface area of a surface patch. Thus the equation for the surface area involves a square root and a numerical integration for all but the simplest surfaces.

Since the volume of a pyramid is equal to one third of the base area times the perpendicular height, we see that the volume subtended by ABCD at the origin O is $\frac{1}{3}\,\mathbf{r}.\left(\frac{\partial r}{\partial u} \times \frac{\partial r}{\partial v}\right)\delta u\, \delta v$, so that the total volume V subtended by a region R of the u,v plane is approximately

$$V = \frac{1}{3}\iint_R \mathbf{r}.\left(\frac{\partial r}{\partial u} \times \frac{\partial r}{\partial v}\right) \mathrm{d}u\, \mathrm{d}v. \tag{4.48}$$

4.3 IMPLICIT EQUATIONS OF SURFACES AND CURVES IN THREE DIMENSIONS

4.3.1 Implicit equations of a surface

The equation $f(x,y,z) = 0$ in general describes a surface in three-dimensional space in much the same way as the equation $f(x,y) = 0$ describes a plane curve.

As we shall see in the next section, continuity of $f(x,y,z)$ is necessary if the surface is to have a well-defined normal at (x,y,z). In this case we can also use the value of f as a measure of proximity to the surface.

Simple examples of such surface equations are the plane

$$ax + by + cz + d = 0$$

(see Section 3.4.3) and the sphere

$$x^2 + y^2 + z^2 - a^2 = 0.$$

The general quadratic equation

$$ax^2 + by^2 + cz^2 + 2hxy + 2gzx + 2fyz + 2ux + 2vy + 2wz + d = 0$$

represents a general **quadric surface**, embracing spheres, cylinders, cones, ellipsoids, paraboloids and hyperboloids. A full treatment of these simple surfaces is given in the books on classical three-dimensional analytical geometry such as Sommerville (1934).

As with the conic sections, some normalisation is necessary to produce a unique implicit representation of a particular quadric surface.

4.3.2 The normal vector to the surface $f(x,y,z) = 0$

To obtain the normal vector we note that this vector is perpendicular to any curve $\mathbf{r} = \mathbf{r}(s)$ lying on the surface, and thus the normal vector is perpendicular to the tangent vector $\mathbf{T} = \dfrac{d\mathbf{r}}{ds}$ for any such curve (see Figure 4.25).

Also $f = 0$ at all points on curves on the surface, so that $\dfrac{df}{ds} = 0$ on these curves. Using the chain rule for the differentiation of a function of three variables, we obtain

$$\frac{df}{ds} = \frac{\partial f}{\partial x}\frac{dx}{ds} + \frac{\partial f}{\partial y}\frac{dy}{ds} + \frac{\partial f}{\partial z}\frac{dz}{ds} = 0,$$

or

$$\left(\frac{\partial f}{\partial x}\mathbf{i} + \frac{\partial f}{\partial y}\mathbf{j} + \frac{\partial f}{\partial z}\mathbf{k}\right) \cdot \frac{d\mathbf{r}}{ds} = 0.$$

Thus the **gradient vector** $\frac{\partial f}{\partial x}\mathbf{i} + \frac{\partial f}{\partial y}\mathbf{j} + \frac{\partial f}{\partial z}\mathbf{k}$, usually written as ∇f, is perpendicular to $\frac{d\mathbf{r}}{ds}$ for any curve in the surface passing through the point where ∇f is evaluated.

Thus the unit normal vector to the surface is given by $\mathbf{n} = \nabla f/|\nabla f|$. Continuity of all the three partial derivatives is therefore needed if there are to be no sharp creases or cusps in the surface. Singularities may also occur at points where $\nabla f = 0$.

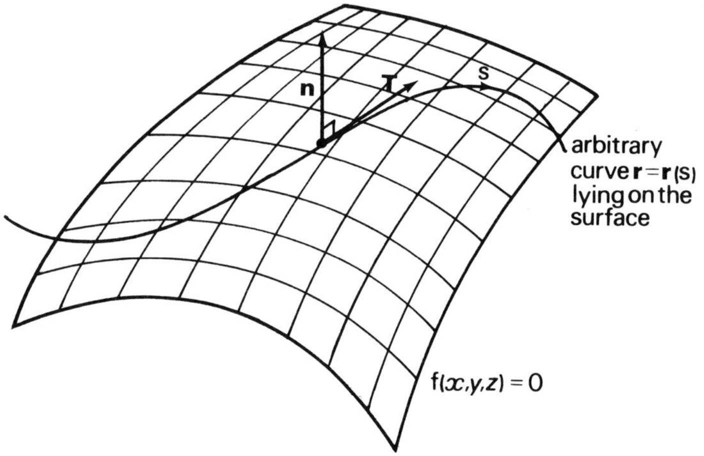

Figure 4.25

Example

The normal to the ellipsoid $x^2 + 2y^2 + 2z^2 = 6$ at the point $(1,-1,1)$ is obtained as follows.

Writing $f(x,y,z) = x^2 + 2y^2 + 2z^2 - 6$, we obtain

$$\nabla f = 2x\mathbf{i} + 4y\mathbf{j} + 4z\mathbf{k}.$$

Thus
$$\nabla f = 2\mathbf{i} - 4\mathbf{j} + 4\mathbf{k},$$

and $|\nabla f| = 6$ at $(1,-1,1)$. Finally, the unit normal at this point is

$$\mathbf{n} = (\mathbf{i} - 2\mathbf{j} + 2\mathbf{k})/3\,.$$

4.3.3 Equations of a space curve

A three-dimensional curve can be described as the intersection of two surfaces (see Figure 4.26), and may be represented by the simultaneous equations

$$f_1(x,y,z) = 0$$

and

$$f_2(x,y,z) = 0.$$

As in two dimensions it is necessary to solve these implicit equations simultaneously in order to determine points on the curve. If the equations are non-linear a numerical technique must usually be employed. This matter is considered further in Chapter 9.

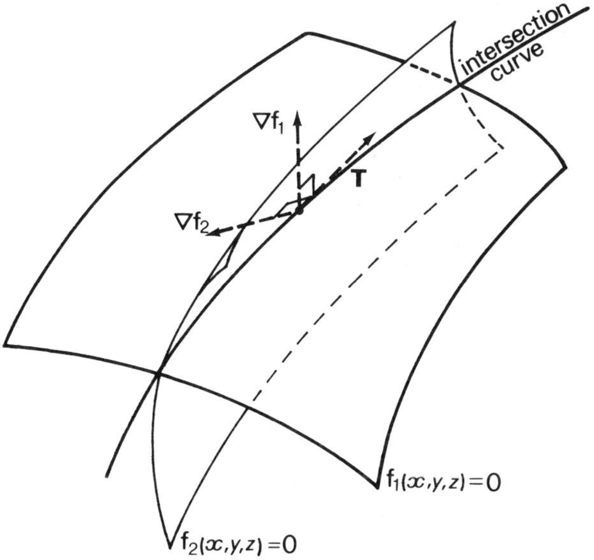

Figure 4.26

4.3.4 Tangent vector to a space curve

We have seen that the tangent vector to any curve in the surface $f(x,y,z) = 0$ is perpendicular to ∇f. Since the space curve described in Section 4.3.3 lies in the two surfaces $f_1(x,y,z) = 0$ and $f_2(x,y,z) = 0$, it follows that its tangent vector **T** is normal to both ∇f_1 and ∇f_2 (see Figure 4.26).

Thus

$$\mathbf{T} = \frac{\pm \nabla f_1 \times \nabla f_2}{|\nabla f_1 \times \nabla f_2|}.$$

Notice that the sense of **T** is not precisely defined when the curve is defined implicitly. The normals and curvature of a space curve may be obtained by consideration of higher derivatives; see for example Willmore (1958).

4.3.5 The APT surface definitions

Although parametrically defined surfaces have recently been added to the APT language for numerical control of machine tools, the major use of the language is based on the classical surfaces of implicit type, such as planes, cylinders and spheres.

Although the surfaces may be generated in many different ways, they are eventually stored in the computer in terms of a comparatively small number of **canonical forms**. These forms are listed below, with comments on their use.

(1) A **plane** $Ax + By + Cz - D = 0$ (note the sign of D) is stored as four coefficients (A,B,C,D), where $A^2 + B^2 + C^2 = 1$. A **line** in two dimensional work is stored as a plane with $C = 0$.

(2) A **sphere** $(x - X)^2 + (y - Y)^2 + (z - Z)^2 = R^2$ is stored as (X,Y,Z,R).

(3) A **circular cylinder**

$$(x - X)^2 + (y - Y)^2 + (z - Z)^2 - [A(x - X) + B(y - Y) + C(z - Z)]^2 = R^2,$$

where (X,Y,Z) is an arbitrary point on the axis, and (A,B,C) is a unit vector along the axis, is stored as (X,Y,Z,A,B,C,R). A **circle** in two dimensional work is stored as a cylinder with $A = B = 0, C = 1$.

(4) A **circular cone**

$$[A(x - X) + B(y - Y) + C(z - Z)]^2 = [(x - X)^2 + (y - Y)^2 + (z - Z)^2] \cos^2 \theta,$$

with vertex at (X,Y,Z), with semi-vertical angle θ, and with axis having the direction of the unit vector (A,B,C), is stored as $(X,Y,Z,A,B,C, \cos \theta)$.

(5) A **quadric** surface

$$Ax^2 + By^2 + Cz^2 + D + 2Fyz + 2Gzx + 2Hxy + 2Px + 2Qy + 2Rz = 0$$

is stored as (A,B,C,D,F,G,H,P,Q,R). Any **conic section** in two dimensions (apart from a straight line or a circle) is stored as a quadric in which $C = F = G = R = 0$; in other words, it is treated as a cylinder.

(6) The **tabulated cylinder** (TABCYL) is a cylinder whose rulings pass through the points of a piecewise cubic curve defined by a set of data points given by the user. The cubic curve is a plane curve in a plane normal to the axis of

the cylinder. Further details are given in Chapter 6.

(7). **Ruled surfaces** are also defined in APT. Although their definition is not parametric, they are effectively surfaces of the kind described in equation 4.39. The two directrices are defined by plane sections of APT surfaces, the planes being defined in terms of points or directions lying in them. On each directrix, the end points P_0 and P_1 are supplied by the user (see Figure 4.27). A direction P_0P_2 must also be supplied which defines a (possibly oblique) axis system in the section plane. The point P on the curve is parametrised linearly along P_0P_1, the points P_0 and P_1 having parameters 0 and 1 respectively. The rulings join points with the same parameter on the two directrices. Although the axes are defined independently on the two directrices, the ruled surfaces are necessarily somewhat restricted.

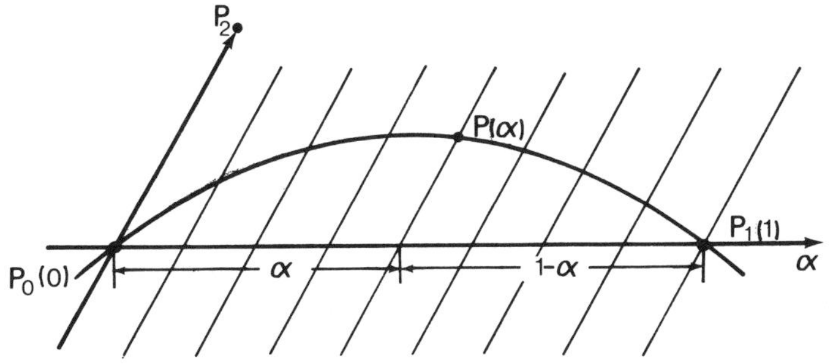

Figure 4.27

In Sections 9.3 and 9.4, we discuss the APT subroutines for calculating cutter positions which maintain given tolerance conditions relative to the surfaces being machined. These routines in turn call subroutines to determine surface normals and line intersections which are individually written for each surface type. The decision to use separate routines for planes, spheres, cylinders and cones, all of which are specialised quadric surfaces, is a compromise between the need for extra routines and the saving to be had from simpler routines when these often-used surfaces are encountered.

Chapter 5
Curve and surface design

5.1 PARAMETRIC CUBIC EQUATIONS FOR CURVE AND SURFACE DESIGN

5.1.1 Introduction

We have mentioned in Section 4.1.5 that parametric equations have advantages in curve and surface design. Since most objects we wish to design do not have simple shapes which can be described by simple analytic functions, we define curves in a piecewise manner, and surfaces in patches. The continuity and smoothness across the joins between the pieces and patches can be built into the parametrisation of the sections on either side. In this chapter we will discuss the properties of individual segments, and leave the discussion of composite curves until Chapter 6.

Because we will wish to determine tangents, normals, curvatures and so on, we will need a parametrisation which makes differentiation easy. Polynomial functions of the parameters are an obvious choice.

Polynomials of high degree can describe complex curves, but they require a large number of coefficients whose physical significance is difficult to grasp. Thus they are an inappropriate tool for the designer. Moreover, the use of high degree polynomials may introduce unwanted oscillations in the curve (see also Appendix 5).

On the other hand, the designer must have sufficient freedom to design without too many pieces or patches. It is found that cubics are a good compromise in most applications, and most design and fitting methods are in fact implemented using cubic parametrisation.

5.1.2 Ferguson's parametric cubic curves

Ferguson (1963) first introduced the use of parametric cubic equations for the definition of curves and surfaces in aircraft design. The segments of these curves are described by the equations of the form

$$\mathbf{r} = \mathbf{r}(u) = \mathbf{a}_0 + u\mathbf{a}_1 + u^2\mathbf{a}_2 + u^3\mathbf{a}_3. \tag{5.1}$$

It can be seen that four vectors (or 12 coefficients) are required to define the segment.

The usual way to determine $\mathbf{a}_0$, $\mathbf{a}_1$, $\mathbf{a}_2$ and $\mathbf{a}_3$ is to specify the values of $\mathbf{r}$ and $\frac{d\mathbf{r}}{du}$ at both ends of the segment. It is usual to assign parameter values $u = 0$ and $u = 1$ to the two ends of the segment, with $0 < u < 1$ in between.

Denoting $\frac{d\mathbf{r}}{du}$ by $\dot{\mathbf{r}}(u)$, we therefore see that

$$\mathbf{a}_0 = \mathbf{r}(0),$$

$$\mathbf{a}_0 + \mathbf{a}_1 + \mathbf{a}_2 + \mathbf{a}_3 = \mathbf{r}(1),$$

$$\mathbf{a}_1 = \dot{\mathbf{r}}(0),$$

and

$$\mathbf{a}_1 + 2\mathbf{a}_2 + 3\mathbf{a}_3 = \dot{\mathbf{r}}(1).$$

Solving for $\mathbf{a}_0$, $\mathbf{a}_1$, $\mathbf{a}_2$ and $\mathbf{a}_3$, we get the expressions

$$\begin{aligned} \mathbf{a}_0 &= \mathbf{r}(0), \\ \mathbf{a}_1 &= \dot{\mathbf{r}}(0), \\ \mathbf{a}_2 &= 3[\mathbf{r}(1) - \mathbf{r}(0)] - 2\dot{\mathbf{r}}(0) - \dot{\mathbf{r}}(1), \\ \text{and} \quad \mathbf{a}_3 &= 2[\mathbf{r}(0) - \mathbf{r}(1)] + \dot{\mathbf{r}}(0) + \dot{\mathbf{r}}(1). \end{aligned} \tag{5.2}$$

By direct substitution in (5.1), we can obtain $\mathbf{r}$ in terms of $\mathbf{r}(0)$, $\mathbf{r}(1)$, $\dot{\mathbf{r}}(0)$ and $\dot{\mathbf{r}}(1)$. Thus

$$\mathbf{r} = \mathbf{r}(u) = \mathbf{r}(0)(1 - 3u^2 + 2u^3) + \mathbf{r}(1)(3u^2 - 2u^3) + \dot{\mathbf{r}}(0)(u - 2u^2 + u^3) + \dot{\mathbf{r}}(1)(-u^2 + u^3).$$

Alternatively, we may write $\mathbf{r} = \mathbf{UCS}$, where $\mathbf{UCS}$ denotes the product of the three matrices given below:

$$\mathbf{r}(u) = \underbrace{[1 \quad u \quad u^2 \quad u^3]}_{\mathbf{U}} \underbrace{\begin{bmatrix} 1 & 0 & 0 & 0 \\ 0 & 0 & 1 & 0 \\ -3 & 3 & -2 & -1 \\ 2 & -2 & 1 & 1 \end{bmatrix}}_{\mathbf{C}} \underbrace{\begin{bmatrix} \mathbf{r}(0) \\ \mathbf{r}(1) \\ \dot{\mathbf{r}}(0) \\ \dot{\mathbf{r}}(1) \end{bmatrix}}_{\mathbf{S}}. \tag{5.3}$$

Note that the elements of the right-hand matrix of this equation are themselves vectors. In all cases in this book where matrices which have *vector* elements are used, only one such matrix will occur in any given product. The product of a matrix of scalar elements with such a matrix follows the rules given in Section A1.3, except that the multiplications of corresponding elements become scalar multiplications of the vector elements concerned. Thus, for example,

$$\begin{bmatrix} 1 & 2 \\ -1 & 4 \end{bmatrix} \begin{bmatrix} \mathbf{a} & \mathbf{b} \\ \mathbf{c} & \mathbf{d} \end{bmatrix} = \begin{bmatrix} \mathbf{a} + 2\mathbf{c} & \mathbf{b} + 2\mathbf{d} \\ 4\mathbf{c} - \mathbf{a} & 4\mathbf{d} - \mathbf{b} \end{bmatrix}.$$

The derivatives $\dot{\mathbf{r}}(0)$ and $\dot{\mathbf{r}}(1)$ are proportional to the unit tangent vectors $\mathbf{T}(0)$ and $\mathbf{T}(1)$ at the ends. Thus, we may write

$$\dot{\mathbf{r}}(0) = \alpha_0 \mathbf{T}(0), \quad \dot{\mathbf{r}}(1) = \alpha_1 \mathbf{T}(1).$$

The significance of the tangent vector magnitudes α_0 and α_1 is as follows. Simultaneous increase of α_0 and α_1 simply gives more fullness to the curve (see Figure 5.1), whereas increasing only α_0 causes the curve to remain close to the direction of $\mathbf{T}(0)$ for a greater part of its length before turning into the direction of $\mathbf{T}(1)$ (see Figure 5.2). For large values of α_0 and α_1 kinks and loops occur. For plane cubic curves, a safe rule is to ensure that the tangent magnitudes α_0 and α_1 do not exceed three times the chord length $|\mathbf{r}(1) - \mathbf{r}(0)|$.

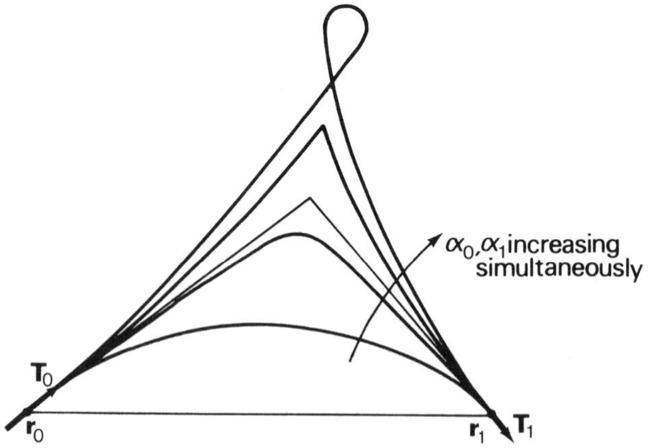

Figure 5.1

In the case of plane curves, the somewhat subjective notion of the 'fullness' of the curve can be replaced by choosing the constants α in such a way as to achieve prescribed values of curvature κ at the ends of any given segment. Then the curvature of two segments can be made continuous at the join, provided that the curvature requirements do not cause undesirable kinks or loops within the segments.

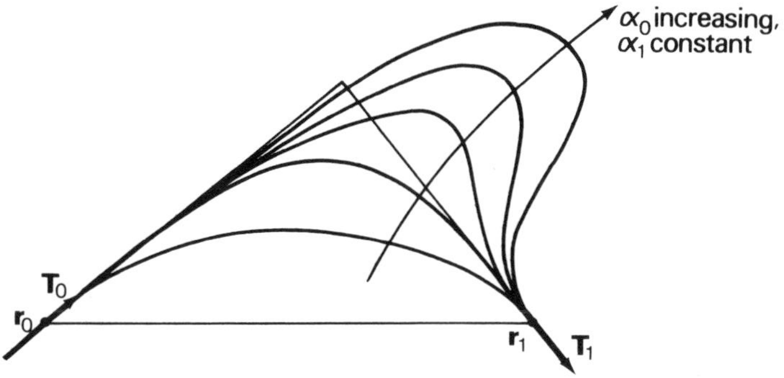

Figure 5.2

Expressions for the curvature are more simply derived in terms of Bézier's notation described in the next section.

5.1.3 Bézier's cubic UNISURF curves

Bézier (1970) has recombined the terms of the Ferguson cubic parametrisation in a way that makes the physical meaning of the vector coefficients more apparent. This is, of course, most important if we wish to design curves, rather than fit them.

In Bézier's form, we write

$$\mathbf{r} = \mathbf{r}(u) = (1-u)^3\mathbf{r}_0 + 3u(1-u)^2\mathbf{r}_1 + 3u^2(1-u)\mathbf{r}_2 + u^3\mathbf{r}_3, \tag{5.4}$$

where again $0 \leqslant u \leqslant 1$ for any given segment.

It can be seen that this is simply a re-arrangement of Ferguson's form, with

$$\mathbf{a}_0 = \mathbf{r}_0,$$

$$\mathbf{a}_1 = 3(\mathbf{r}_1 - \mathbf{r}_0),$$

$$\mathbf{a}_2 = 3(\mathbf{r}_2 - 2\mathbf{r}_1 + \mathbf{r}_0),$$

and $$\mathbf{a}_3 = \mathbf{r}_3 - 3\mathbf{r}_2 + 3\mathbf{r}_1 - \mathbf{r}_0.$$

The important consequences of this re-arrangement are that

$$\begin{aligned} \mathbf{r}(0) &= \mathbf{r}_0, \\ \mathbf{r}(1) &= \mathbf{r}_3, \\ \dot{\mathbf{r}}(0) &= 3(\mathbf{r}_1 - \mathbf{r}_0), \end{aligned} \tag{5.5}$$

and $$\dot{\mathbf{r}}(1) = 3(\mathbf{r}_3 - \mathbf{r}_2).$$

Thus the curve described by Bézier's form passes through the points $\mathbf{r}_0$ and $\mathbf{r}_3$, has a tangent at $\mathbf{r}_0$ in the direction from $\mathbf{r}_0$ to $\mathbf{r}_1$ and at $\mathbf{r}_3$ has a tangent in the direction from $\mathbf{r}_2$ to $\mathbf{r}_3$ (see Figures 5.3, 5.4, 5.5).

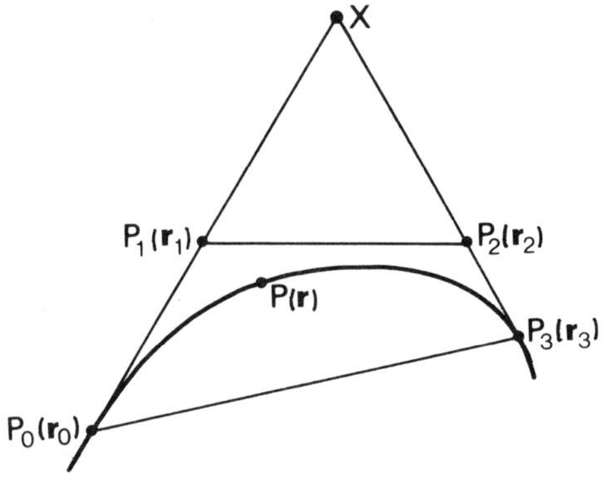

Figure 5.3

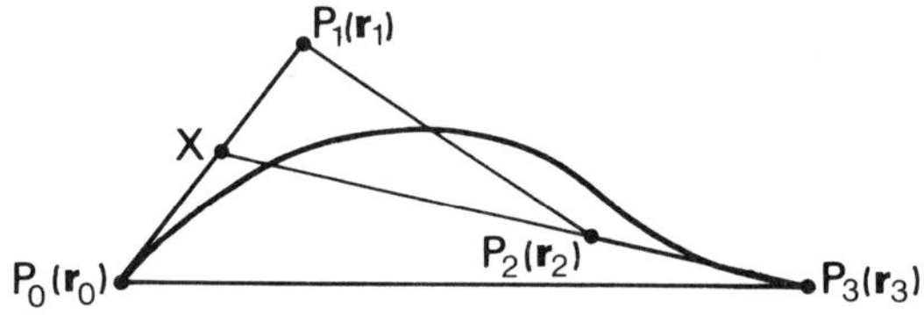

Figure 5.4

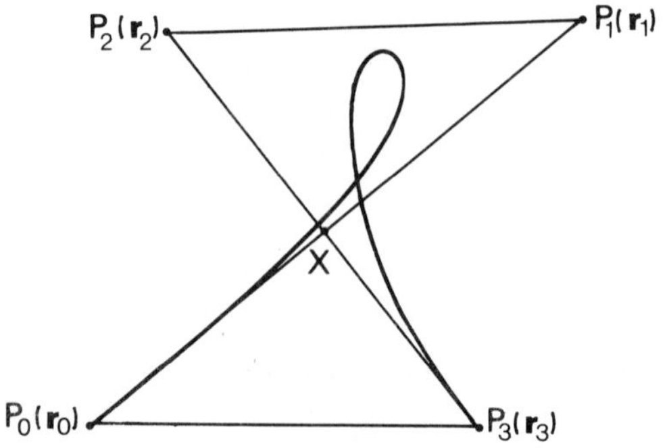

Figure 5.5

The straight lines P_0P_1, P_1P_2 and P_2P_3 form a figure called the characteristic polygon of the curve, although it is not usually a closed figure. In order to design a curve, we therefore choose the points P_0 and P_3 through which we want the curve to pass, and then place P_1 and P_2 on the desired tangents at P_0 and P_3. The lengths of P_0P_1 and P_2P_3 are then adjusted simultaneously to give greater fullness to the curve, or differentially in order to draw the curve nearer to one or other tangent, in the same way as the values of α_0 and α_1 may be used to modify the curves in Figures 5.1 and 5.2.

The matrix form of the Bézier cubic curve is $\mathbf{r} = \mathbf{UMR}$, or in detail

$$\mathbf{r}(u) = \underset{\mathbf{U}}{[1 \quad u \quad u^2 \quad u^3]} \underset{\mathbf{M}}{\begin{bmatrix} 1 & 0 & 0 & 0 \\ -3 & 3 & 0 & 0 \\ 3 & -6 & 3 & 0 \\ -1 & 3 & -3 & 1 \end{bmatrix}} \underset{\mathbf{R}}{\begin{bmatrix} \mathbf{r}_0 \\ \mathbf{r}_1 \\ \mathbf{r}_2 \\ \mathbf{r}_3 \end{bmatrix}}. \tag{5.6}$$

The curvature of the Bézier curve is obtained by noting that $\dot{\mathbf{r}} \times \ddot{\mathbf{r}} = \dot{s}^3\kappa\mathbf{B}$ (see equations 4.17 and 4.20), so that $\kappa = |\dot{\mathbf{r}} \times \ddot{\mathbf{r}}|/\dot{s}^3$.

$$\begin{aligned} \text{Now} \quad & \ddot{\mathbf{r}}(0) = 6(\mathbf{r}_0 - 2\mathbf{r}_1 + \mathbf{r}_2) = 6(\mathbf{r}_2 - \mathbf{r}_1) - 6(\mathbf{r}_1 - \mathbf{r}_0) \\ \text{and} \quad & \ddot{\mathbf{r}}(1) = 6(\mathbf{r}_1 - 2\mathbf{r}_2 + \mathbf{r}_3) = 6(\mathbf{r}_3 - \mathbf{r}_2) - 6(\mathbf{r}_2 - \mathbf{r}_1). \end{aligned} \tag{5.7}$$

$$\begin{aligned} \text{Thus} \quad & \kappa(0) = \frac{2|(\mathbf{r}_1 - \mathbf{r}_0) \times (\mathbf{r}_2 - \mathbf{r}_1)|}{3|\mathbf{r}_1 - \mathbf{r}_0|^3} \\ \text{and} \quad & \kappa(1) = \frac{2|(\mathbf{r}_2 - \mathbf{r}_1) \times (\mathbf{r}_3 - \mathbf{r}_2)|}{3|\mathbf{r}_3 - \mathbf{r}_2|^3}. \end{aligned} \tag{5.8}$$

For a plane curve, the curvature is always towards the chord P_0P_3 if the points P_1 and P_2 lie within the segments P_0X and XP_3, where X is the intersection of P_0P_1 and P_2P_3 (see Figure 5.3). In Figure 5.4, the point P_1 is beyond X on P_0P_1, so that the curvature at P_3 is away from the chord. When $P_0P_1 > P_0X$ and $P_2P_3 > XP_3$, the curve may even have a loop, as shown in Figure 5.5. If $P_0P_1 = \alpha P_0X$ and $P_2P_3 = \beta XP_3$, then the curve loops if $(\alpha - \frac{4}{3})(\beta - \frac{4}{3}) > \frac{4}{9}$, and the loop occurs between $u = 0$ and $u = 1$ if $\alpha > 1$ and $\beta > 1$. Forrest (1968) suggests that P_0P_1 and P_2P_3 should never exceed the chord P_0P_3, and this ensures that α and β are less than unity. It follows that any loops occurring in curves obeying this rule will always lie outside the range $0 \leqslant u \leqslant 1$.

The Bézier cubic curve may alternatively be written in the form

$$\mathbf{r} = \mathbf{r}_0 + 3u(1-u)^2(\mathbf{r}_1 - \mathbf{r}_0) + 3u^2(1-u)(\mathbf{r}_2 - \mathbf{r}_0) + u^3(\mathbf{r}_3 - \mathbf{r}_0).$$

Thus $\mathbf{r} = \mathbf{r}_0 + \alpha(\mathbf{r}_1 - \mathbf{r}_0) + \beta(\mathbf{r}_2 - \mathbf{r}_0) + \gamma(\mathbf{r}_3 - \mathbf{r}_0)$, where $0 < \alpha < 1, 0 < \beta < 1$ and $0 < \gamma < 1$ whenever $0 < u < 1$.

If we regard α, β and γ as oblique co-ordinates for $\mathbf{r}$ in a system with origin $\mathbf{r}_0$ and axes in the directions of $\mathbf{r}_1 - \mathbf{r}_0$, $\mathbf{r}_2 - \mathbf{r}_0$ and $\mathbf{r}_3 - \mathbf{r}_0$, then these inequalities show that the curve lies entirely inside the tetrahedron $P_0P_1P_2P_3$ (see Figure 5.6) when $0 < u < 1$. In fact the curve is even more closely constrained, since we can show that $0 < \alpha < 4/9$ and $0 < \beta < 4/9$.

This property of the curve in relation to its characteristic polygon is called the **hull convexity property**, and is a most desirable feature for any curve defined in terms of a 'polygon'.

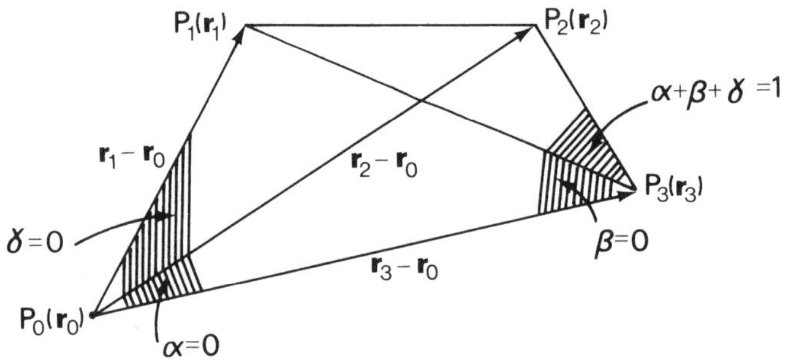

Figure 5.6

5.1.4 The Bernstein-Bézier polynomial curve

Bézier (1970) describes a more general polynomial curve given by

$$\mathbf{r} = \mathbf{r}(u) = \sum_{i=0}^{n} \frac{n!}{(n-i)!i!} u^i(1-u)^{n-i}\mathbf{r}_i, \tag{5.9}$$

where $\mathbf{r}_0, \mathbf{r}_1, \mathbf{r}_2, \ldots, \mathbf{r}_n$ are the position vectors of the $n+1$ vertices $P_0, P_1, \ldots, P_n$ of a generalised characteristic 'polygon'. The Bézier cubic curve is an example of this general curve, where $n = 3$.

It can then be shown that

$$\mathbf{r}(0) = \mathbf{r}_0, \qquad \mathbf{r}(1) = \mathbf{r}_n,$$

$$\dot{\mathbf{r}}(0) = n(\mathbf{r}_1 - \mathbf{r}_0), \qquad \dot{\mathbf{r}}(1) = n(\mathbf{r}_n - \mathbf{r}_{n-1}). \tag{5.10}$$

Thus the general polynomial curve passes through the points P_0 and P_n, where the tangents are in the directions of the vectors P_0P_1 and $P_{n-1}P_n$. It is an extension to vector-valued functions of the Bernstein polynomials defined in Appendix 3.

The advantage of higher order curves is that several orders of continuity can be achieved between segments of composite curves. For example, a fifth order polynomial curve will permit the specification of end points, end tangents and curvature $\kappa\mathbf{B}$ at both ends, still leaving the constants α_0 and α_1 available for shape adjustment.

However, although the shape of the characteristic polygon still gives some indication of the shape of the associated curve, the relationship becomes weaker as the order of the curve is increased.

The first order curve $\mathbf{r} = (1 - u)\mathbf{r}_0 + u\mathbf{r}_1$ is the straight line equation described in Section 3.3.5.

The second order curve has the equation

$$\mathbf{r} = (1 - u)^2\mathbf{r}_0 + 2u(1 - u)\mathbf{r}_1 + u^2\mathbf{r}_2. \tag{5.11}$$

The point with parameter u on this curve can be constructed as follows. The point A which divides P_0P_1 in the ratio $u{:}1 - u$ is joined to the point B which divides P_1P_2 in the same ratio (see Figure 5.7). The segment AB is now divided in the same ratio at the point C.

Now $\quad \mathbf{r}_A = (1 - u)\mathbf{r}_0 + u\mathbf{r}_1$

and $\quad \mathbf{r}_B = (1 - u)\mathbf{r}_1 + u\mathbf{r}_2.$

Then
$$\begin{aligned}\mathbf{r}_C &= (1 - u)\mathbf{r}_A + u\mathbf{r}_B\\ &= (1 - u)^2\mathbf{r}_0 + 2u(1 - u)\mathbf{r}_1 + u^2\mathbf{r}_2.\end{aligned}$$

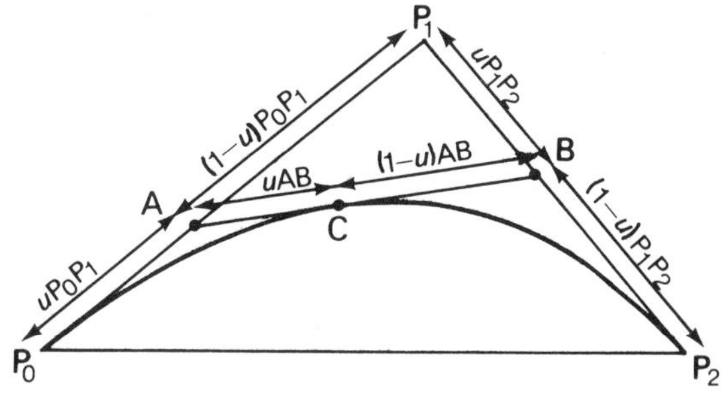

Figure 5.7

Thus C is the point with parameter u on the parametric quadratic curve (5.11). Bézier (1970) shows that this method of construction can be extended to higher order Bernstein-Bézier curves.

Alternatively, we can show that the curve represented by (5.11) is the parabolic 'proportional curve' described in Section 1.2.3. The derivation is omitted here because a more general proof will be given in Section 5.2.1 for the general conic segment.

5.1.5 Special cases of the Bézier cubic curve

The following particular cases of the Bézier cubic curve are of practical importance.

(1) *Straight line*
Whenever the vertices of the characteristic polygon are collinear, the curve becomes a straight line. In particular, if we set $\mathbf{r}_1 = (2\mathbf{r}_0 + \mathbf{r}_3)/3$ and $\mathbf{r}_2 = (\mathbf{r}_0 + 2\mathbf{r}_3)/3$ in equation (5.4), then we obtain the standard uniform parametrisation $\mathbf{r} = (1-u)\mathbf{r}_0 + u\mathbf{r}_3$.

(2) *Parabola*
The curve $\mathbf{r} = (1-u)^2\mathbf{r}_0 + 2u(1-u)\mathbf{r}^* + u^2\mathbf{r}_3$, representing a parabolic arc whose end points are $\mathbf{r}_0$ and $\mathbf{r}_3$ and whose end tangents intersect at $\mathbf{r}^*$, is obtained by setting $\mathbf{r}_1 = (\mathbf{r}_0 + 2\mathbf{r}^*)/3$ and $\mathbf{r}_2 = (\mathbf{r}_3 + 2\mathbf{r}^*)/3$.

(3) *Circle*
A close approximation to the circular arc $\mathbf{r} = \cos\theta\mathbf{i} + \sin\theta\mathbf{j}$ for $0 \leqslant \theta \leqslant \pi/2$ is obtained by writing $\mathbf{r}_0 = \mathbf{i}$, $\mathbf{r}_2 = \mathbf{i} + k\mathbf{j}$, $\mathbf{r}_3 = k\mathbf{i} + \mathbf{j}$ and $\mathbf{r}_4 = \mathbf{j}$, where $k = 4(\sqrt{2} - 1)/3$. The radius of the approximate arc varies between 1 and 1.00027 in this case, the maximum deviation from the mean radius being ±0.13%. A more general formula for circular arcs is given in Gossling (1976).

5.1.6 Ferguson's cubic surface patches

The surface fitting procedure described by Ferguson (1963) extends the curve definition of Section 5.1.2 by allowing the vectors $\mathbf{a}_0$, $\mathbf{a}_1$, $\mathbf{a}_2$ and $\mathbf{a}_3$ to be functions of a second parameter v. Using a similar cubic parametrisation, we write

$$\mathbf{a}_i = \mathbf{a}_{i0} + v\mathbf{a}_{i1} + v^2\mathbf{a}_{i2} + v^3\mathbf{a}_{i3}, \quad i = 0, 1, 2, 3.$$

Then the curve $\mathbf{r} = \mathbf{r}(u)$ moves and changes its shape as v progresses from 0 to 1, and the variable curve sweeps out the surface given by

$$\mathbf{r} = \mathbf{r}(u,v) = \mathbf{a}_{00} + v\mathbf{a}_{01} + v^2\mathbf{a}_{02} + v^3\mathbf{a}_{03}$$
$$+ u\mathbf{a}_{10} + uv\mathbf{a}_{11} + uv^2\mathbf{a}_{12} + uv^3\mathbf{a}_{13} + \ldots.$$

$$= \sum_{i=0}^{3} \sum_{j=0}^{3} \mathbf{a}_{ij} u^i v^j \; .$$

In matrix notation, $\mathbf{r}(u,v) = \mathbf{UAV}$. In full,

$$\mathbf{r}(u,v) = [1 \quad u \quad u^2 \quad u^3] \begin{bmatrix} \mathbf{a}_{00} & \mathbf{a}_{01} & \mathbf{a}_{02} & \mathbf{a}_{03} \\ \mathbf{a}_{10} & \mathbf{a}_{11} & \mathbf{a}_{12} & \mathbf{a}_{13} \\ \mathbf{a}_{20} & \mathbf{a}_{21} & \mathbf{a}_{22} & \mathbf{a}_{23} \\ \mathbf{a}_{30} & \mathbf{a}_{31} & \mathbf{a}_{32} & \mathbf{a}_{33} \end{bmatrix} \begin{bmatrix} 1 \\ v \\ v^2 \\ v^3 \end{bmatrix} . \tag{5.13}$$

It can be seen that there are 16 vector coefficients to be determined by the user. This requires, for example, the specification of the values of $\mathbf{r}$, $\dfrac{\partial \mathbf{r}}{\partial u}$, $\dfrac{\partial \mathbf{r}}{\partial v}$ and $\dfrac{\partial^2 \mathbf{r}}{\partial u \partial v}$ at the four corners. (We will adopt the shorthand $\dfrac{\partial \mathbf{r}}{\partial u} = \mathbf{r}_u$, $\dfrac{\partial \mathbf{r}}{\partial v} = \mathbf{r}_v$ and $\dfrac{\partial^2 \mathbf{r}}{\partial u \partial v} = \mathbf{r}_{uv}$ in what follows).

The vectors $\mathbf{r}_u$ and $\mathbf{r}_v$ are proportional to the tangent vectors $\mathbf{T}_u$ and $\mathbf{T}_v$ to the parametric curves v = constant and u = constant. The vector $\mathbf{r}_{uv}$ has been said to measure the 'twist' in the surface, or the rate of change of $\mathbf{T}_u$ with v or $\mathbf{T}_v$ with u (see Figure 5.8). Thus $\mathbf{r}_{uv}$ is called the **twist vector.** In fact, $\mathbf{r}_{uv}$ is a property of the *parametrisation* as well as of the surface itself, and may be non-zero even on plane surfaces. It should really be described as the **cross-derivative vector.**

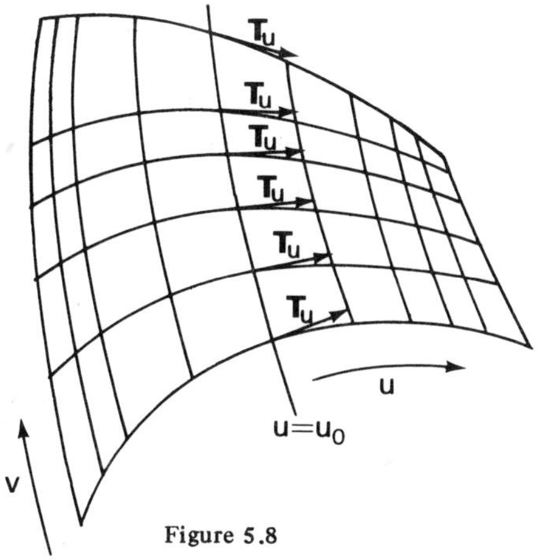

Figure 5.8

5.1.7 Bézier UNISURF surface patches

We may generalise the UNISURF curve in much the same way as the Ferguson curve by specifying the path traced by each vertex of the characteristic polygon. The final result is symmetric in u and v in the same way as that for the Ferguson surface patch. Then

$$\mathbf{r} = \sum_{i=0}^{3} \sum_{j=0}^{3} \mathbf{r}_{ij} \frac{3!}{(3-i)!i!} \frac{3!}{(3-j)!j!} u^i(1-u)^{3-i} v^j (1-v)^{3-j} \tag{5.14}$$

where $\mathbf{r}_{ij}$ are the vertices of a characteristic polyhedron, as shown in Figure 5.9. Only the vertices $\mathbf{r}_{00}$, $\mathbf{r}_{03}$, $\mathbf{r}_{30}$ and $\mathbf{r}_{33}$ lie on the surface; the peripheral vertices form the characteristic polygons of the curves forming the edges of the patch, and the four interior vertices influence the cross derivatives at the four corners, as we shall see below.

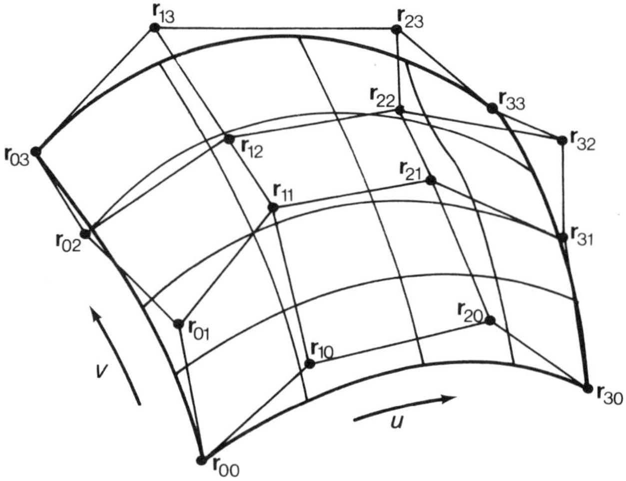

Figure 5.9

In matrix form, we may describe the Bézier surface patch by the equation

$$\mathbf{r} = \mathbf{r}(u,v) = [1 \quad u \quad u^2 \quad u^3]\ \mathbf{MBM}^T \begin{bmatrix} 1 \\ v \\ v^2 \\ v^3 \end{bmatrix} \tag{5.15}$$

where $\mathbf{M}$ and $\mathbf{B}$ are given respectively by

$$\mathbf{M} = \begin{bmatrix} 1 & 0 & 0 & 0 \\ -3 & 3 & 0 & 0 \\ 3 & -6 & 3 & 0 \\ -1 & 3 & -3 & 1 \end{bmatrix}, \quad \mathbf{B} = \begin{bmatrix} \mathbf{r}_{00} & \mathbf{r}_{01} & \mathbf{r}_{02} & \mathbf{r}_{03} \\ \mathbf{r}_{10} & \mathbf{r}_{11} & \mathbf{r}_{12} & \mathbf{r}_{13} \\ \mathbf{r}_{20} & \mathbf{r}_{21} & \mathbf{r}_{22} & \mathbf{r}_{23} \\ \mathbf{r}_{30} & \mathbf{r}_{31} & \mathbf{r}_{32} & \mathbf{r}_{33} \end{bmatrix}.$$

The values of **r** and its derivatives at the patch corners can be evaluated and expressed as follows:

$$\begin{bmatrix} \mathbf{r}(0,0) & \mathbf{r}(0,1) & \mathbf{r}_v(0,0) & \mathbf{r}_v(0,1) \\ \mathbf{r}(1,0) & \mathbf{r}(1,1) & \mathbf{r}_v(1,0) & \mathbf{r}_v(1,1) \\ \mathbf{r}_u(0,0) & \mathbf{r}_u(0,1) & \mathbf{r}_{uv}(0,0) & \mathbf{r}_{uv}(0,1) \\ \mathbf{r}_u(1,0) & \mathbf{r}_u(1,1) & \mathbf{r}_{uv}(1,0) & \mathbf{r}_{uv}(1,1) \end{bmatrix} =$$

$$\begin{bmatrix} 1 & 0 & 0 & 0 \\ 0 & 0 & 0 & 1 \\ -3 & 3 & 0 & 0 \\ 0 & 0 & -3 & 3 \end{bmatrix} \begin{bmatrix} \mathbf{r}_{00} & \mathbf{r}_{01} & \mathbf{r}_{02} & \mathbf{r}_{03} \\ \mathbf{r}_{10} & \mathbf{r}_{11} & \mathbf{r}_{12} & \mathbf{r}_{13} \\ \mathbf{r}_{20} & \mathbf{r}_{21} & \mathbf{r}_{22} & \mathbf{r}_{23} \\ \mathbf{r}_{30} & \mathbf{r}_{31} & \mathbf{r}_{32} & \mathbf{r}_{33} \end{bmatrix} \begin{bmatrix} 1 & 0 & -3 & 0 \\ 0 & 0 & 3 & 0 \\ 0 & 0 & 0 & -3 \\ 0 & 1 & 0 & 3 \end{bmatrix} \quad (5.16)$$

Thus, for example, the values of **r** and its derivatives at the corner $u = v = 0$ are given by

$$\mathbf{r}(0,0) = \mathbf{r}_{00},$$

$$\mathbf{r}_u(0,0) = 3(\mathbf{r}_{10} - \mathbf{r}_{00}),$$

$$\mathbf{r}_v(0,0) = 3(\mathbf{r}_{01} - \mathbf{r}_{00}),$$

and

$$\mathbf{r}_{uv}(0,0) = 9(\mathbf{r}_{00} - \mathbf{r}_{01} - \mathbf{r}_{10} + \mathbf{r}_{11}).$$

5.2 RATIONAL PARAMETRIC CURVES AND SURFACES

5.2.1 Rational quadratic parametrisation of a segment of a conic section

In order to describe the segment $P_0P'P_2$ of the conic section shown in Figure 5.10, we construct the tangents at the points P_0, P' and P_2. The points P_1, A and B are the intersections of these tangents. If we regard P_0, P_1 and P_2 as fixed, then A and B vary as P'moves along the segment. The ratios $g_1 = P_1A/P_0A$ and $g_2 = P_1B/P_2B$ then depend on P' and may be used to provide a parameter for P'.

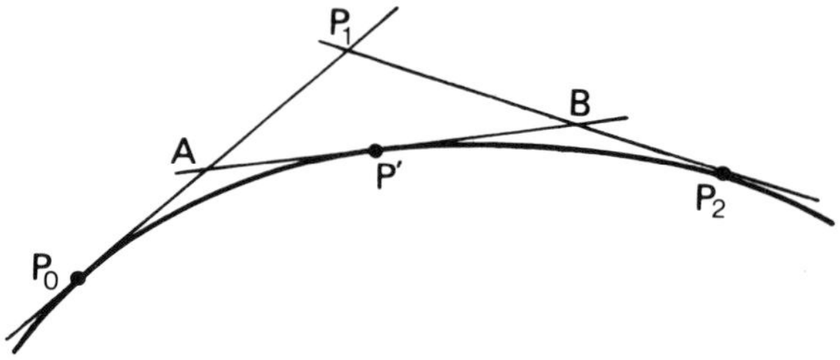

Figure 5.10

In Section 1.2.2, we showed that the conics through P_0 and P_2 with tangents P_0P_1 and P_1P_2 form a pencil. If P_0P_1, P_1P_2 and P_0P_2 have the equations $\ell_1 = 0$, $\ell_2 = 0$ and $\ell_3 = 0$ respectively, the pencil has the equation

$$(1 - \lambda)\ell_1\ell_2 + \lambda\ell_3{}^2 = 0.$$

Whereas in this equation the conics are set in the *Oxy* plane using rectangular co-ordinates, it is convenient here to use oblique co-ordinates (α,β) in the plane containing P_0, P_1 and P_2, taking P_1 as the origin, P_0 as the point (1,0) and P_2 as the point (0,1). As shown in Figure 5.11, the general point $P(\alpha,\beta)$ has position vector $\mathbf{r}$ given by

$$\mathbf{r} = \mathbf{r}_1 + \alpha(\mathbf{r}_0 - \mathbf{r}_1) + \beta(\mathbf{r}_2 - \mathbf{r}_1) \tag{5.17}$$

The lines P_0P_1, P_1P_2 and P_0P_2 have the equations $\beta = 0$, $-\alpha = 0$ and $\alpha + \beta - 1 = 0$, so that the pencil of conics is given by

$$S(\alpha,\beta) = (1 - \lambda)\alpha\beta - \lambda(\alpha - \beta - 1)^2 = 0 \tag{5.18}$$

in oblique co-ordinates. We have chosen to describe P_1P_2 by $-\alpha = 0$ rather than $+\alpha = 0$ since we would like those conics of the pencil which pass inside the triangle $P_0P_1P_2$ to correspond to $0 < \lambda < 1$.

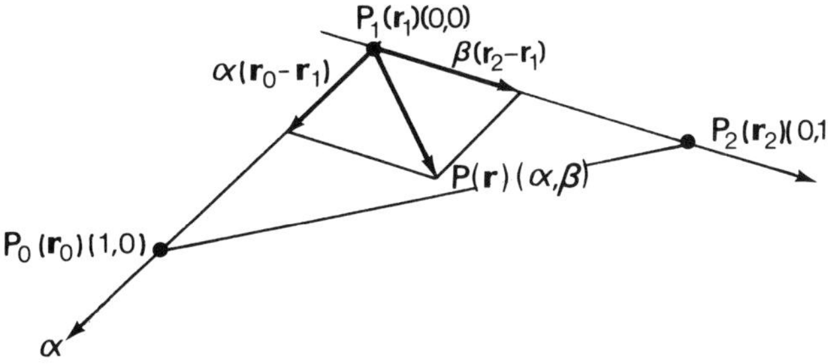

Figure 5.11

The tangent to this conic at $P'(\alpha',\beta')$ may be obtained by using equation (1.13) in oblique co-ordinates:

$$S_\alpha(\alpha',\beta')(\alpha - \alpha') + S_\beta(\alpha',\beta')(\beta - \beta') = 0.$$

This tangent intersects P_0P_1 at $(\alpha,0)$, where

$$\alpha = \frac{\alpha' S_\alpha(\alpha',\beta') + \beta' S_\beta(\alpha',\beta')}{S_\alpha(\alpha',\beta')}$$

$$= \frac{2(1-\lambda)\alpha'\beta' - 2\lambda(\alpha'+\beta')(\alpha'+\beta'-1)}{(1-\lambda)\beta' - 2\lambda(\alpha'+\beta'-1)}$$

$$= \frac{-2\lambda(\alpha'+\beta'-1)}{(1-\lambda)\beta' - 2\lambda(\alpha'+\beta'-1)},$$

since $S(\alpha',\beta') = 0$. Thus

$$g_1 = \frac{\alpha}{1-\alpha} = \frac{-2\lambda(\alpha'+\beta'-1)}{(1-\lambda)\beta'}. \qquad (5.19)$$

Similarly $$g_2 = \frac{\beta}{1-\beta} = \frac{-2\lambda(\alpha'+\beta'-1)}{(1-\lambda)\alpha'}. \qquad (5.20)$$

Hence $$g_1g_2 = \frac{4\lambda^2(\alpha'+\beta'-1)^2}{(1-\lambda)^2\alpha'\beta'} = \frac{4\lambda}{1-\lambda},$$

by using (5.18), so that g_1g_2 is constant for any given conic of the pencil, and is positive for those conics passing through the triangle $P_0P_1P_2$. We accordingly write

$$g_1g_2 = k^2 \qquad (5.21)$$

for such conics.

The point of contact of the tangent can be expressed in terms of g_1 and g_2 by solving (5.19) and (5.20) for α' and β'. We obtain

$$\alpha' = \frac{g_1}{g_1 + 2 + g_2}, \quad \beta' = \frac{g_2}{g_1 + 2 + g_2}. \tag{5.22}$$

Thus the point on the conic section where the tangent intersects P_0P_1 and P_1P_2 in ratios $g_1:1$ and $g_2:1$ respectively has position vector $\mathbf{r}$ given by

$$\mathbf{r} = \mathbf{r}_1 + \frac{g_1(\mathbf{r}_0 - \mathbf{r}_1)}{g_1 + 2 + g_2} + \frac{g_2(\mathbf{r}_2 - \mathbf{r}_1)}{g_1 + 2 + g_2}$$

$$= \frac{g_1\mathbf{r}_0 + 2\mathbf{r}_1 + g_2\mathbf{r}_2}{g_1 + 2 + g_2}. \tag{5.23}$$

We now write $g_1 = \dfrac{w_0(1-u)}{w_1 u}$ and $g_2 = \dfrac{w_2 u}{w_1(1-u)}$, so that $g_1 g_2 = k^2 = \dfrac{w_0 w_2}{w_1{}^2}$

where w_0, w_1 and w_2 are constants known as **weights**. Thus finally

$$\mathbf{r} = \frac{w_0\mathbf{r}_0(1-u)^2 + 2w_1\mathbf{r}_1 u(1-u) + w_2\mathbf{r}_2 u^2}{w_0(1-u)^2 + 2w_1 u(1-u) + w_2 u^2}, \tag{5.24}$$

which is a **rational quadratic parametric equation** for the conic segment with tangents P_0P_1 and P_1P_2 and having $g_1 g_2 = \dfrac{w_0 w_2}{w_1{}^2}$.

The parametrisation of the curve can be altered by changing the individual weights whilst keeping the value $(w_0w_2)/w_1{}^2$ constant. Note that equation 5.24 is unaltered if all three weights are multiplied by the same constant. It is therefore the ratios $w_0{:}w_1{:}w_2$ which determine the nature of the curve and its parametrisation. A more general parametrisation is given by $g_1 = \dfrac{w_0 c}{w_1}\dfrac{(b-u')}{(u'-a)}$ and $g_2 = \dfrac{w_2}{w_1 c}\dfrac{(u'-a)}{(b-u')}$ which, when substituted into (5.23) still produces a rational quadratic equation. The shape of the curve is unchanged since $g_1 g_2 = \dfrac{w_0 w_2}{w_1{}^2}$ as before. The resulting transformation between u and u' is of the bilinear form $Auu' + Bu + Cu' + D = 0$. The range $0 \leqslant u \leqslant 1$ corresponds to $a \leqslant u' \leqslant b$, and the constant c determines the distribution of the parameter values along the length of the curve.

Rational parametric curves are most simply expressed in terms of **homogenous co-ordinates** (xw,yw,zw,w) (see Chapter 3).

The equations

$$xw = x_0w_0(1 - u)^2 + 2x_1w_1u(1 - u) + x_2w_2u^2$$

$$yw = y_0w_0(1 - u)^2 + 2y_1w_1u(1 - u) + y_2w_2u^2$$

$$zw = z_0w_0(1 - u)^2 + 2z_1w_1u(1 - u) + z_2w_2u^2$$

and

$$w = w_0(1 - u)^2 + 2w_1u(1 - u) + w_2u^2$$

are equivalent to equations (5.24), as can be seen by writing $x = \dfrac{xw}{w}$, for example.

Denoting the vectors $\begin{bmatrix} xw \\ yw \\ zw \\ w \end{bmatrix}$ and $\begin{bmatrix} x_iw_i \\ y_iw_i \\ z_iw_i \\ w_i \end{bmatrix}$ by $\mathbf{P}$ and $\mathbf{P}_i, i = 0, 1, 2$, we may therefore write equation (5.24) in the form

$$\mathbf{P} = (1 - u)^2\mathbf{P}_0 + 2u(1 - u)\mathbf{P}_1 + u^2\mathbf{P}_2, \tag{5.25}$$

which is similar to the equation of the quadratic Bézier curve. However, the rational quadratic curve may represent any conic segment, whereas the Bézier curve represents a parabola only. Then

$$\mathbf{P}(0) = \mathbf{P}_0,$$

$$\mathbf{P}(1) = \mathbf{P}_2,$$

$$\dot{\mathbf{P}}(0) = 2(\mathbf{P}_1 - \mathbf{P}_0), \tag{5.26}$$

and

$$\dot{\mathbf{P}}(1) = 2(\mathbf{P}_2 - \mathbf{P}_1),$$

where $\dot{\mathbf{P}}(u) \equiv \frac{d\mathbf{P}}{du}$. Now $\dot{\mathbf{P}} \equiv \frac{d}{du}(w\mathbf{R})$, where $\mathbf{R} = \begin{bmatrix} x \\ y \\ z \\ 1 \end{bmatrix} = \begin{bmatrix} \mathbf{r} \\ \hline 1 \end{bmatrix}$.

Thus $$\dot{\mathbf{P}} = \dot{w}\mathbf{R} + w\dot{\mathbf{R}},$$

and hence $$\dot{\mathbf{R}} = \frac{1}{w}(\dot{\mathbf{P}} - \dot{w}\mathbf{R}). \qquad (5.27)$$

Finally, we may use (5.26) and (5.27) to obtain the Cartesian relationships

$$\mathbf{r}(0) = \mathbf{r}_0$$

$$\mathbf{r}(1) = \mathbf{r}_2$$

$$\dot{\mathbf{r}}(0) = \frac{2w_1}{w_0}(\mathbf{r}_1 - \mathbf{r}_0) \qquad (5.28)$$

and $$\dot{\mathbf{r}}(1) = \frac{2w_1}{w_2}(\mathbf{r}_2 - \mathbf{r}_1).$$

Thus the rational quadratic represents, for $0 \leqslant u \leqslant 1$, a conic segment with end points P_0 and P_2, and with tangents P_0P_1 and P_1P_2 at these points.

We showed in Section 5.1.3 that a Bézier cubic curve segment remains inside the tetrahedron whose vertices are the points of its characteristic polygon. In a similar fashion, we can show that the conic segment remains inside triangle $P_0P_1P_2$ for $0 < u < 1$ when the weights are positive.

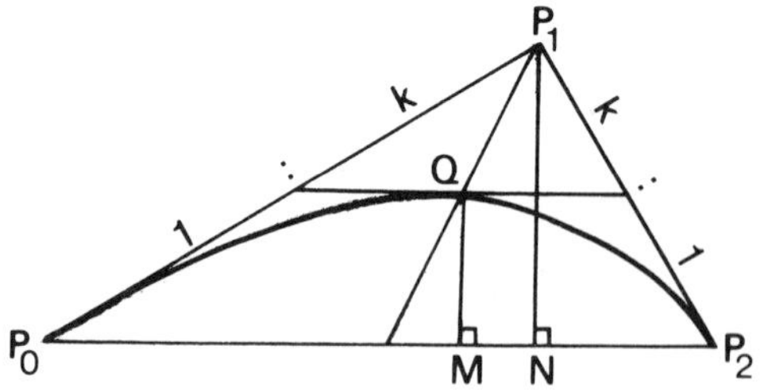

Figure 5.12

Moreover, referring to Figure 5.12, it is evident that the maximum distance QM from the conic segment to the chord P_0P_2 occurs at the point of contact Q of the tangent having equal ratios $g_1 = g_2 = k$ on the defining tangents. Then the maximum distance is $QM = P_1N/(1 + k)$, where P_1N is the perpendicular from P_1 onto P_0P_2. Thus the fullness of the curve is determined by k. In the particular parametrisation where $w_0 = w_2 = 1 - p$, and $w_1 = p$, we have $QM = pP_1N$, so that the parameter p determines the curve fullness. We will return to this again in considering the rational cubic curve.

When $w_0 = w_1 = w_2$ (or $k = 1$), the denominator becomes unity, and we obtain the Bézier quadratic parametrisation of a parabola. Thus the parabola has $k = 1$, whereas $k > 1$ gives an ellipse and $k < 1$ a hyperbola.

The user can specify the fullness of the curve and its parametrisation most conveniently *either* by means of a given tangent and the parameter u^* at its point of contact, *or* by providing a third point $P^*(\mathbf{r}^*)$ on the curve and the corresponding parameter u^*.

In the first case, the tangent is specified by the ratios $g_1{}^*$ and $g_2{}^*$. Then

$$\frac{w_0}{w_1} = \frac{g_1{}^*u^*}{1 - u^*}, \qquad \frac{w_2}{w_1} = \frac{g_2{}^*(1 - u^*)}{u^*}.$$

If, on the other hand, we are given $\mathbf{r}^* = \mathbf{r}(u^*)$, then, provided P^* lies inside the triangle $P_0P_1P_2$, we may use (5.24) to obtain the equation

$$w_0(1 - u^*)^2\,(\mathbf{r}^* - \mathbf{r}_0) + 2w_1u^*(1 - u^*)\,(\mathbf{r}^* - \mathbf{r}_1) + w_2u^{*2}(\mathbf{r}^* - \mathbf{r}_2) = \mathbf{0}\,. \tag{5.29}$$

Provided that $\mathbf{r}_0$, $\mathbf{r}_1$ and $\mathbf{r}_2$ are not collinear, we may define the normal $\mathbf{n}$ to their plane, and then (5.29) may be solved for the ratios $w_0 : w_1$ and so on in the form

$$\frac{w_0(1 - u^*)^2}{\mathbf{n}.(\mathbf{r}^* - \mathbf{r}_1)\times(\mathbf{r}^* - \mathbf{r}_2)} = \frac{2w_1u^*(1 - u^*)}{\mathbf{n}.(\mathbf{r}^* - \mathbf{r}_2)\times(\mathbf{r}^* - \mathbf{r}_0)} = \frac{w_2u^{*2}}{\mathbf{n}.(\mathbf{r}^* - \mathbf{r}_0)\times(\mathbf{r}^* - \mathbf{r}_1)}.$$

An alternative to providing a third point or tangent is to maintain continuity of curvature with the segment to the left.

We may show that the curvature at $u = 0$ is given by

$$\kappa(0) = \frac{w_0w_2}{2w_1{}^2}\,\frac{|(\mathbf{r}_1 - \mathbf{r}_0)\times(\mathbf{r}_2 - \mathbf{r}_1)|}{|\mathbf{r}_1 - \mathbf{r}_0|^3},$$

and at $u = 1$ by $\qquad \kappa(1) = |\mathbf{r}_1 - \mathbf{r}_0|^3\kappa(0)/|\mathbf{r}_2 - \mathbf{r}_1|^3.$

Since $\mathbf{r}_0$, $\mathbf{r}_1$ and $\mathbf{r}_2$ are fixed by continuity and tangency requirements, these expressions give a relationship between the constant $k^2 = \dfrac{w_0w_2}{w_1{}^2}$ and the curvature at the end points. Moreover, we see from the second equation that the *ratio* of the curvatures at each end is fixed by $\mathbf{r}_0$, $\mathbf{r}_1$ and $\mathbf{r}_2$. It is therefore only possible to match curvature at one end of a segment or chain of segments.

Examples

(1) The homogeneous equation of the straight line through the points $\mathbf{r}_0$ and $\mathbf{r}_2$ may be obtained by substituting

$$w_1\mathbf{r}_1 = (w_0\mathbf{r}_0 + w_2\mathbf{r}_2)/2, \qquad w_1 = (w_0 + w_2)/2$$

in equation (5.24). The point $\mathbf{r}_1$ is then collinear with $\mathbf{r}_0$ and $\mathbf{r}_2$, and the line has the equation

$$\mathbf{r} = \frac{w_0\mathbf{r}_0(1 - u) + w_2\mathbf{r}_2 u}{w_0(1 - u) + w_2 u}.$$

The ratio $w_0{:}w_2$ determines the parametrisation of the line, which is uniform when $w_0 = w_2$, when we obtain the usual parametric equation

$$\mathbf{r} = \mathbf{r}_0(1 - u) + \mathbf{r}_2 u.$$

(2) The equation of the segment of the circle $x^2 + y^2 = 1, z = 0$ in the first quadrant is obtained by setting $\mathbf{r}_0 = \mathbf{i}$, $\mathbf{r}_1 = \mathbf{i} + \mathbf{j}$, $\mathbf{r}_2 = \mathbf{j}$; $w_0 = \sqrt{2}, w_1 = 1$ and $w_2 = \sqrt{2}$. From equation (5.24), it can then be verified that $x^2 + y^2 = 1$. An alternative parametrisation, with $w_0 = w_1 = 1$ and $w_2 = 2$, gives the form

$$\mathbf{r} = \frac{1 - u^2}{1 + u^2}\mathbf{i} + \frac{2u}{1 + u^2}\mathbf{j},$$

which was illustrated in Figure 4.3, Section 4.1.1.

5.2.2 Rational cubic curves

The rational cubic curve is a natural extension of the rational quadratic formulation of conics described in Section 5.2.1. The classical theory of the twisted cubic curve is to be found in Semple and Kneebone (1952), and its application to design has been described by Coons (1967) and Forrest (1968).

The points of the curve are again described by homogeneous coordinates (xw,yw,zw,w) where xw, yw, zw and w are cubic functions of the parameter u, so that x, y and z are rational cubic functions of u.

We may now adopt a formulation of the rational cubic segment which is similar to Bézier's treatment of the standard cubic curve. Thus

$$\mathbf{P} = (1 - u)^3\mathbf{P}_0 + 3u(1 - u)^2\mathbf{P}_1 + 3u^2(1 - u)\mathbf{P}_2 + u^3\mathbf{P}_3,$$

where
$$\mathbf{P} = \begin{bmatrix} xw \\ yw \\ zw \\ w \end{bmatrix}, \qquad \mathbf{P}_i = \begin{bmatrix} x_i w_i \\ y_i w_i \\ z_i w_i \\ w_i \end{bmatrix}, i = 0, 1, 2, 3. \tag{5.31}$$

The vectors $\mathbf{P}_i$ are homogeneous co-ordinates of the vertices of a characteristic polygon having a similar meaning to that of the Bézier polygon. For any given segment, the parameter u again lies in the interval $0 \leqslant u \leqslant 1$. The values and derivatives of $\mathbf{P}$ at the ends of the segment are given by

$$\mathbf{P}(0) = \mathbf{P}_0,$$

$$\mathbf{P}(1) = \mathbf{P}_3,$$

$$\mathbf{P}(0) = 3(\mathbf{P}_1 - \mathbf{P}_0),$$

and
$$\dot{\mathbf{P}}(1) = 3(\mathbf{P}_3 - \mathbf{P}_2), \tag{5.32}$$

which are directly equivalent to the Equations (5.7).

Finally, we may use 5.32 and 5.27 to obtain the Cartesian coordinate relationships

$$\mathbf{r}(0) = \mathbf{r}_0,$$

$$\mathbf{r}(1) = \mathbf{r}_3,$$

$$\dot{\mathbf{r}}(0) = \frac{3w_1}{w_0}(\mathbf{r}_1 - \mathbf{r}_0),$$

and
$$\dot{\mathbf{r}}(1) = \frac{3w_2}{w_3}(\mathbf{r}_3 - \mathbf{r}_2). \tag{5.33}$$

Thus the polygon defined by $\mathbf{r}_0, \mathbf{r}_1, \mathbf{r}_2, \mathbf{r}_3$ has much the same significance as Bézier's characteristic polygon, except that we now have freedom to adjust the parametrisation of the curve as well as its shape.

The curvatures at $u = 0$ and $u = 1$ can be shown to be given by

$$\kappa(0) = \frac{2w_0 w_2}{3w_1^2} \frac{|(\mathbf{r}_1 - \mathbf{r}_0) \times (\mathbf{r}_2 - \mathbf{r}_1)|}{|\mathbf{r}_1 - \mathbf{r}_0|^3}$$

$$\kappa(1) = \frac{2w_1 w_3}{3w_2^2} \frac{|(\mathbf{r}_2 - \mathbf{r}_1) \times (\mathbf{r}_3 - \mathbf{r}_2)|}{|\mathbf{r}_3 - \mathbf{r}_2|^3}.$$

The numbers $\dfrac{w_0 w_2}{w_1{}^2}$ and $\dfrac{w_1 w_3}{w_2{}^2}$ control the curvature for a given polygon in a similar manner to the constant $k^2 = \dfrac{w_0 w_2}{w_1{}^2}$ in the case of the conic segment.

It should be noted that the parametrisation of the rational cubic curve may be altered using a bilinear transformation $Auu' + Bu + Cu' + D = 0$ in the same way as may be done for the rational quadratic. For there to be no change in the *range* of parametrisation, the end values of the range must be the self-corresponding points of the transformation. Thus, for example, we may wish to change the parametrisation whilst retaining the range (0,1). Then, setting $u' = u$, we see that $Au^2 + (B + C)u + D \equiv Au(u - 1) = 0$ must have solutions $u = 0$ and $u = 1$, so that $D = 0$ and $B + C = -A$. It follows that any bilinear transformation of the form

$$Auu' + Bu - (A + B)u' = 0$$

retains the curve shape and the parameter range (0,1). More details of the invariance of curve shape are given in Forrest (1968).

The rational cubic curve has the advantage of incorporating both conics and parametric cubic curves as special cases. Thus if the weights w_i are all equal, we recover the Bézier cubic curve, and if $\mathbf{P}_0 - 3\mathbf{P}_1 + 3\mathbf{P}_2 - \mathbf{P}_3 = 0$, the curve becomes a rational quadratic, that is a conic. Another special case is the straight line in its homogenous form, To obtain this, we set $\mathbf{P}_1 = (2\mathbf{P}_0 + \mathbf{P}_3)/3$ and $\mathbf{P}_2 = (\mathbf{P}_0 + 2\mathbf{P}_3)/3$ and obtain $\mathbf{P} = (1 - u)\mathbf{P}_0 + u\mathbf{P}_3$.

The rational cubic curve forms the basis of the CONSURF surface lofting program (Ball 1974, 1975), except that CONSURF uses a slightly different polynomial basis. We have chosen to retain the Bernstein polynomials for the sake of uniformity.

The CONSURF program uses two specialisations of the rational cubic curve which are of special interest. The first is the **linear parameter segment**, in which a fixed axis direction $\mathbf{n}$ is given. Then the position vector $\mathbf{r}$ of a point on the curve is required to have a component $\mathbf{r.n}$ which increases linearly with the curve parameter u. Thus $\mathbf{r.n} = k_1(1 - u) + k_2 u$, where k_1 and k_2 are constants ($k_1 \neq k_2$). Then

$$\mathbf{P.N} = w(u)\,[k_1(1 - u) + k_2 u]\ ,$$

where $\mathbf{P.N} = P_1 n_1 + P_2 n_2 + P_3 n_3$ is a scalar product of the homogeneous vectors $\mathbf{P} = (P_1, P_2, P_3, P_4)$ and $\mathbf{N} = (n_1, n_2, n_3, 0)$.

By writing down the cubic expressions for $\mathbf{P}$ and w, expanding the right-hand side and equating coefficients, the following conditions are obtained for the linear parameter segment:

$$w_0 - 3w_1 + 3w_2 - w_3 = 0,$$

$$3w_1(\mathbf{r}_1 - \mathbf{r}_0).\mathbf{n} = w_0(\mathbf{r}_3 - \mathbf{r}_0).\mathbf{n}, \tag{5.34}$$

and

$$3w_2(\mathbf{r}_2 - \mathbf{r}_3).\mathbf{n} = w_3(\mathbf{r}_0 - \mathbf{r}_3).\mathbf{n}.$$

To simplify the use of this segment, a **simple linear parameter segment** is defined in which the user provides values for the end point vectors $\mathbf{r}_0$ and $\mathbf{r}_3$, corresponding tangent vectors $\mathbf{T}_0$ and $\mathbf{T}_3$, and the p-**ratio** described below. In terms of these,

$$\left.\begin{aligned} \mathbf{r}_1 &= \mathbf{r}_0 + \alpha_0\mathbf{T}_0, \\ \mathbf{r}_2 &= \mathbf{r}_3 - \alpha_3\mathbf{T}_3, \\ w_0 &= w_3 = p, \\ w_1 &= w_2 = 1 - p, \end{aligned}\right\} \tag{5.35}$$

where

$$\alpha_0 = \frac{p(\mathbf{r}_3 - \mathbf{r}_0).\mathbf{n}}{3(1 - p)\mathbf{T}_0.\mathbf{n}}$$

and

$$\alpha_3 = \frac{p(\mathbf{r}_3 - \mathbf{r}_0).\mathbf{n}}{3(1 - p)\mathbf{T}_3.\mathbf{n}}.$$

We see that increasing the p-ratio moves $\mathbf{r}_1$ and $\mathbf{r}_2$ farther along the tangents, and thus gives greater fullness to the segment.

The second specialisation of the rational cubic curves is the **generalised conic segment**. As its name implies, this is a generalisation of the rational quadratic plane conic segment. The new segment has the advantages of allowing an inflexion point, and need not be a plane curve. Whereas the rational quadratic curve is bounded by the defining triangle (see Section 5.2.1), the generalised segment is bounded by the two right circular cones we shall now describe.

As with the linear parameter segment, the end point vectors $\mathbf{r}_0$ and $\mathbf{r}_3$, tangent vectors $\mathbf{T}_0$ and $\mathbf{T}_3$ and a p-ratio are defined. Then $\mathbf{r}_1 = \mathbf{r}_0 + \lambda\mathbf{T}_0$ and $\mathbf{r}_2 = \mathbf{r}_3 - \mu\mathbf{T}_3$, are points lying on the circle of intersection of the cones obtained by rotating the vectors $\mathbf{T}_0$ and $\mathbf{T}_3$ about the line joining the end-points (see Figure 5.13).

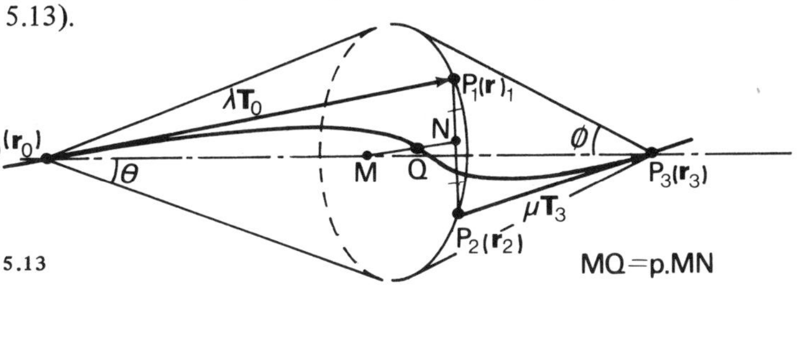

Figure 5.13

Let the semi-angles of the cones with vertices P_0 and P_3 be θ and ϕ respectively.

Then $$\lambda \cos\theta + \mu\cos\phi = d\,,$$

and $$\lambda \sin\theta - \mu \sin\phi = 0, \qquad (5.36)$$

where $$d = |\mathbf{r}_3 - \mathbf{r}_0|.$$

Hence $$\lambda = \frac{d\sin\phi}{\sin(\theta+\phi)}\,,\quad \mu = \frac{d\sin\theta}{\sin(\theta+\phi)}\,.$$

Ball (1975) gives the derivation of λ and μ in terms of $\mathbf{r}_0$, $\mathbf{r}_3$, $\mathbf{T}_0$ and $\mathbf{T}_3$.

In the case of a generalised conic segment, we define $w_1 = w_2 = \dfrac{p}{3}$ and $w_0 = w_3 = 1 - p$ in order to ensure that increasing values of p increase the fullness of the curve in the same way as in the linear parameter segment.

Then when $u = ½$,

$$\mathbf{r}(½) = (1-p)\left(\frac{\mathbf{r}_0+\mathbf{r}_3}{2}\right) + p\left(\frac{\mathbf{r}_1+\mathbf{r}_2}{2}\right),$$

so that the curve crosses the line joining the midpoint M $\left(\dfrac{\mathbf{r}_0+\mathbf{r}_3}{2}\right)$ of P_0P_3 and the midpoint N$\left(\dfrac{\mathbf{r}_1+\mathbf{r}_2}{2}\right)$ of P_1P_2 at a point which divides MN in the ratio $p{:}1{-}p$ (see Figure 5.13). We see that this is a straightforward generalisation of the p-ratio used to specify the fullness of the plane conic curve in Section 5.2.1.

5.2.3 Rational surface patches

The rational quadratic and cubic curves can be generalised to give rational surfaces in much the same way as Bézier's cubic curves are the basis for his bicubic UNISURF patches. Thus the rational bicubic patch takes the same form as equation (5.12), except that the vectors are in homogeneous co-ordinates throughout.

Hence $$\mathbf{P} = \mathbf{P}(u,v) = [1 \;\; u \;\; u^2 \;\; u^3]\;\mathbf{M}\mathbf{B}^*\mathbf{M}^T\begin{bmatrix}1\\ v\\ v^2\\ v^3\end{bmatrix} \qquad (5.37)$$

where $\mathbf{M}$ is the Bézier coefficient matrix of equation (5.6) and $\mathbf{B}^*$ is given by

$$\mathbf{B}^* = \begin{bmatrix} \mathbf{P}_{00} & \mathbf{P}_{01} & \mathbf{P}_{02} & \mathbf{P}_{03} \\ \mathbf{P}_{10} & \mathbf{P}_{11} & \mathbf{P}_{12} & \mathbf{P}_{13} \\ \mathbf{P}_{20} & \mathbf{P}_{21} & \mathbf{P}_{22} & \mathbf{P}_{23} \\ \mathbf{P}_{30} & \mathbf{P}_{31} & \mathbf{P}_{32} & \mathbf{P}_{33} \end{bmatrix}.$$

The rational biquadratic patch is defined in a similar manner. It should be noted that, although both the u and v parametric curves are conic sections, these patches are not in general parts of quadric surfaces. However, every quadric surface may be parametrised in this way. In fact, the rational biquadratic form includes all Steiner surfaces (see Sommerville, 1934), which in turn include all quadric surfaces.

5.3 PARAMETER TRANSFORMATIONS FOR POLYNOMIAL AND RATIONAL PARAMETRIC CURVES AND SURFACES

The polynomial curve segments in Section 5.1 were all parametrised with the end points have parameters $u = 0, u = 1$.

The properties of these curves have therefore been considered in relation to this particular parametrisation. If we wish to select a *portion* of such a segment for separate examination or modification, it is convenient to transform the parameters in such a way that the end points $u = u_0$ and $u = u_1$ have new parameters $u' = 0$ and $u' = 1$, so that we can apply our standard results to the sub-segment $u_0 \leqslant u \leqslant u_1$.

We will discuss the curves in the Bézier form, and therefore write the transformation in the same spirit.

If the degree of the polynomial is to be unchanged, the transformation must be linear, and takes the form

$$u = (1 - u')u_0 + u'u_1. \qquad (5.38)$$

It can be seen that the parameters $u' = 0{,}1$ correspond to $u = u_0, u_1$ as required.

If we write (5.38) in the form $u = u_0 + (u_1 - u_0)u'$, and denote $u_1 - u_0$ by Δu_0, we obtain the transformations

$$u^n = \sum_{r=0}^{n} {}^nC_r u_0^r (u'\Delta u_0)^{n-r} \qquad (5.39)$$

so that, taking the cubic Bézier curve as an example,

$$[1\ u\ u^2\ u^3] = [1\ u'\ u'^2\ u'^3]\begin{bmatrix}1 & u_0 & u_0^2 & u_0^3\\ 0 & \Delta u_0 & 2u_0\Delta u_0 & 3u_0^2\Delta u_0\\ 0 & 0 & \Delta u_0^2 & 3u_0\Delta u_0^2\\ 0 & 0 & 0 & \Delta u_0^3\end{bmatrix}$$

or

$$\mathbf{U} = \mathbf{U}'\mathbf{T}. \qquad (5.40)$$

Then equation (5.6) may be written in terms of the new parameter u' as

$$\mathbf{r}(u) = \mathbf{UMR} = \mathbf{U}'\mathbf{TMR} = \mathbf{U}'\mathbf{MR}' = \mathbf{r}'(u'),$$

where the modified polygon points are given by the equation $\mathbf{MR}' = \mathbf{TMR}$, or

$$\mathbf{R}' = \mathbf{M}^{-1}\mathbf{TMR}. \qquad (5.41)$$

In detail,

$$\mathbf{r}_0' = (1 - u_0)^3\mathbf{r}_0 + 3u_0(1 - u_0)^2\mathbf{r}_1 + 3u_0^2(1 - u_0)\mathbf{r}_2 + u_0^3\mathbf{r}_3,$$

$$\mathbf{r}_1' = (1 - u_0)^2(1 - u_1)\mathbf{r}_0 + (1 - u_0)(2u_0 + u_1 - 3u_0u_1)\mathbf{r}_1$$
$$+ u_0(2u_1 + u_0 - 3u_0u_1)\mathbf{r}_2 + u_0^2u_1\mathbf{r}_3, \qquad (5.42)$$

$$\mathbf{r}_2' = (1 - u_0)(1 - u_1)^2\mathbf{r}_0 + (1 - u_1)(2u_1 + u_0 - 3u_0u_1)\mathbf{r}_1$$
$$+ u_1(2u_0 + u_1 - 3u_0u_1)\mathbf{r}_2 + u_1^2u_0\mathbf{r}_3,$$

$$\mathbf{r}_3' = (1 - u_1)^3\mathbf{r}_0 + 3u_1(1 - u_1)^2\mathbf{r}_1 + 3u_1^2(1 - u_1)\mathbf{r}_2 + u_1^3\mathbf{r}_3.$$

Equations (5.40) and (5.41) can easily be generalised to higher order polynomial curves using (5.39) to obtain the coefficients of **T**.

The corresponding transformation for a rational curve, which again produces a rational equation of the same degree, is given by

$$u = \frac{(1 - u')\omega_0u_0 + u'\omega_1u_1}{(1 - u')\omega_0 + u'\omega_1}. \qquad (5.43)$$

The parameters $u' = 0,1$ again correspond to $u = u_0, u_1$ but we may adjust the parametrisation within the segment by means of the ratio $\omega_0:\omega_1$. In particular, we may leave the end-points unchanged, but adjust the uniformity of parametrisation, by setting $u_0 = 0, u_1 = 1$ in (5.43) and obtain

$$(\omega_1 - \omega_0)uu' + \omega_0 u - \omega_1 u' = 0. \tag{5.44}$$

Equations of the form of (5.43) and (5.44) represent bilinear transformations between u and u'. A general theory of such transformations can be found in books on projective geometry such as Semple and Kneebone (1952).

For a polynomial surface of the form $\mathbf{r}(u,v) = \mathbf{UMBM}^T\mathbf{V}$ described by equation (5.15), we may select a sub-patch $u_0 \leqslant u \leqslant u_1$, $v_0 \leqslant v \leqslant v_1$ by simultaneous transformation of u and v, to obtain the following generalisation of (5.4.1):

$$\mathbf{B}' = (\mathbf{M}^{-1}\mathbf{T}_u\mathbf{M})\mathbf{B}(\mathbf{M}^{-1}\mathbf{T}_v\mathbf{M})^T \tag{5.45}$$

in which $\mathbf{T}_u$ is the matrix $\mathbf{T}$ in (5.40), and $\mathbf{T}_v$ is the corresponding matrix in v_0 and v_1.

5.4 AREA SUBTENDED BY A PLANE BÉZIER CURVE

In the cross-sectional design of ducts, for example, the cross-sectional area is of particular interest (see later in Chapter 8).

If a plane Bézier curve is expressed in terms of a coordinate system whose origin lies in the plane of the curve, we may use equation (4.46) to obtain the area subtended by the curve at the origin. This enables us to calculate the cross-sectional area of any design whose normal cross-sections are composed of plane Bézier segments.

For a Bézier cubic curve, integration using (4.46) gives the result

$$\mathbf{A} = \frac{3}{10}(\mathbf{r}_1 - \mathbf{r}_0) \times (\mathbf{r}_2 - \mathbf{r}_1) + \frac{3}{10}(\mathbf{r}_2 - \mathbf{r}_1) \times (\mathbf{r}_3 - \mathbf{r}_2) - \frac{9}{20}(\mathbf{r}_3 - \mathbf{r}_2) \times (\mathbf{r}_1 - \mathbf{r}_0) \tag{5.46}$$

for the area $\mathbf{A}$ *between the curve and the chord* P_0P_3 in terms of the polygon points P_0, P_1, P_2, and P_3.

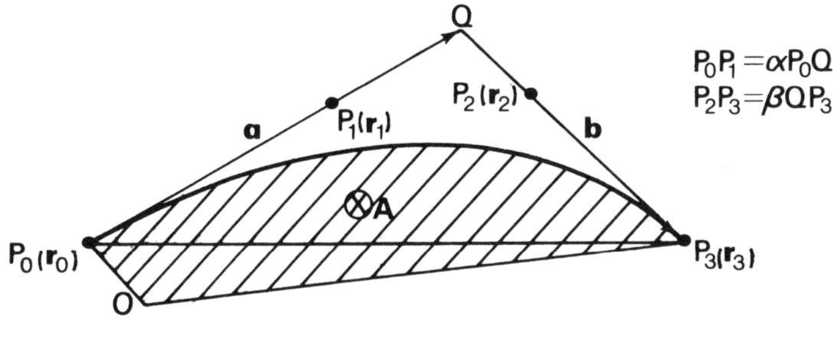

Figure 5.14

For these plane cubic curves, we may write $\mathbf{r}_1 = \mathbf{r}_0 + \alpha\mathbf{a}$, $\mathbf{r}_2 = \mathbf{r}_3 - \beta\mathbf{b}$, $\mathbf{r}_3 = \mathbf{r}_0 + \mathbf{a} + \mathbf{b}$ (see Figure 5.14). In these terms the total vector area $\mathbf{A}^*$ subtended at the origin by the segment is

$$\mathbf{A}^* = \frac{3}{20}\,[2\alpha + 2\beta - \alpha\beta]\,(\mathbf{a} \times \mathbf{b}) + \mathbf{A}_0, \tag{5.47}$$

where $\mathbf{A}_0$ is the vector area of triangle OP_0P_3.

For example, the approximate circle described in Section 5.1.5 has $\alpha = \beta = 4(\sqrt{2} - 1)/3$, so that $|\mathbf{A}| = 0.285618$, compared with 0.285398 for a true circular arc. The areal error is 0.00022, and the average radial error $2\delta A/\pi = 0.00014$.

Chapter 6

Composite curves and splines

6.1 INTRODUCTION

We are concerned in this book with computer-aided methods for the design and manufacture of some artefact whose geometry involves curved surfaces. Many advantages arise, as pointed out in the Introduction, from setting up mathematical representations of these curved surfaces. Sometimes this is easy, for instance when a body is composed entirely of simple geometrical shapes such as spheres, cylinders or cones. But very often this is not so; car bodies and the fuselages and aerodynamic surfaces of aircraft are obvious examples. The procedure used for the mathematical representation of more complex surfaces such as these commonly runs roughly as follows:

(i) Two notional sets of lines are envisaged to lie in the surface, one set running fore and aft, the other transversely. This network of lines defines a number of 'topologically rectangular' patches, each of which, for a smooth surface, will be bounded by four smooth continuous curves.

(ii) The coordinates of the intersections of this notional mesh are measured from a model or a set of cross-sectional drawings of the surface.

(iii) An interpolatory technique is used to establish firm mathematical specifications of the two sets of lines comprising the network.

(iv) Each mesh cell of the network now has four well-defined boundaries, and the interior of the cell is 'filled in' using two-dimensional interpolation.

In this chapter we describe a number of methods which may be used to define the framework of curves on which this procedure is based. Their extension to the complete definition of a surface in three dimensions is dealt with in Chapter 7.

The mathematical details of the process described are simplest when the curved surface is defined with respect to a flat plane, the (x,y)-plane, the mesh being composed of lines of constant x and of constant y. All the mesh curves are then plane curves. We will accordingly examine the fitting of plane curves first, turning later in the chapter to the more general problem of fitting curves in three dimensions.

In what follows we shall distinguish between derivatives with respect to spatial coordinates and derivatives with respect to curve parameters by using primes for the former but dots for the latter. For example, we will write the derivative of a function $f(x)$ in Cartesian coordinates as $f'(x)$, but the derivative of $\mathbf{r}(s)$, parametrised in terms of arc length s, will be written as $\dot{\mathbf{r}}(s)$ as in previous chapters.

6.2 THE FITTING OF PLANE CURVES

6.2.1 Classical Methods

First, we examine the application of the classical methods of numerical analysis which are summarised in Appendix 5.

We assume that we wish to construct a surface of the form $z = \dot{z}(x,y)$. A line of constant y in this surface will be a line $z = z(x)$, and we require that this line fits the set of data points (x_0,z_0), (x_1,z_1), . . . , (x_n,z_n), where $x_0 < x_1 < \ldots < x_n$.

Since we have $(n + 1)$ data points, we can form the Lagrangean interpolating polynomial $p_L(x)$ of degree n which passes through these points. However, if n is large we are likely to encounter undesirable oscillations in $p_L(x)$, which may possess as many as $(n - 1)$ maxima and minima if all the zeros of $p_L{}'(x)$ happen to be real. Clearly this oscillatory tendency will become greater as the degree of $p_L(x)$ increases, that is as the number of points to be interpolated increases.

One way of avoiding this problem is to construct a **composite curve** by fitting successive low-degree polynomials (cubics, for example) to successive groups of data points. The resulting **piecewise polynomial** function will be continuous, but will in general have discontinuities in slope at the joins between the successive curve segments. In most applications this will be unacceptable, because if the final surface is to be smooth we must require the mesh lines from which it is constructed to be smooth.

We might consider the use of a single low-degree least-squares polynomial over the whole range, but this would not give a sufficiently good fit in most cases. Increasing the degree of the polynomial to obtain a better fit would be likely to lead to oscillatory problems once again.

A further possibility is Hermite interpolation. For this we need to know the value of $z'(x)$ as well as that of z at all the data points. Unfortunately, although the position of a point on a surface may be measured fairly accurately from a model or drawing, the accurate measurement of a gradient is more difficult. Consequently the gradient values are likely to be less reliable than the z-values, and since Hermite's formula is rather sensitive to changes in derivative values we will probably obtain only a poor representation of the desired curve. Oscillation problems will again arise if a high degree is chosen for the interpolating polynomial, and for this reason piecewise Hermite interpolation using low-degree polynomials (as described above for Lagrangean interpolation) will generally be preferable.

Let us examine this last suggestion in more detail. If we consider a single interval $x_i \leqslant x \leqslant x_{i+1}$, with both z and $z'(x)$ specified at each end, we can construct a cubic interpolating polynomial over that interval. If this is done over every interval, the resulting piecewise cubic function will have continuity of slope at all the data points. A surface constructed from a mesh of lines of this type may therefore be smooth everywhere. It should be noted that second derivatives (and hence curvature) will change discontinuously at data points, however.

The attractions of piecewise Hermite interpolation, then, are that it can give first-derivative continuity without severe oscillation problems. Its main disadvantage is the requirement for gradient values. There is a natural tendency, when measured gradient values are not available, to use values estimated from the z-values at the data points using classical methods. However, this practice should be avoided if possible, because the estimation of derivatives is an inherently inaccurate process, as pointed out in Appendix 5. The estimates obtained are too crude to warrant the use of the comparatively accurate Hermite formula.

Before leaving the classical methods, we remark that all of them may only be used, in their basic forms, for interpolating single-valued functions. Multiple-valued functions (corresponding to cross-sections of folded surfaces) may be dealt with using piecewise polynomials as suggested earlier. An illustration is given in Figure 6.1:

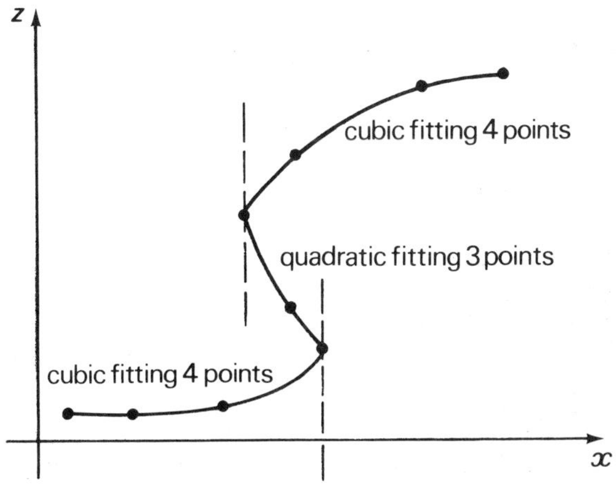

Figure 6.1 – Piecewise Lagrangean Interpolation of a Multivalued Function

Since no polynomial ever has an infinite gradient, comparatively large errors will occur in the vicinity of any vertical tangent of the surface.

6.2.2 Polynomial Splines

Each of the methods suggested in the last section has one at least of the following major disadvantages:

(i) oscillatory tendencies,

(ii) inability to produce a smooth curve, or

(iii) a requirement for gradient values at the data points.

We now show how it is possible to interpolate the data smoothly using low-degree polynomials to reduce problems of oscillation, yet with the demand for gradient values reduced to an absolute minimum.

Consider once again a set of data points (x_0,z_0), (x_1,z_1), . . . , (x_n,z_n), with $x_0 < x_1 < \ldots < x_n$. We seek to fit these points with a composite function $\phi(x)$ having the following properties:

(i) over each subinterval $x_{i-1} \leqslant x \leqslant x_i$, $i = 1, 2, \ldots, n$, $\phi(x)$ is a cubic polynomial.

(ii) $\phi(x)$ and its first *and second* derivatives are continuous at the data points.

The resulting smooth piecewise cubic curve is called a **cubic spline**. Splines of higher degree result when derivatives of the third or higher orders are also required to be continuous at each of the data points; for instance, a fifth-degree spline has a continuous fourth derivative. Splines of even degree are not often used, since they have certain characteristics making them less suitable for our present purpose (see Chapter 3 of Ahlberg, Nilson and Walsh, 1967).

The term **spline** arises by analogy with a draughtsman's aid of that name, a thin metal or wooden strip, which is bent elastically so as to pass through certain points of constraint. The curve taken up by a physical spline is one which minimises its internal strain energy. Mathematically, the curve is the one whose mean squared curvature is a minimum; in this sense, it is the 'smoothest' curve which passes through the fixed points. In terms of Cartesian coordinates, the minimum energy curve is that which minimises the integral, taken between its end-points,

$$\int \frac{z''^2 \, dx}{(1 + z'^2)^{5/2}}$$

The determination of the appropriate curve $z = z(x)$ from this integral is a variational problem with no elementary solution. However, if it is assumed that $z' \ll 1$ along the entire length of the spline, we obtain the simpler problem of minimising

$$\int z''^2 \, dx \quad ,$$

whose solution proves to be a piecewise cubic function, continuous up to its second derivatives, namely a cubic spline in the mathematical sense. In practice, large slopes may be dealt with by local redefinition of axes in appropriate regions, as in the diagram below:

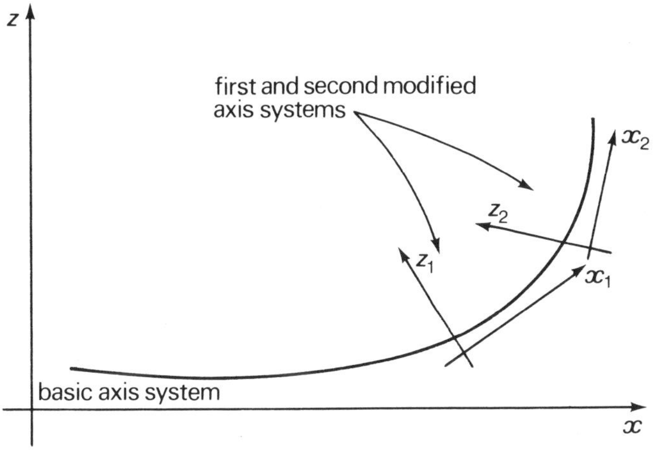

Figure 6.2

In this way it may be ensured that slopes are always small in terms of the local system of coordinates. The APT interpolation procedure TABCYL (see Section 6.4) uses this stratagem. Alternatively, a parametric spline may be used (see Section 6.3.4), which has the advantage of being axis-independent.

In passing, we note that the KURGLA curve-fitting procedure used by Mehlum (1969) in his AUTOKON ship-designing system is based upon the minimisation of the original strain-energy integral rather than the simplified approximation. Mehlum does not approximate until later in the algorithm, so that no problems arise in the treatment of large slopes (see Section 6.4).

We now consider the generation of a cubic spline $\phi(x)$ which fits the data points (x_i,z_i), $i = 0, 1, 2, \ldots, n$. We recall that

(i) $\phi(x)$ is a polynomial of degree $\leqslant 3$ in each interval $x_{i-1} \leqslant x \leqslant x_i$,

(ii) $\phi(x_i) = z_i$,

(iii) $\phi'(x)$ and $\phi''(x)$ are continuous at the points $x_1, x_2, \ldots, x_{n-1}$.

In the terminology of splines the points x_i are known as **knots** and each interval $x_{i-1} \leqslant x \leqslant x_i$ is called a **span**.

On the first span, $x_0 \leqslant x \leqslant x_1$, we know that $\phi(x_0) = z_0$ and $\phi(x_1) = z_1$. In order to determine uniquely the four coefficients of an interpolating cubic over this span we need two additional items of information. Suppose for the moment then, that $\phi'(x_0) = z_0'$ and $\phi''(x_0) = z_0''$ are also prescribed at x_0. This enables us to construct a cubic on the first span which assumes the specified function and derivative values at the end points.

Once this has been done, we may determine $\phi'(x_1)$ and $\phi''(x_1)$ from the cubic. But these two derivatives are to be continuous at x_1, and so we have the four values $\phi(x_1)$, $\phi'(x_1)$, $\phi''(x_1)$ and $\phi(x_2) = z_2$ which determine the cubic function comprising the second span of the spline. The remaining segments of the spline may be successively calculated in this way until the last data point is reached.

The argument above shows that the complete cubic interpolating spline may be constructed by using the $(n + 1)$ function values z_i and just two other items of information. In practice splines are not usually computed by this method, for two reasons. Firstly, the accumulation of rounding errors leads to significant deviations from the true spline if many spans are to be calculated. Secondly, it is more often convenient to specify one derivative value at each end of the spline than to specify two at the same end as we have done here. However, Nutbourne (1973) has described a practical implementation of the process outlined above. We now give the method which is more usually used in practice, and which gives reasonable accuracy for fairly uniform knot spacings. We consider initially only a single span of the spline, whose width is $x_i - x_{i-1} = h_i$, as shown in Figure 6.3.

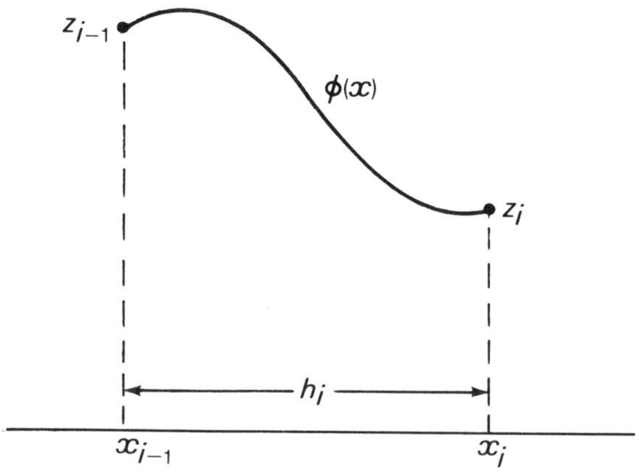

Figure 6.3

Since $\phi(x)$ is cubic, $\phi'(x)$ is quadratic and $\phi''(x)$ linear in x over this span. If $\phi''(x_{i-1}) = s_{i-1}$ and $\phi''(x_i) = s_i$ then $\phi''(x)$ is given at any point on the span by the linear interpolation formula

$$\phi''(x) = \frac{s_{i-1}(x_i - x) + s_i(x - x_{i-1})}{h_i} .$$

Two integrations then result in

$$\phi(x) = \frac{s_{i-1}(x_i - x)^3 + s_i(x - x_{i-1})^3}{6h_i} + c_1 x + c_2 ,$$

where the constants of integration c_1 and c_2 may be evaluated from the end conditions $\phi(x_{i-1}) = z_{i-1}, \phi(x_i) = z_i$ to give

$$\phi(x) = \frac{s_{i-1}(x_i - x)^3}{6h_i} + \frac{s_i(x - x_{i-1})^3}{6h_i} \tag{6.1}$$

$$+ \left\{\frac{z_{i-1}}{h_i} - \frac{s_{i-1}h_i}{6}\right\}(x_i - x) + \left\{\frac{z_i}{h_i} - \frac{s_ih_i}{6}\right\}(x - x_{i-1}) \,.$$

This expresses the interpolating cubic over the span $x_{i-1} \leqslant x \leqslant x_i$ in terms of two known quantities, z_{i-1}, z_i, and two unknown quantities, s_{i-1}, s_i.

In order to determine the values of s_{i-1} and s_i we use the property of first-derivative continuity at the knots of the spline. On differentiating (6.1) with respect to x and setting $x = x_i$ we find, after some simplification,

$$\phi'(x_i) = \frac{z_i - z_{i-1}}{h_i} + \frac{s_ih_i}{3} + \frac{s_{i-1}h_i}{6} \,. \tag{6.2}$$

If now we replace i by i+1 throughout (6.1) we obtain the cubic which interpolates the next span, $x_i \leqslant x \leqslant x_{i+1}$. Differentiating this and setting $x = x_i$ we get

$$\phi'(x_i) = \frac{z_{i+1} - z_i}{h_{i+1}} - \frac{s_ih_{i+1}}{3} - \frac{s_{i+1}h_{i+1}}{6} \,, \tag{6.3}$$

where $h_{i+1} = x_{i+1} - x_i$. Since $\phi'(x)$ is to be continuous across $x = x_i$, the right-hand sides of (6.2) and (6.3) must be equal. This gives

$$h_is_{i-1} + 2(h_i + h_{i+1})s_i + h_{i+1}s_{i+1} = \frac{6(z_{i+1} - z_i)}{h_{i+1}} - \frac{6(z_i - z_{i-1})}{h_i} \,, \tag{6.4}$$

which is linear in the three unknowns s_{i-1}, s_i and s_{i+1}. We can set up an equation of this type for each of the $(n-1)$ internal knots $x_1, x_2, \ldots, x_{n-1}$. But we have $(n + 1)$ values of s to calculate, namely s_0, $s_1, \ldots, s_n$, and we therefore need precisely two additional relations to complete the system and enable us to compute the spline unambiguously. The additional relations chosen will depend upon physical or other considerations in any specific application. The possibilities include

(i) free ends: no curvature of the spline at x_0 and x_n, that is $s_0 = s_n = 0$ (this gives a **natural spline**);

(ii) built-in ends: specified first derivatives at x_0 and x_n. Setting $i = 0$ and $\phi'(x_0) = g_0$ in (6.3) we find

$$2h_1s_0 + h_1s_1 = \frac{6(z_1 - z_0)}{h_1} - 6g_0 ,$$

and we may use (6.2) to obtain a similar relation at x_n if $\phi'(x_n) = g_n$ is specified:

$$h_ns_{n-1} + 2h_ns_n = 6g_n - \frac{6(z_n - z_{n-1})}{h_n} ;$$

(iii) quadratic end spans: since a quadratic has a constant second derivative, our two extra constraint equations are then $s_0 = s_1$ and $s_{n-1} = s_n$.

This last assumption is one fairly obvious method of absolving the user from having to specify any derivative values. A more satisfactory way of doing this, however, is to set up a single cubic over the double span $x_0 \leqslant x \leqslant x_2$ which is also required to interpolate the point (x_1,z_1), and similarly at the other end of the spline. The resulting equations are

$$h_2s_0 - (h_1 + h_2)s_1 + h_1s_2 = 0 ,$$

$$h_ns_{n-2} - (h_{n-1} + h_n)s_{n-1} + h_{n-1}s_n = 0 .$$

It is possible of course, to use different constraints at the two ends of the spline.

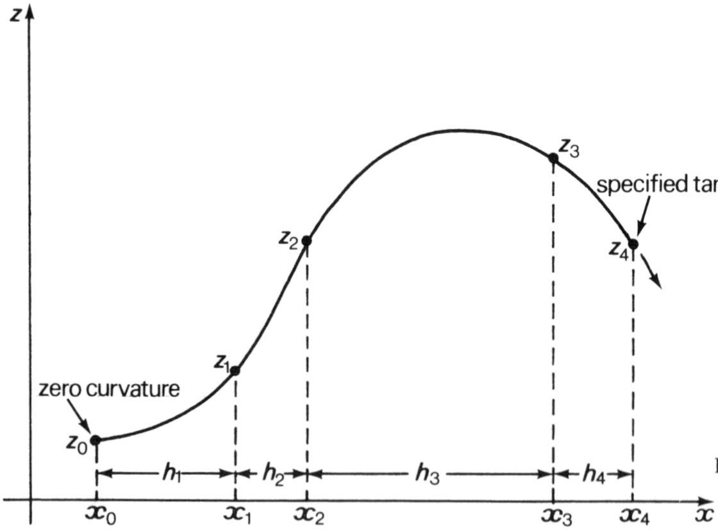

Figure 6.4

To illustrate the computation of a spline curve, we quote the system of equations for a case with $n = 4$ and with end conditions $s_0 = 0$ and $h_4s_3 + 2h_4s_4 = 6g_4 - \dfrac{6(z_4 - z_3)}{h_4}$, corresponding to one free end and one end with specified gradient g_4 (see Figure 6.4). The equations for the unknown second derivatives, written in matrix form, are

$$\begin{bmatrix} 2(h_1+h_2) & h_2 & & \\ h_2 & 2(h_2+h_3) & h_3 & \\ & h_3 & 2(h_3+h_4) & h_4 \\ & & h_4 & 2h_4 \end{bmatrix} \begin{bmatrix} s_1 \\ s_2 \\ s_3 \\ s_4 \end{bmatrix}$$

$$= 6 \begin{bmatrix} \frac{1}{h_1} z_0 - \left(\frac{1}{h_1} + \frac{1}{h_2}\right) z_1 + \frac{1}{h_2} z_2 \\ \frac{1}{h_2} z_1 - \left(\frac{1}{h_2} + \frac{1}{h_3}\right) z_2 + \frac{1}{h_3} z_3 \\ \frac{1}{h_3} z_2 - \left(\frac{1}{h_3} + \frac{1}{h_4}\right) z_3 + \frac{1}{h_4} z_4 \\ g_4 + \frac{1}{h_4} z_3 - \frac{1}{h_4} z_4 \end{bmatrix}.$$

We note that the matrix of coefficients of this system is **tridiagonal** (having non-zero elements only on the leading diagonal and immediately to either side of it). A solution to such a system may be computed accurately and efficiently using well-established techniques, as explained in Chapter 3 of Goult *et al* (1974). Once all the s-values have been calculated, (6.1) gives the particular cubic which fits each individual span of the spline.

The gradients $g_i = \phi'(x_i)$ may be calculated, if desired, from (6.2). Alternatively, the roles of s_i and g_i in the derivation of the formulae may be reversed. The algebra is more complicated, but a similar system of equations results, and the values of the g_i rather than the s_i are then found explicitly in the course of the computation. The equations which replace (6.1) and (6.4) are respectively

$$\phi(x) = \frac{g_{i-1}(x_i - x)^2(x - x_{i-1})}{h_i^2} - \frac{g_i(x - x_{i-1})^2(x_i - x)}{h_i^2}$$

$$+ \frac{z_{i-1}(x_i - x)^2[2(x - x_{i-1}) + h_i]}{h_i^3} + \frac{z_i(x - x_{i-1})^2[2(x_i - x) + h_i]}{h_i^3}$$

and

$$h_{i+1}g_{i-1} + 2(h_i + h_{i+1})g_i + h_i g_{i+1} = \frac{3h_{i+1}(z_i - z_{i-1})}{h_i} + \frac{3h_i(z_{i+1} - z_i)}{h_{i+1}}.$$

The use of these equations would be appropriate when values of first (but not second) derivatives are required at the knots once the spline has been constructed.

From the fact that the spline may be constructed in this alternative manner, we see that cubic spline interpolation is an application of the piecewise Hermite interpolation discussed in Section 6.2.1. In effect, the whole of the data set is first surveyed to determine the values which must be chosen for the first derivatives at the knots so that the piecewise Hermite interpolant also has continuous second derivatives there.

As mentioned earlier, it is possible to compute interpolating splines of higher than cubic degree. Details of the fifth-degree spline are given, for instance, in Späth (1974), which also contains some useful computational subroutines. Detailed treatments of the general theory of polynomial splines and their applications are given by Ahlberg, Nilson and Walsh (1967) and Greville (1969).

The **spline in tension** (Schweikert, 1966) is a mathematical analogue of the mechanical spline subject to a uniform tension between its ends. This type of curve is attracting increasing interest because it is possible to adjust the tension parameter to avoid unwanted oscillations or inflections which polynomial splines sometimes exhibit. Cline (1974) and Nielson (1974) discuss the computational aspects of such curves.

It should by now be apparent that splines are smooth yet flexible curves having many applications in curve fitting and design. They have the great advantages that second-order continuity is attained (so that curvature is continuous) and that little *a priori* derivative information is required in their construction. They do have limitations, however, amongst which are

(i) a local modification involves the recomputation of the entire splines;

(ii) a spline as treated here will not cope with a vertical tangent;

(iii) oscillation problems may arise in the approximation of a curve with a discontinuity in its second derivative, for example the continuation of a straight line by a circular arc;

(iv) splines have certain aesthetic objections. The curvature of a surface constructed from spline curves sometimes varies in a rather non-uniform manner so that the pattern of reflections from, say, a car body might contain odd-looking distortions.

Of these four problems, the first can be avoided by use of B-splines, as explained in the next section, while the second can be overcome and the third alleviated by the use of parametric splines (Section 6.3.4). The fourth (the 'highlight' problem) is probably less acute for the splines in tension mentioned in the last paragraph but one.

6.2.3 B-splines

In the last section we saw how a cubic spline can be constructed over n spans. By virtue of the continuity conditions imposed, the only data requirements are the function values z_i to be interpolated at the $(n + 1)$ knots, together with just two extra items of information, making $(n + 3)$ items of data in all.

With this in mind, we consider the construction of a cubic spline $\phi(x)$ which has the properties $\phi(x) = \phi'(x) = \phi''(x) =$ at each end. These six items of data should enable us to compute a unique spline over three spans. This is indeed so, but the spline in question proves to be $\phi(x) \equiv 0$, which undeniably fits the end conditions and has the required continuity at the knots. The case of four spans is much more fruitful; now we need an additional item of data, and can therefore specify a non-zero function value at an internal knot, which ensures that the spline is not identically zero. The general form of the resulting curve is as shown below:

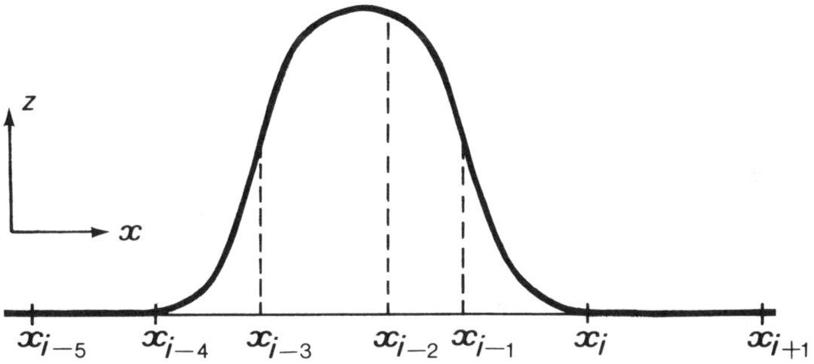

Figure 6.5

In this diagram the spline has been extended from its end-points x_{i-4}, x_i by straight lines along the x-axis. The result is a cubic spline over an indefinite number of spans, but which departs from zero only over precisely four of those spans. Such a function is called a **B-spline** (or **fundamental spline**) of order 4 (or degree 3). A B-spline is said to be a spline of minimal support, its **support** being the number of spans over which a spline is non-zero.

Note that a cubic B-spline is completely determined by the knot set on which it is defined and just one specified value of z. More generally, the B-spline $M_{mi}(x)$ of order m (or degree $m-1$) on a given knot set is zero everywhere except over the m successive spans $x_{i-m} < x < x_i$. Again, $M_{mi}(x)$ is determined by the knot set and one value of z. It is customary to remove this last degree of freedom and to fix the amplitude of the B-spline by standardising in some way. In what follows, we shall standardise according to

$$\int_{x_{i-m}}^{x_i} M_{mi}(x)\mathrm{d}x = m^{-1}$$

as proposed by Cox (1972) and de Boor (1972). Computationally the use of the **normalised** B-spline $N_{mi}(x)$, related to $M_{mi}(x)$ by

$$N_{mi}(x) = (x_i - x_{i-m})M_{mi}(x) ,$$

is often convenient. We will show in the next section that $M_{mi}(x)$, and hence also $N_{mi}(x)$, is everywhere $\geqslant 0$.

The practical importance of B-splines is due to the property (Curry and Schoenberg, 1966) that any spline of order m on a set of knots $x_0, x_1, \ldots, x_n$ can be expressed as a sum of multiples of B-splines defined on the same knot set extended by $(m-1)$ additional knots at each end of the range, which may be chosen arbitrarily: $x_{-m+1}, x_{-m+2}, \ldots, x_{-1}$ and $x_{n+1}, \ldots, x_{n+m-1}$. It is possible to construct $m + n - 1$ successive B-splines on the extended knot set, each of which is non-zero over just m consecutive spans. We can therefore write

$$\phi(x) = \sum_{i=1}^{m+n-1} c_i M_{mi}(x) , \tag{6.5}$$

where $\phi(x)$ is any spline of degree $(m - 1)$ on the original knot set and $M_{mi}(x)$ is the B-spline on the extended knot set which is non-zero for $x_{i-m} < x < x_i$. The c_i are numerical coefficients. We could equally well write (6.5) in terms of the normalised B-splines $N_{mi}(x)$ by adjusting these coefficients appropriately.

This result is very significant because the B-spline is only locally non-zero, as we have seen. If a spline of degree $m - 1$ is expressed in terms of B-splines, changing the coefficient of one of the B-splines alters precisely m spans of the curve, without affecting its continuity properties. By this means we may make local modifications to the curve without having to recompute it completely. Such a feature is desirable in any design system. Moreover, representations of spline curves in terms of B-splines have certain advantages concerned with computational stability (Varah, 1977), and the future of these functions in computational geometry therefore seems assured.

6.2.4 Further aspects of B-splines

Space limitations permit no detailed discussion of the practical implementation of B-splines. However, the present section outlines some of the underlying ideas, and is intended as an introduction to the specialist literature on the subject.

First we introduce a useful notation, defining the truncated power function t_+^k for any variable t and positive integer k by

$$t_+^k = \begin{cases} t^k, & t \geqslant 0 \\ 0, & t < 0. \end{cases}$$

Next we consider the function $(x - x_i)_+^{m-1}$, whose graph has the following form:

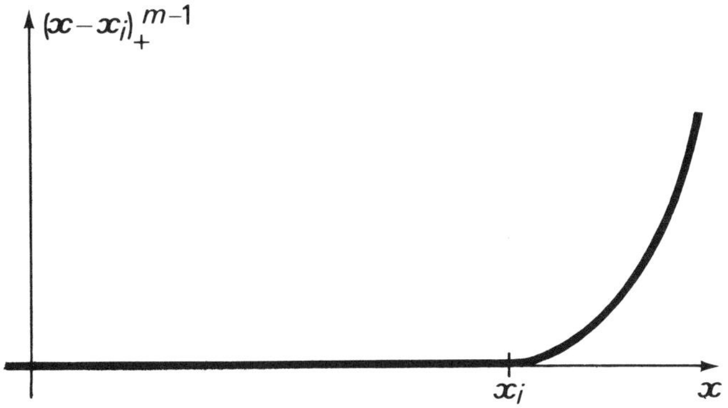

Figure 6.6

This function is continuous at $x = x_i$, and so are its derivatives up to order $(m - 2)$. Then $(x - x_i)_+^{m-1}$ is a spline of order m (degree $m - 1$), and consequently if we add it to any other spline of order m on a knot set including x_i, their sum will have the same nature. In fact it is possible to express any spline of order m on the knot set $x_0, x_1, \ldots, x_n$ as

$$\phi(x) = p(x) + \sum_{r=1}^{n-1} \alpha_r (x - x_r)_+^{m-1} , \tag{6.6}$$

where $p(x)$ is a polynomial of degree m–1 and the α_r are constants. This was first established by Schoenberg and Whitney (1953). The terms up to $r = k-1$ in the summation determine the spline as far as x_k. The addition of the kth term does not violate the spline conditions at x_k, but introduces a discontinuity in $\phi^{(m-1)}(x)$ there, thus modifying the polynomial fitting the span $x_{k-1} \leqslant x \leqslant x_k$ to give the polynomial fitting $x_k \leqslant x \leqslant x_{k+1}$.

One computational approach to B-splines is based on divided differences (see Appendix 5) of a truncated power function. We illustrate for the cubic case ($m = 4$), though generalisation to splines of higher degree is straightforward. Consider, then, the function of two variables

$$F[x;y] = (y - x)_+^3 .$$

Taking the value of x to be fixed initially, we form the fourth divided difference of this function with respect to y, based on the set of knots $y = x_{i-4}, x_{i-3}, \ldots, x_i$. It is a standard result from the theory of divided differences that for any function $f[x]$,

$$f[x_j, x_{j+1}, \ldots, x_{j+k}] = \sum_{r=0}^{k} \frac{f(x_{j+r})}{\omega'(x_{j+r})} ,$$

where $\omega(x) = (x - x_j)(x - x_{j+1}) \ldots (x - x_{j+k})$ and the prime denotes differentiation with respect to x (see, for example, Hildebrand (1956), p. 39). In terms of the function of present interest, this gives

$$F[x; x_{i-4}, x_{i-3}, x_{i-2}, x_{i-1}, x_i] = \sum_{r=0}^{4} \frac{(x_{i+r-4} - x)_+^3}{\omega'(x_{i+r-4})} . \tag{6.7}$$

From what has gone before we recognise the right-hand side as a cubic spline $\psi(x)$ in the variable x, based on the knots $x_{i-4}, x_{i-3}, \ldots, x_i$. In fact $\psi(x)$ is the cubic B-spline $M_{4i}(x)$, for the following reasons:

(i) When $x \geqslant x_i$ in (6.7) all terms in the summation are zero, and $\psi(x)$ is identically zero.
(ii) When $x \leqslant x_{i-4}$, $F[x;y]$ is a pure cubic polynomial in y over $x_{i-4} \leqslant y \leqslant x_i$. Now a fourth divided difference of any cubic is zero, as shown in Appendix 5, and hence $\psi(x)$ is zero for $x \leqslant x_{i-4}$.

As given by (6.7), the B-spline proves to be standardised so that its integral over $x_{i-4} \leqslant x \leqslant x_i$ is ¼.

From the foregoing, we see that there are two straightforward ways of computing a B-spline for a given value of x. The first is by evaluating the sum in (6.7), and the second by forming a conventional divided difference table and computing the left-hand side of (6.7). Unfortunately, both these procedures can be computationally unsatisfactory with regard to the growth of numerical errors. The present standard method avoids these problems by making use of the following recurrence relation, due independently to Cox (1972) and de Boor (1972):

$$M_{\mu i}(x) = \frac{(x - x_{i-\mu})M_{\mu-1,i-1}(x) + (x_i - x)M_{\mu-1,i}(x)}{x_i - x_{i-\mu}} . \tag{6.8}$$

This enables $M_{4i}(x)$ to be computed from the array

$$\begin{array}{ccccccc}
M_{1,i-3}(x) & & & & & & \\
 & & M_{2,i-2}(x) & & & & \\
M_{1,i-2}(x) & & & & M_{3,i-1}(x) & & \\
 & & M_{2,i-1}(x) & & & & M_{4,i}(x) , \\
M_{1,i-1}(x) & & & & M_{3,i}(x) & & \\
 & & M_{2,i}(x) & & & & \\
M_{1,i}(x) & & & & & &
\end{array}$$

in which the columns contain respectively splines of order 1 (piecewise constant), of order 2 (piecewise linear), of order 3 (piecewise quadratic) and of order 4 (piecewise cubic). Now a B-spline of order 1 will be non-zero over just one span, where its value will be constant. If it is standardised in the usual way, its integral over that span will be unity, and we therefore set

$$M_{1j}(x) = \begin{cases} (x_j - x_{j-1})^{-1}, & x_{j-1} \leqslant x < x_j \\ 0, & \text{otherwise.} \end{cases}$$

In practice the foregoing array may be simplified because the last relation shows that some of its elements are zero. For example, if $x_{i-2} \leqslant x < x_{i-1}$ we need only compute the rhomboidal array

$$\begin{array}{cccc} & & M_{3,i-1}(x) & \\ & M_{2,i-1}(x) & & M_{4,i}(x)\,, \\ M_{1,i-1}(x) & & M_{3,i}(x) & \\ & M_{2,i}(x) & & \end{array}$$

since all the remaining elements in the original triangular array are zero. Note that if the value of the *normalised* B-spline $N_{4i}(x) = (x_i - x_{i-4})M_{4i}$ is required, it is only necessary to omit the division by $(x_i - x_{i-4})$ in the last step.

Since entries in the first column of the array are non-negative and successive columns are calculated by adding together non-negative quantities, the statement that $M_{mi}(x) \geqslant 0$ for all x, made in the last section, is clearly justified. The fact that addition is the basic operation in the use of (6.8) is the actual reason for the computational advantage of the Cox-de Boor method. The procedures based on (6.7) involve subtraction, an operation which can lead to drastic loss of accuracy if the difference between the two numbers involved is small.

Cox (1972) gives a detailed error analysis of this method and of the divided difference method based on (6.7) for computing B-spline values. The method described here proves both more efficient and more accurate. In particular, it gives much better accuracy in the computation of splines of high degree and/or with highly non-uniform knot spacing.

Finally, we turn to the problem of interpolation using B-splines, with function values specified at the knots $x_0, x_1, \ldots, x_n$. Again we illustrate the cubic case for simplicity. First, three additional knots must be chosen at each end of the range to give the extended knot set. Next an obvious extension of the method just outlined is used to compute the value at each original knot $x_0, x_1, \ldots, x_n$ of the three B-splines which are non-zero there. For instance, at the knot x_i the non-zero B-splines are $M_{4,i+1}$, $M_{4,i+2}$ and $M_{4,i+3}$ (see Figure 6.7). Their values at x_i may all be found from an array such as the following:

0			
	0		
0		0	
	0		$M_{4,i+1}(x_i)$
0		$M_{3,i+1}(x_i)$	
	$M_{2,i+1}(x_i)$		$M_{4,i+2}(x_i)$.
$M_{1,i+1}(x_i)$		$M_{3,i+2}(x_i)$	
	$M_{2,i+2}(x_i)$		$M_{4,i+3}(x_i)$
0		$M_{3,i+3}(x_i)$	
	0		
0			

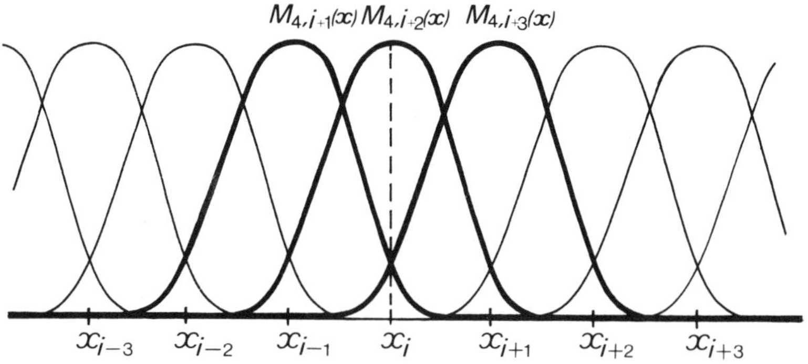

Figure 6.7

If the value to be interpolated at x_i is z_i, then we have, from (6.5),

$$z_i = \phi(x_i) = c_{i+1}M_{4,i+1}(x_i) + c_{i+2}M_{4,i+2}(x_i) + c_{i+3}M_{4,i+3}(x_i) ,$$

in which the c's are unknown. There will be one such linear equation for each of the $(n + 1)$ knots $x_0, x_1, \ldots, x_n$. But there are $(n + 3)$ B-splines on the extended knot set, and hence $(n + 3)$ coefficients $c_1, c_2, \ldots, c_{n+3}$ to be determined. We therefore need, as previously, just two extra items of information to determine $\phi(x)$ completely. These will usually be derivative values in practice. For instance, if $z_0'' = \phi''(x_0) = 0$ is specified, then

$$z_0'' = \phi''(x_0) = c_1M''_{41}(x_0) + c_2M''_{42}(x_0) + c_3M''_{43}(x_0).$$

If a similar condition is given at x_n we then have the two extra equations needed to complete the system. The B-spline derivatives in equations such as these may be evaluated as suggested by de Boor (1972) or by a method due to Butterfield (1976) which has better numerical stability.

Once all the coefficients in (6.5) have been determined, the resulting linear combination of B-splines may be evaluated for any desired value of x. This can be done straightforwardly, by using a table based on (6.8) to evaluate the four B-splines which are non-zero at x and using (6.5) directly. A more elegant alternative is the 'method of convex combinations' (de Boor, 1972), but Cox (1976) has recently shown that the straightforward approach is comparable both in efficiency and accuracy.

For further reading on the subject of B-splines, the papers of Cox (1972) and de Boor (1972) are fundamental. A generalisation of the interpolation process outlined here is given in Cox (1975). Other papers dealing with the use of B-splines in interpolation and curve-fitting problems are those of Schumaker (1969), Cox and Hayes (1973), Hayes (1974), Hayes and Halliday (1974), Cox (1976) and de Boor (1976).

6.3 PARAMETRIC COMPOSITE CURVES

The remainder of this chapter is largely concerned with the problem of constructing a general curve in three dimensions. From the classical point of view, such a curve is defined as the line of intersection of two surfaces. While this approach has applications in the field of numerical control, it is not usually a very helpful viewpoint for the designer, who will usually wish to specify the curve first and only subsequently enlist its aid in constructing one or more surfaces.

Another way of defining a curve in Cartesian space is by means of its projections onto the (x,y)-plane and the (x,z)-plane. In fitting a set of points (x_i, y_i, z_i), $i = 0, 1, \ldots, n$, we construct one plane composite curve fitting the points (x_i, y_i) and another fitting the points (x_i, z_i). Then for any given x we can calculate the corresponding values of y and z. This approach is satisfactory for certain limited applications, but complications arise if the x_i do not satisfy $x_0 < x_1 < \ldots < x_n$ (for instance if the curve turns back on itself). We also run into problems if $\dfrac{dy}{dx}$ or $\dfrac{dz}{dx}$ become infinite anywhere on the desired curve, since the techniques given earlier in this chapter are not well suited for coping with such situations.

These disadvantages may be overcome by the use of parametric methods, whose virtues with regard to curve and surface design were explained earlier in Section 4.1.5. Although we deal with space curves in what follows, parametric methods have corresponding advantages for the treatment of plane curves, and the necessary specialisation to two dimensions is immediate.

We first consider the construction of composite curves from the various types of parametric curve segments introduced in Chapter 5.

Suppose, then, that we wish to join a segment $\mathbf{r}^{(1)}(u_1)$, $0 \leqslant u_1 \leqslant 1$ to a segment $\mathbf{r}^{(2)}(u_2)$, $0 \leqslant u_2 \leqslant 1$. We will usually require the curve to be continuous at the join, and to have continuity of slope there. Then we must have

$$\mathbf{r}^{(1)}(1) = \mathbf{r}^{(2)}(0)\ , \tag{6.9}$$

and

$$\left.\begin{aligned} \dot{\mathbf{r}}^{(1)}(1) &= \alpha_1 \mathbf{T} \\ \dot{\mathbf{r}}^{(2)}(0) &= \alpha_2 \mathbf{T} \end{aligned}\right\}\ , \tag{6.10}$$

using the notation of Section 5.1.2, where a dot denotes differentiation with respect to the appropriate parameter, $\mathbf{T}$ is the unit vector in the common tangent direction at the join and α_1, α_2 are scalar constants which, as shown earlier, influence the fullness of the curve segments.

The attainment of curvature continuity is less straightforward. In terms of arc length s, the curvature κ at any point on the curve is given by

$$\frac{\mathrm{d}^2\mathbf{r}}{\mathrm{d}s^2} = \kappa \mathbf{N}\ ,$$

where $\mathbf{N}$ is the unit vector in the direction of the principal normal, namely the direction of the line joining the point under consideration to the centre of curvature. If the centre of curvature is to move continuously as we cross the join, both κ and $\mathbf{N}$ must be continuous there. If $\mathbf{N}$ and $\mathbf{T}$ are continuous, so is the binormal vector $\mathbf{B} = \mathbf{T} \times \mathbf{N}$, and in terms of this vector we have the result, derived in Section 4.2.5,

$$\kappa \mathbf{B} = \frac{\dot{\mathbf{r}} \times \ddot{\mathbf{r}}}{|\dot{\mathbf{r}}|^3}\ . \tag{6.11}$$

This is convenient for curvature analysis in that it involves derivatives with respect to the curve parameter alone, and not with respect to s. For curvature continuity at the join, then, we require

$$\frac{\dot{\mathbf{r}}^{(2)}(0) \times \ddot{\mathbf{r}}^{(2)}(0)}{|\dot{\mathbf{r}}^{(2)}(0)|^3} = \frac{\dot{\mathbf{r}}^{(1)}(1) \times \ddot{\mathbf{r}}^{(1)}(1)}{|\dot{\mathbf{r}}^{(1)}(1)|^3}\ ,$$

which becomes, on taking (6.10) into account,

$$\mathbf{T} \times \ddot{\mathbf{r}}^{(2)}(0) = \left(\frac{\alpha_2}{\alpha_1}\right)^2 \mathbf{T} \times \ddot{\mathbf{r}}^{(1)}(1)\ .$$

This relation is satisfied by

$$\ddot{\mathbf{r}}^{(2)}(0) = \left(\frac{\alpha_2}{\alpha_1}\right)^2 \ddot{\mathbf{r}}^{(1)}(1) + \mu\dot{\mathbf{r}}^{(1)}(1) , \qquad (6.12)$$

where μ is an arbitrary scalar. However, most practical curve-defining systems take μ to be zero, sacrificing a degree of flexibility in the interests of simplicity. As will be seen in what follows, different methods deal in various ways with the tangent vector magnitudes α_1 and α_2. A general discussion of these magnitudes and their significance is given later, in Section 6.3.6.

6.3.1 Composite Ferguson Curves

It was shown in Section 5.1.2 that a Ferguson curve segment may be written in terms of its end points and end tangents as

$$\mathbf{r}(u) = \mathbf{r}(0)(1 - 3u^2 + 2u^3) + \mathbf{r}(1)(3u^2 - 2u^3) + \dot{\mathbf{r}}(0)(u - 2u^2 + u^3) + \dot{\mathbf{r}}(1)(-u^2 + u^3) , \quad 0 \leqslant u \leqslant 1 . \qquad (6.13)$$

Ferguson (1963, 1964) confines his attention to the obvious method for obtaining curvature continuity, and matches $\mathbf{r}$, $\dot{\mathbf{r}}$ and $\ddot{\mathbf{r}}$ across joins between segments, so that (6.9) holds, $\alpha_1 = \alpha_2$ in (6.10) and also $\ddot{\mathbf{r}}^{(1)}(1) = \ddot{\mathbf{r}}^{(2)}(0)$. The second derivatives are easily evaluated from (6.13), when this last condition becomes

$$6\mathbf{r}^{(1)}(0) - 6\mathbf{r}^{(1)}(1) + 2\dot{\mathbf{r}}^{(1)}(0) + 4\dot{\mathbf{r}}^{(1)}(1) = -6\mathbf{r}^{(2)}(0) + 6\mathbf{r}^{(2)}(1) - 4\dot{\mathbf{r}}^{(2)}(0) - 2\dot{\mathbf{r}}^{(2)}(1) .$$

This simplifies because of the assumptions that $\mathbf{r}^{(1)}(1)=\mathbf{r}^{(2)}(0)$ and $\dot{\mathbf{r}}^{(1)}(1)=\dot{\mathbf{r}}^{(2)}(0)$, though the result is best expressed in an alternative notation. If we regard ourselves as fitting a composite Ferguson curve through a set of points $\mathbf{r}_0$, $\mathbf{r}_1$, . . . , $\mathbf{r}_n$, and the tangents at those points are $\mathbf{t}_0$, $\mathbf{t}_1$, . . . , $\mathbf{t}_n$, the last equation gives

$$\mathbf{t}_{i-1} + 4\mathbf{t}_i + \mathbf{t}_{i+1} = 3(\mathbf{r}_{i+1} - \mathbf{r}_{i-1}) , \qquad i = 1, 2, \ldots, n-1.$$

This is a recurrence relation between the tangents at three successive points. It is only necessary to specify $\mathbf{t}_0$ and $\mathbf{t}_n$ to obtain a system of equations from which all the remaining tangents may be determined in terms of positional data alone. Taking these values for the tangents will then assure curvature continuity for the composite curve. Clearly, this procedure is closely akin to the one given in Section 6.2.2 for calculating cubic splines. In fact a composite Ferguson curve is a particular case of a **parametric spline**, a type of curve which we discuss more generally in Sections 6.3.4 and 6.3.6. As explained there, Ferguson's method is well suited to the automatic fitting of a curve through a specified sequence of data points, provided their physical spacing is not very uneven.

6.3.2 Composite Bézier Curves

The cubic Bézier segment, introduced in Section 5.1.3, is given by

$$\mathbf{r}(u) = (1-u)^3\mathbf{r}_0 + 3u(1-u)^2\mathbf{r}_1 + 3u^2(1-u)\mathbf{r}_2 + u^3\mathbf{r}_3 \ , \quad 0 \leqslant u \leqslant 1.$$

Here $\mathbf{r}_0$, $\mathbf{r}_1$, $\mathbf{r}_2$ and $\mathbf{r}_3$ are the vertices of the 'characteristic polygon' to which the curve is, in a sense, an approximation. Bézier is less restrictive than Ferguson in his choice of inter-segment continuity conditions. Whereas Ferguson's method is used primarily for curve fitting, Bézier's approach offers the extra freedom needed for curve design.

Suppose we wish to construct a curve segment $\mathbf{r}^{(2)}(u_2)$ which has slope and curvature continuity at its join with an existing segment $\mathbf{r}^{(1)}(u_1)$. The first requirement, since $\mathbf{r}(0) = \mathbf{r}_0$ and $\mathbf{r}(1) = \mathbf{r}_3$, is

$$\mathbf{r}_3{}^{(1)} = \mathbf{r}_0{}^{(2)} \ . \tag{6.14}$$

Also $\dot{\mathbf{r}}(0) = 3(\mathbf{r}_1 - \mathbf{r}_0)$ and $\dot{\mathbf{r}}(1) = 3(\mathbf{r}_3 - \mathbf{r}_2)$, whence for continuity of tangent direction we require

$$\frac{3}{\alpha_1}(\mathbf{r}_3{}^{(1)} - \mathbf{r}_2{}^{(1)}) = \frac{3}{\alpha_2}(\mathbf{r}_1{}^{(2)} - \mathbf{r}_0{}^{(2)}) = \mathbf{T} \ , \tag{6.15}$$

where α_1, α_2 are the magnitudes of $\dot{\mathbf{r}}^{(1)}(1)$ and $\dot{\mathbf{r}}^{(2)}(0)$ respectively, which we now permit to be different. Note that if (6.14) and (6.15) both hold, the three points $\mathbf{r}_2^{(1)}$, $\mathbf{r}_3^{(1)} = \mathbf{r}_0^{(2)}$ and $\mathbf{r}_1^{(2)}$ must be collinear. We also find

$$\left.\begin{aligned} \ddot{\mathbf{r}}^{(1)}(1) &= 6(\mathbf{r}_1^{(1)} - 2\mathbf{r}_2^{(1)} + \mathbf{r}_3^{(1)}) \\ \text{and} \qquad \ddot{\mathbf{r}}^{(2)}(0) &= 6(\mathbf{r}_0^{(2)} - 2\mathbf{r}_1^{(2)} + \mathbf{r}_2^{(2)}) \end{aligned}\right\} . \tag{6.16}$$

The curvature continuity requirement that

$$\kappa\mathbf{B} = \frac{\dot{\mathbf{r}} \times \ddot{\mathbf{r}}}{|\dot{\mathbf{r}}|^3}$$

shall be continuous across the join then becomes, from (6.12),

$$\ddot{\mathbf{r}}^{(2)}(0) = \lambda^2\ddot{\mathbf{r}}^{(1)}(1) + \mu\dot{\mathbf{r}}^{(1)}(1) \ ,$$

where $\lambda = \alpha_2/\alpha_1$. Then (6.15) and (6.16) give

$$6(\mathbf{r}_0{}^{(2)} - 2\mathbf{r}_1{}^{(2)} + \mathbf{r}_2{}^{(2)}) = 6\lambda^2(\mathbf{r}_1{}^{(1)} - 2\mathbf{r}_2{}^{(1)} + \mathbf{r}_3{}^{(1)}) + 3\mu(\mathbf{r}_3{}^{(1)} - \mathbf{r}_2{}^{(1)}).$$

We may now eliminate $\mathbf{r}_1^{(2)}$ using (6.15) and $\mathbf{r}_0^{(2)}$ using (6.14) to obtain

$$\mathbf{r}_2^{(2)} = \lambda^2 \mathbf{r}_1^{(1)} - (2\lambda^2 + 2\lambda + \tfrac{1}{2}\mu)\mathbf{r}_2^{(1)} + (\lambda^2 + 2\lambda + 1 + \tfrac{1}{2}\mu)\mathbf{r}_3^{(1)}. \tag{6.17}$$

This determines $\mathbf{r}_2^{(2)}$ in terms of $\mathbf{r}_1^{(1)}$, $\mathbf{r}_2^{(1)}$, $\mathbf{r}_3^{(1)}$ and the chosen values of λ and μ. The vertices $\mathbf{r}_0^{(2)}$ and $\mathbf{r}_1^{(2)}$ have already been fixed by the conditions for positional and gradient continuity, leaving only the fourth vertex $\mathbf{r}_3^{(2)}$ of the characteristic polygon of $\mathbf{r}^{(2)}(u_2)$ to be freely chosen if curvature continuity is to result. If $\mathbf{r}_3^{(1)}$ is subtracted from both sides of (6.17) the right-hand side may be expressed as a combination of $(\mathbf{r}_3^{(1)} - \mathbf{r}_2^{(1)})$ and $(\mathbf{r}_2^{(1)} - \mathbf{r}_1^{(1)})$. This shows that $\mathbf{r}_2^{(2)}$ must be coplanar with $\mathbf{r}_1^{(1)}$, $\mathbf{r}_2^{(1)}$, $\mathbf{r}_3^{(1)} = \mathbf{r}_0^{(2)}$ and $\mathbf{r}_1^{(2)}$, as shown in Figure 6.8.

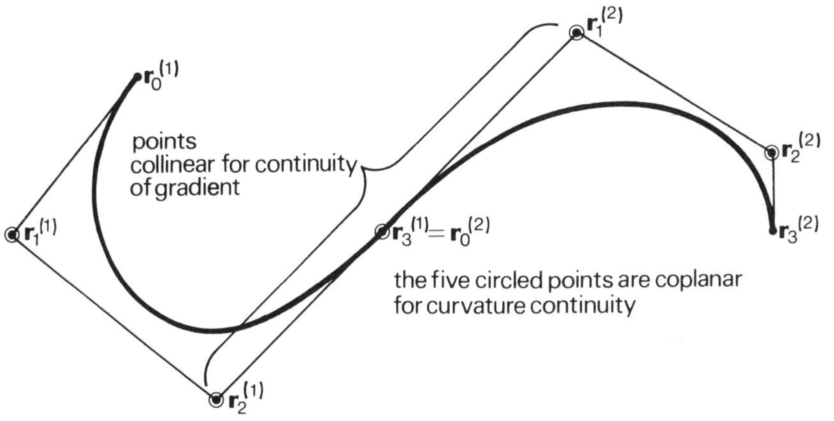

Figure 6.8 – Curvature Continuity between Bézier Segments.

Using these results it is possible to build up a composite Bézier curve with positional, gradient and curvature continuity. We start at one end, adding one segment at a time. For each new segment only λ, μ and the vertex $\mathbf{r}_3$ may be freely chosen for shape determination, because, as we have seen, the vertices $\mathbf{r}_0$, $\mathbf{r}_1$ and $\mathbf{r}_2$ are all immediately specified in terms of vertices from the preceding segment once λ and μ have been chosen.

Only in the case of a plane curve does the possibility arise of constructing a cubic segment with specified positions, tangent directions and curvatures at both end points. In the plane, a position vector is specified by two scalars, the direction of a unit tangent by one, and a radius of curvature by one. Then we have eight scalars, which suffice in principle to determine the four two-dimensional position vectors defining the characteristic polygon. The segment will be completely specified if we can determine appropriate magnitudes for the end tangents, and the equations for these magnitudes prove to be quartic (fourth-degree) polynomial equations. Such an equation does not necessarily have any real solutions, and so a segment having the required properties may or may not exist. Further details are given by Bézier (1972, p. 123 ff.). The possibility does not arise in three dimensions, where a position vector and a curvature vector each require three scalars and a tangent direction two for their specification. Then we have sixteen conditions to satisfy but only twelve degrees of freedom, three scalars for each of the four vectors defining the characteristic polygon.

It is clear that first-order continuity of composite Bézier curves is easily obtained, but that second-order continuity requires rather restrictive conditions. More freedom is available if the higher order Bernstein-Bézier curve segments of Section 5.1.4 are used; the necessary analysis follows the same lines as that given here. The reader is referred to Forrest (1972b) and Gordon and Riesenfeld (1974b) for further discussion of Bézier techniques.

6.3.3 Composite Rational Quadratic and Cubic Curves

The equation of a rational quadratic curve segment is

$$\mathbf{r}(u) = \frac{w_0\mathbf{r}_0(1-u)^2 + 2w_1\mathbf{r}_1u(1-u) + w_2\mathbf{r}_2u^2}{w_0(1-u)^2 + 2w_1u(1-u) + w_2u^2}, \quad 0 \leqslant u \leqslant 1. \tag{6.18}$$

As explained in Section 5.2.1, this is a plane conic curve which approximates the 'polygon' whose vertices are the points $\mathbf{r}_0$, $\mathbf{r}_1$ and $\mathbf{r}_2$. The first and second derivatives with respect to u at $u = 0$ and $u = 1$ are

$$\dot{\mathbf{r}}(0) = 2\frac{w_1}{w_0}(\mathbf{r}_1 - \mathbf{r}_0), \qquad \dot{\mathbf{r}}(1) = 2\frac{w_1}{w_2}(\mathbf{r}_2 - \mathbf{r}_1),$$

$$\ddot{\mathbf{r}}(0) = \left(4\frac{w_1}{w_0} + 2\frac{w_2}{w_0} - 8\frac{w_1^2}{w_0^2}\right)(\mathbf{r}_1 - \mathbf{r}_0) + 2\frac{w_2}{w_0}(\mathbf{r}_2 - \mathbf{r}_1)$$

and

$$\ddot{\mathbf{r}}(1) = \left(2\frac{w_0}{w_2} + 4\frac{w_1}{w_2} - 8\frac{w_1^2}{w_2^2}\right)(\mathbf{r}_1 - \mathbf{r}_2) + 2\frac{w_0}{w_2}(\mathbf{r}_0 - \mathbf{r}_1).$$

Using (6.11), we find that the curvatures at the end points are given by the following equations, where $k^2 = \dfrac{w_0w_2}{w_1^2}$ as in Section 5.2.1:

$$\kappa(0)\mathbf{B}(0) = \tfrac{1}{2}k^2\,\frac{(\mathbf{r}_1 - \mathbf{r}_0) \times (\mathbf{r}_2 - \mathbf{r}_1)}{|\mathbf{r}_1 - \mathbf{r}_0|^3},$$

$$\kappa(1)\mathbf{B}(1) = \tfrac{1}{2}k^2 \frac{(\mathbf{r}_2 - \mathbf{r}_1) \times (\mathbf{r}_0 - \mathbf{r}_1)}{|\mathbf{r}_2 - \mathbf{r}_1|^3} .$$

As with Bézier curves, continuity of the curve and its gradient is achieved between segments $\mathbf{r}^{(1)}(u_1)$ and $\mathbf{r}^{(2)}(u_2)$ by making the three points $\mathbf{r}_1{}^{(1)}$, $\mathbf{r}_2{}^{(1)} = \mathbf{r}_0{}^{(2)}$ and $\mathbf{r}_1{}^{(2)}$ collinear. Curvature continuity requires $\kappa^{(1)}(1)\mathbf{B}^{(1)}(1)$ and $\kappa^{(2)}(0)\mathbf{B}^{(2)}(0)$ to be equal, whence

$$(k^{(1)})^2 \frac{(\mathbf{r}_2{}^{(1)} - \mathbf{r}_1{}^{(1)}) \times (\mathbf{r}_0{}^{(1)} - \mathbf{r}_1{}^{(1)})}{|\mathbf{r}_2{}^{(1)} - \mathbf{r}_1{}^{(1)}|^3} = (k^{(2)})^2 \frac{(\mathbf{r}_1{}^{(2)} - \mathbf{r}_0{}^{(2)}) \times (\mathbf{r}_2{}^{(2)} - \mathbf{r}_1{}^{(2)})}{|\mathbf{r}_1{}^{(2)} - \mathbf{r}_0{}^{(2)}|^3} . \quad (6.19)$$

For (6.19) to hold as regards direction, the factors in the cross products must all lie in a plane perpendicular to the direction of **B** at the join. Thus the five points $\mathbf{r}_0{}^{(1)}$, $\mathbf{r}_1{}^{(1)}$, $\mathbf{r}_2{}^{(1)} = \mathbf{r}_0{}^{(2)}$, $\mathbf{r}_1{}^{(2)}$ and $\mathbf{r}_2{}^{(2)}$ must be coplanar, which is not surprising since each segment is a plane curve. The magnitudes of the left-hand sides of (6.19) may be matched by adjusting the k's.

In conclusion, then, it is possible to construct a composite rational quadratic curve having continuity of slope and curvature. It will be a plane curve, and each new segment added must have its value of k adjusted so that (6.19) is satisfied at the join.

We turn now to the rational cubic, a single curve segment of which may be expressed by

$$\mathbf{r}(u) = \frac{w_0\mathbf{r}_0(1-u)^3 + 3w_1\mathbf{r}_1u(1-u)^2 + 3w_2\mathbf{r}_2u^2(1-u) + w_3\mathbf{r}_3u^3}{w_0(1-u)^3 + 3w_1u(1-u)^2 + 3w_2u^2(1-u) + w_3u^3}, \quad 0 \leqslant u \leqslant 1. \quad (6.20)$$

From the discussion in Section 5.2.2, we know that this segment approximates the 'polygon' with vertices $\mathbf{r}_0$, $\mathbf{r}_1$, $\mathbf{r}_2$ and $\mathbf{r}_3$. At $u = 0$ and $u = 1$ we find

$$\dot{\mathbf{r}}(0) = 3\frac{w_1}{w_0}(\mathbf{r}_1 - \mathbf{r}_0) , \qquad \dot{\mathbf{r}}(1) = 3\frac{w_2}{w_3}(\mathbf{r}_3 - \mathbf{r}_2) ,$$

$$\ddot{\mathbf{r}}(0) = (6\frac{w_1}{w_0} + 6\frac{w_2}{w_0} - 18\frac{w_1^2}{w_0^2})(\mathbf{r}_1 - \mathbf{r}_0) + 6\frac{w_2}{w_0}(\mathbf{r}_2 - \mathbf{r}_1) ,$$

and $$\ddot{\mathbf{r}}(1) = (6\frac{w_1}{w_3} + 6\frac{w_2}{w_3} - 18\frac{w_2^2}{w_3^2})(\mathbf{r}_2 - \mathbf{r}_3) + 6\frac{w_1}{w_3}(\mathbf{r}_1 - \mathbf{r}_2) .$$

Continuity of the curve and its gradient at a join requires collinearity of the three points $\mathbf{r}_2{}^{(1)}$, $\mathbf{r}_3{}^{(1)} = \mathbf{r}_0{}^{(2)}$ and $\mathbf{r}_1{}^{(2)}$. Following the reasoning given for the rational quadratic, we find that curvature continuity at the join requires

$$K_2^{(1)}\frac{(\mathbf{r}_2^{(1)}-\mathbf{r}_1^{(1)})\times(\mathbf{r}_3^{(1)}-\mathbf{r}_2^{(1)})}{|\mathbf{r}_3^{(1)}-\mathbf{r}_2^{(1)}|^3}=K_1^{(2)}\frac{(\mathbf{r}_1^{(2)}-\mathbf{r}_0^{(2)})\times(\mathbf{r}_2^{(2)}-\mathbf{r}_1^{(2)})}{|\mathbf{r}_1^{(2)}-\mathbf{r}_0^{(2)}|^3} \tag{6.21}$$

where $K_1 = w_0w_2/w_1^2$ and $K_2 = w_1w_3/w_2^2$. This implies that $\mathbf{r}_1^{(1)}$, $\mathbf{r}_2^{(1)}$, $\mathbf{r}_3^{(1)} = \mathbf{r}_0^{(2)}$, $\mathbf{r}_1^{(2)}$ and $\mathbf{r}_2^{(2)}$ must be coplanar, though $\mathbf{r}_0^{(1)}$ and $\mathbf{r}_3^{(2)}$ can lie out of the plane so that the composite curve is now not necessarily a plane curve. The magnitude of the two sides of (6.21) may be matched by adjusting the K's. We note that K_1 and K_2 may be separately adjusted for each segment by variation of the weights w_0, w_1, w_2 and w_3. This permits a specified curvature to be attained at both ends of a segment, and it is now therefore possible to interpose a segment in a gap between two existing segments and match position, gradient and curvature at both resulting joins.

A curve design method based on rational cubics is given by Ahuja (1968).

6.3.4 Parametric Splines

Spline functions were introduced in Section 6.2.2, in a two-dimensional context. In three dimensions, the adoption of a parametric approach enables us to use the techniques introduced earlier with little modification. Given a set of points (x_i,y_i,z_i), $i=0,1,\ldots,n$, we may generate three separate spline curves

$$x = x(u)\,, \qquad y = y(u)\,, \qquad z = z(u)\,,$$

in terms of a parameter u. These curves interpolate the sets of points (u_i,x_i), (u_i,y_i) and (u_i,z_i) respectively. As previously, we need some additional information concerning derivatives at the ends of the curve. This point is discussed in Section 6.3.6. Once the three spline functions have been computed, any particular value of the parameter u gives a point on the required space curve. The use of splines gives a curve whose curvature is continuous, with the added advantage that no derivative information is needed at the internal data points. The splines may either be computed directly, by the method of Section 6.2.2, or in terms of a B-spline basis as in Section 6.2.4.

An obvious choice for the parameter u would be accumulated arc length, but this choice leads to difficulties in practice: we cannot establish the arc length between data points until we have computed the curve, but we need the arc length in order to compute the curve. The most natural alternative is to use accumulated *chord* length, setting $u_0 = 0$ and

$$u_{i+1} = u_i + d_{i+1}\,, \qquad i = 0,1,2,\ldots,n-1\,,$$

where $d_{i+1} = [(x_{i+1} - x_i)^2 + (y_{i+1} - y_i)^2 + (z_{i+1} - z_i)^2]^{1/2}$.

Computationally, this is much more convenient. We may start from a curve parametrised in this way and use an iterative technique to obtain a curve parametrised with respect to arc length, but the differences are usually insignificant in practical terms. Another possibility is simply to assign successive integer (or other equally spaced) values to u at the knots, but this can lead to problems if the physical spacing of the knots is very uneven, as explained in Section 6.3.6.

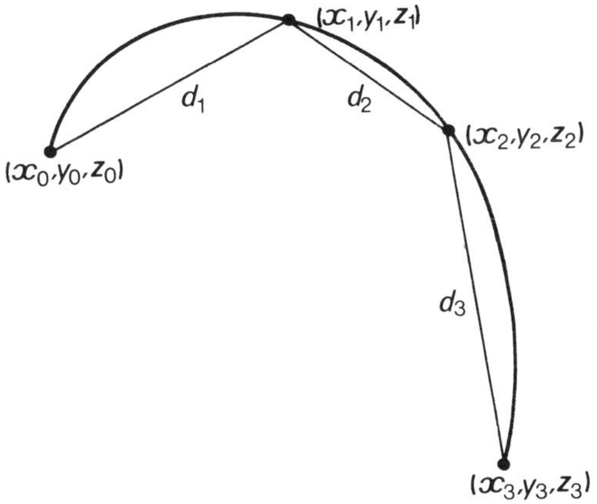

Figure 6.9 – Chord-length Parametrisation of a Spline Curve.

The parametric approach may also be used to define a plane curve, when only $x = x(u)$ and $y = y(u)$ have to be generated. In either two or three dimensions, no problems arise in dealing with vertical tangents when a parametric method is used. It is therefore possible to generate closed curves and curves containing loops. Ahlberg, Nilson and Walsh (1967) quote results concerning the approximation of the unit circle, using chord-length parametrisation of periodic cubic splines $x(u)$ and $y(u)$. A periodic spline is one for which the value of the function and its first two derivatives are matched at the end points, in which case no additional end conditions are needed. With data points equidistantly spaced on the circle's periphery, the error bounds on the radius were found to be as follows:

No. of spans	$\|\delta r\|$
4	<0.01
8	0.00112
12	0.000165

This example shows that parametric splines can cope very effectively with vertical tangents.

In fact it is possible to improve upon the figures given above if curvature matching at the knots is based on equation (6.12) rather than on simple Ferguson-type matching of second derivatives. An example is provided by the circular arc approximation given in Example 3 of Section 5.1.5. That curve, when reflected first in the x-axis and then in the y-axis, gives a still better curvature-continuous approximation to the unit circle. It was obtained, in effect, by taking $\alpha_1 = \alpha_2$ in (6.12) and then finding the value of μ which minimises the maximum deviation from the circle, whereas the results quoted by Ahlberg, Nilson and Walsh correspond to the case $\mu = 0$.

The same authors also report their experience that the oscillatory tendencies which arise in the fitting of splines to functions with curvature discontinuities (see p. 162) are far less acute for parametric than for non-parametric splines.

6.3.5 B-Spline Curves

We now make a brief study of a class of curves, based upon B-splines, which have an affinity with Bézier curves and possess corresponding advantages in the curve design area.

We confine our attention to the cubic case here; a more detailed and general discussion is given by Gordon and Riesenfeld (1974a). Using the notation of Section 6.2.3, we denote by $N_{4i}(u)$ the normalised cubic B-spline which is defined on an appropriate extension of the integer knot set $u_i = i$, $i = 0, 1, \ldots, n$ and which is non-zero for $u_{i-4} < u < u_i$. We will assume that $n \geqslant 3$. The most convenient way to extend the original knot set is to choose $u_{-3} = u_{-2} = u_{-1} = 0$ and $u_{n+1} = u_{n+2} = u_{n+3} = n$. Then the extra spans introduced all have length zero, and the original initial and final knots become **multiple knots**; in the present case their multiplicity is 4. The Cox - de Boor algorithm for calculating B-splines (Section 6.2.4) will cope with multiple knots provided that the convention $0/0 = 0$ is adopted in its use.

A mathematical investigation shows that the effect of introducing a multiple knot is to decrease the order of continuity of the spline at that knot. Thus whereas a cubic spline is continuous up to its second derivatives at a simple knot, it will generally be discontinuous in its second derivatives at a double knot, discontinuous also in its first derivative at a triple knot and discontinuous in itself at a quadruple knot. These effects are illustrated in Figure 6.10 below, in which we portray the complete set of normalised cubic B-splines which can be constructed on the extended knot set 0, 0, 0, 0, 1, 2, 3, 4, 5, 5, 5, 5. Note that $\ddot{N}_{43}(0)$ is non-zero, $\ddot{N}_{42}(0)$ and $\dot{N}_{42}(0)$ are non-zero, $\ddot{N}_{41}(0)$, $\dot{N}_{41}(0)$ and $N_{41}(0)$ are non-zero, while the same type of behaviour occurs at the multiple knot $u = 5$. We have here employed dots to denote derivatives, since u will be used as a curve parameter in what follows.

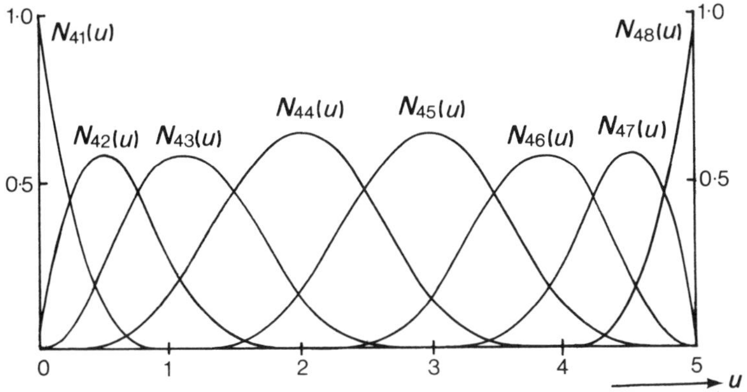

Figure 6.10 – B-splines on $0 \leqslant u \leqslant 5$.

We recall from Section 6.2.3 that any cubic spline $\phi(u)$ on the knot set $0, 1, 2, \ldots, n$ can be expressed in terms of cubic B-splines by

$$\phi(u) = \sum_{i=1}^{n+3} c_i N_{4i}(u),$$

where the c_i are scalar coefficients. The introduction of multiple knots does not invalidate this result. If we have a set of vectors $\mathbf{r}_0, \mathbf{r}_1, \ldots, \mathbf{r}_n$, we may use them to construct a vector-valued analogue of the last equation by defining

$$\mathbf{r}(u) = \sum_{i=0}^{n} \mathbf{r}_i N_{4,i+1}(u). \tag{6.22}$$

Since we have $(n+1)$ vector coefficients we need a set of $(n+1)$ B-splines, which we will define on the extended integer knot set $0, 0, 0, 0, 1, 2, \ldots, (n-3), (n-2), (n-2), (n-2), (n-2)$.

For $0 \leqslant u \leqslant n-2$, (6.22) is the equation of a cubic **B-spline curve**. Some of the simpler properties of such a curve follow from the result

$$\sum_{i=0}^{n} N_{4,i+1}(u) = 1, \quad 0 \leqslant u \leqslant n-2 \tag{6.23}$$

(see Gordon and Riesenfeld, 1974a). At $u = 0$, (6.22) gives

$$\mathbf{r}(0) = \sum_{i=0}^{n} \mathbf{r}_i N_{4,i+1}(0).$$

Now it is obvious from Figure 6.10 that the only B-spline which is non-zero at $u = 0$ is $N_{41}(u)$; further, (6.23) then implies that $N_{41}(0) = 1$. Thus

$$\mathbf{r}(0) = \mathbf{r}_0 . \tag{6.24}$$

We also see from (6.22) that

$$\dot{\mathbf{r}}(0) = \sum_{i=0}^{n} \mathbf{r}_i \dot{N}_{4,i+1}(0) .$$

On differentiating (6.23) we obtain

$$\sum_{i=0}^{n} \dot{N}_{4,i+1}(u) = 0 , \quad 0 \leqslant u \leqslant n-2 ,$$

and since in this summation only $\dot{N}_{41}(u)$ and $\dot{N}_{42}(u)$ are non-zero at $u = 0$ (see Figure 6.10), we conclude that

$$\dot{N}_{41}(0) = -\dot{N}_{42}(0) .$$

It follows that

$$\dot{\mathbf{r}}(0) = \dot{N}_{42}(0)\,(\mathbf{r}_1 - \mathbf{r}_0) . \tag{6.25}$$

From (6.24) and (6.25) we see that the cubic B-spline curve has properties in common with the cubic Bézier curve. For instance, if we regard $\mathbf{r}_0, \mathbf{r}_1, \ldots, \mathbf{r}_n$ as the vertices of an open polygon, the B-spline curve commences at $\mathbf{r}_0$, and its tangent at that point is in the direction of $(\mathbf{r}_1 - \mathbf{r}_0)$. We find analogous behaviour at the other end of the curve, and in fact the curve may be regarded as an approximation to the polygon in much the same way as a Bézier curve approximates its characteristic polygon. The B-spline curve has the advantage that a change in one vertex only alters four spans of the curve, so that local adjustments can be made without disturbing the rest of the curve. Further, we can construct a cubic B-spline curve to approximate a polygon with as many sides as we like, the properties of the B-splines ensuring continuity up to second derivatives everywhere on the curve. By contrast, if the polygon has more than three sides the cubic Bézier technique will require a composite curve, with the associated problem of matching curvature at joins between segments if second-order continuity is required.

Cubic B-spline curves possess some interesting geometrical properties. Since for any non-integer value of u only four of the terms in (6.22) can be non-zero, each span of the curve is determined by at most four consecutive vertices of its defining polygon. Now the coefficients of the four corresponding vectors in (6.22) are positive and sum to unity, according to (6.23). Thus for any u, $\mathbf{r}(u)$ is a weighted average of these four vectors, and the entire span must lie inside what is called their **convex hull.** Two two-dimensional cases are shown below in Figure 6.11. In three dimensions the convex hull will be the tetrahedron defined by the four vertices. As explained in Section 5.1.3, Bézier curve segments have a similar property.

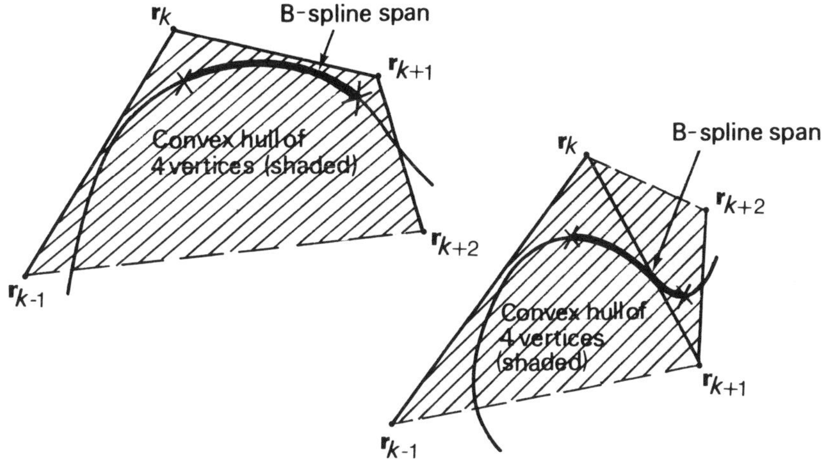

Figure 6.11 – The Convex Hull Property of a Cubic B-spline Curve.

An important consequence of this convex hull property is that if four consecutive vertices of the polygon are collinear, their convex hull degenerates to a straight line; the resulting span of the B-spline curve must therefore be linear. Then a B-spline curve may have locally linear sections embedded in it, which is a useful feature from the design point of view.

In two dimensions, it is quite easy to sketch the cubic B-spline curve for any given polygon. As we have seen, the end points of the curve and of the polygon coincide, and the tangents to the curve at its end points are in the directions of the initial and final sides of the polygon. Two further useful facts are (i) the curve passes close to the mid-point of each side of the polygon with the exception of the first and the last, and (ii) the curve passes through the points

$$\frac{1}{6}\mathbf{r}_{k-1} + \frac{2}{3}\mathbf{r}_k + \frac{1}{6}\mathbf{r}_{k+1} \doteq \frac{2}{3}\mathbf{r}_k + \frac{1}{3}\left[\frac{1}{2}(\mathbf{r}_{k-1} + \mathbf{r}_{k+1})\right],$$

for $k = 2, 3, \ldots, n-2$. These points are one third the way along the straight line joining $\mathbf{r}_k$ to the mid-point of the line joining $\mathbf{r}_{k-1}$ and $\mathbf{r}_{k+1}$, as shown in Figure 6.12.

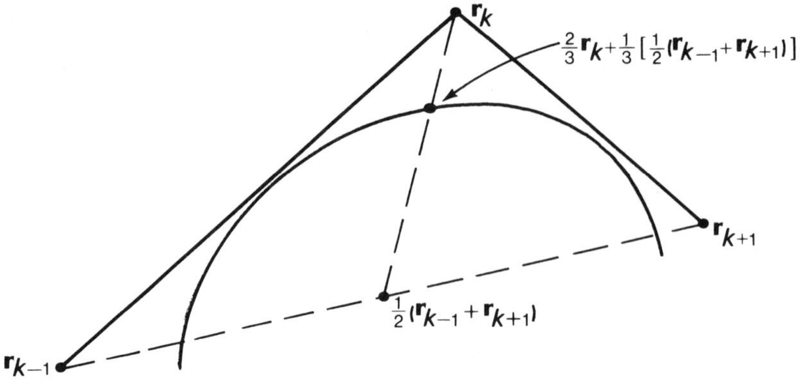

Figure 6.12

Armed with this information and the convex hull property, we can quickly sketch the curve which approximates a given polygon, as in Figure 6.13 below. Here the crosses mark points whose approximate locations are given by the two properties just cited.

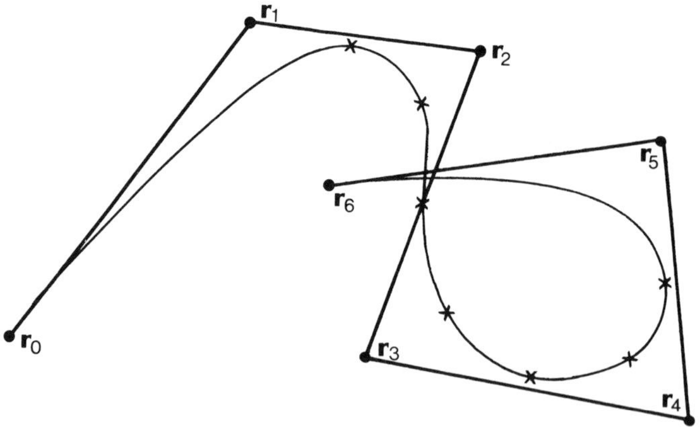

Figure 6.13

There are a number of fruitful modifications and generalisations of the type of B-spline curve we have studied. First, we can construct *closed* B-spline curves, based on closed polygons. The knot set is now cyclic (that is $u_n = u_0$, $u_{n+1} = u_1$, and so on) and the use of multiple knots is unnecessary. The curve can be constructed entirely in terms of B-splines of the basic type such as $N_{44}(u)$ or $N_{45}(u)$ in Figure 6.10, translated by integer distances along the u-axis. A second possibility is to base the B-splines on a knot set with unequal spacing. Gordon and Riesenfeld (1974a) experimented along these lines, relating distances between knots to the lengths of the corresponding sides of the polygon. The resulting

non-uniform B-spline curves closely resembled the corresponding uniform curves except in cases of considerable disparity between the lengths of the polygon sides. It was found that discontinuities in the curvature or direction of the curve could be obtained by allowing two or three polygon vertices to coalesce, which provides added flexibility for the designer. A third extension, of course, is to use B-splines of degree higher than cubic, though even the cubic case gives great versatility.

The design procedure proposed by Gordon and Riesenfeld contains the following steps:

(i) the designer specifies a set of points which lie approximately on the desired curve;
(ii) the computer responds with an interpolating spline, and then uses an inversion procedure to determine the polygon which corresponds to the interpolated curve;
(iii) the designer now works with the polygon, moving individual vertices to perturb the initial interpolated curve into what he considers to be a more satisfactory shape.

6.3.6 Parametrisation and Tangent Vectors

Most of the parametric curves we have studied are constructed from segments with parameter u running from 0 to 1 between their end-points. The exceptions are the parametric spline and the B-spline curve, which are parametrised so that u is zero at one end of the curve as a whole, taking increasing values at the knots or joins between segments as we move towards the other end. We now examine the relations between curves parametrised in these two different ways.

First we consider a single parametric cubic segment, whose behaviour typifies that of parametric polynomial segments in general, and which we will write in the Ferguson form

$$\mathbf{r}(u) = \sum_{j=0}^{3} \mathbf{a}_j u^j , \qquad 0 \leqslant u \leqslant 1 .$$

We will say that this segment has 'parametric length' 1. As we saw in Section 5.3, the most general parameter transformation which preserves degree is a linear one, which we will here write as $u' = bu + c$, where b and c are constants and we will assume in what follows that $b > 0$. In terms of the new parameter we have

$$\mathbf{r}(u') = \sum_{j=0}^{3} \mathbf{a}_j \left(\frac{u' - c}{b}\right)^j = \sum_{j=0}^{3} \mathbf{a}'_j u'^j ,$$

where the $\mathbf{a}'_j$ are linear combinations of the $\mathbf{a}_j$. The degree is still cubic, but the range of u' corresponding to $0 \leqslant u \leqslant 1$ is $c \leqslant u' \leqslant b + c$, so that the 'parametric length' of the segment in terms of u' is b. Since we can choose b and c as we wish, we can thus express a Ferguson segment as a cubic segment with arbitrary parameter limits. It is not difficult to see that the converse operation is also possible.†

A side effect of our parameter change concerns the magnitudes of tangent vectors. We observe that

$$\dot{\mathbf{r}}(u') = \frac{du}{du'}\,\dot{\mathbf{r}}(u) = \frac{1}{b}\dot{\mathbf{r}}(u) \;,$$

where a dot denotes differentiation with respect to the indicated variable. Then in multiplying the 'parametric length' of the segment by b we have multiplied the tangent vector by b^{-1}. Since also $\ddot{\mathbf{r}}(u') = b^{-2}\ddot{\mathbf{r}}(u)$ the second derivative is multiplied by b^{-2}, and so on. In particular, the parameter transformation changes the derivative magnitudes at the end-points, and so affects the inter-segment continuity conditions of composite curves.

Let us examine what happens when we reparametrise a composite curve which has continuity of direction and magnitude for its first and second derivatives. We first remark that when $b = 1$ we have the simple translation of parameter $u' = u + c$, which leaves derivative magnitudes unaltered. Then we may set $u' = u$, $u' = u + 1$, $u' = u + 2$ in the first, second and third segments, and so on, to express a composite curve of n segments in terms of a single parameter u' which increases from 0 to n along the curve, taking successive integer values at the knots. Because we have maintained the original derivative magnitudes we have retained our former inter-segment continuity conditions, and the curve is now seen to be a parametric spline of the type discussed in Section 6.3.4, but defined on the integer knot parameter set $u' = 0, 1, 2, \ldots, n$.

Matters are different if we reparametrise our original Ferguson curve more generally. On setting $w = b_i u + c_i$ on the ith span or segment, we may obtain continuity of w along the curve by equating its values at the end of each segment and the beginning of the next. This gives

$$b_i + c_i = c_{i+1} \;, \qquad i = 1, 2, \ldots, n{-}1 \;.$$

With these conditions imposed, the reparametrised curve has parameter values $c_1, c_2, c_3, \ldots$ at the knots, while the 'parametric length' of the ith segment is $b_i = c_{i+1} - c_i$. This being so, the original first and second derivatives on that segment are multiplied by b_i^{-1} and b_i^{-2} respectively, and it follows that continunity of derivative magnitudes will be destroyed unless the b_i are the same for each segment. If we wish to express the original curve in terms of a single parameter taking specified *non-uniformly* spaced values at the knots, then, we may retain continuity of direction but not of magnitude for the derivatives. The ratio

†It was also mentioned in Section 5.3 that a *rational* parametric curve segment preserves its nature under a bilinear transformation such as $u' = (b_1 u + c_1)/(b_2 u + c_2)$. We do not pursue this matter here, but Forrest (1968) gives further details.

of the magnitudes of $\dot{\mathbf{r}}(w)$ and $\ddot{\mathbf{r}}(w)$ across the join between the ith and the $(i + 1)$th span will be (b_{i+1}/b_i) and $(b_{i+1}/b_i)^2$ respectively.

For our given set of knots, it is only curves which have this precise relationship between the derivative magnitude ratios and the parametric lengths of their segments which are equivalent to the original Ferguson curve. For instance, the parametric spline fitting the same set of knots but with knot parameter values $w = c_1, c_2, c_3, \ldots$ has a tangent magnitude ratio of unity across all knots. It therefore does not conform to the required pattern, and this curve will differ geometrically from the earlier one.

It is apparent from the illustrations which have been given that continuity of derivative magnitudes between the segments of a composite curve has more to do with the parametrisation than the actual geometry of the curve. In fact an insistence on this kind of continuity is geometrically rather pointless, since a composite curve with continuity of tangent direction only can always be re-parametrised to make the tangent magnitudes continuous also. The original and reformulated curves will be geometrically identical. From the practical point of view, however, a system which matches derivatives both in direction and magnitude at the knots does absolve the user from having to specify tangent magnitude ratios, and thus has the virtue of simplicity.

The true geometrical significance of tangent vector magnitudes is revealed only by comparison with some reference length. For instance, suppose that $\mathbf{r}(u)$, $0 \leqslant u \leqslant 1$, is a curve segment, and let s denote arc length along the curve. Then

$$\dot{\mathbf{r}}(u) = \frac{\mathrm{d}\mathbf{r}}{\mathrm{d}s}\,\frac{\mathrm{d}s}{\mathrm{d}u} ,$$

and since $\dfrac{\mathrm{d}\mathbf{r}}{\mathrm{d}s}$ is a unit vector (see Section 4.2.3) it follows that $|\dot{\mathbf{r}}(u)| = \dfrac{\mathrm{d}s}{\mathrm{d}u}$. Then if S is the total arc length of the segment, we have

$$S = \int_0^S \mathrm{d}s = \int_0^1 \frac{\mathrm{d}s}{\mathrm{d}u}\,\mathrm{d}u = \int_0^1 |\dot{\mathbf{r}}(u)|\,\mathrm{d}u .$$

The trapezium rule then gives

$$S \simeq \tfrac{1}{2}[\,|\dot{\mathbf{r}}(0)| + |\dot{\mathbf{r}}(1)|\,]$$

as a rough estimate of the total arc length.

Now suppose that $\mathbf{r}(0)$ and $\mathbf{r}(1)$ are fixed, as also are the directions of $\dot{\mathbf{r}}(0)$ and $\dot{\mathbf{r}}(1)$. We will initially take as our reference length $S_0 = 1$, the approximate arc length when $|\dot{\mathbf{r}}(0)| = |\dot{\mathbf{r}}(1)| = 1$. It is clear that multiplying both tangent vector magnitudes by $\beta > 1$ will increase the approximate arc length by the factor β. Since the end-points are fixed, this implies an increase in the fullness of the curve. If β is large, there is a danger that the curve will describe a loop in accommodating the much increased arc length (see the diagrams in Section 5.1.2). On the other hand, if $\beta < 1$ the arc length will be less than S_0 and the curve configuration will become closer to the chord line $\mathbf{r}(1) - \mathbf{r}(0)$.

In fact, the length of this chord line is a more convenient reference than S_0 as used above. It is possible to show that loops will not occur in a plane cubic segment provided that the projections of $\dot{\mathbf{r}}(0)$ and $\dot{\mathbf{r}}(1)$ onto the chord direction each do not exceed three times the chord length. For a twisted segment the notion of a loop is less clear-cut; there will always be a loop in some projection of such a curve. Nevertheless, it seems likely that problems will arise in the definition of a surface patch if the foregoing limits are exceeded in any of its boundary curves.

Next, suppose that $|\dot{\mathbf{r}}(0)|$ and $|\dot{\mathbf{r}}(1)|$ are changed independently of each other. Since curvature κ is given by

$$\kappa(u)\mathbf{B}(u) = \frac{\dot{\mathbf{r}}(u) \times \ddot{\mathbf{r}}(u)}{|\dot{\mathbf{r}}(u)|^3},$$

where $\mathbf{B}$ is the binormal vector, we may expect small curvature at $u = 0$ if $|\dot{\mathbf{r}}(0)|$ is large, and *vice versa*. Then increasing $|\dot{\mathbf{r}}(0)|$ while $|\dot{\mathbf{r}}(1)|$ is held constant may be expected to straighten out the portion of the curve segment near $\mathbf{r}(0)$. A corresponding result will occur near $\mathbf{r}(1)$ if $|\dot{\mathbf{r}}(1)|$ is increased while $|\dot{\mathbf{r}}(0)|$ is held constant. These predictions are confirmed by experience (again, see the diagrams in Section 5.1.2).

To illustrate some of these findings, we consider a spline curve $\mathbf{r}(u)$, parametrised so that $u = 0, 1, \ldots, n$ at the knots, which are physically very non-uniformly spaced. In particular, we examine what happens when the ith and $(i+1)$th spans have chord-lengths d_i and d_{i+1} such that $d_i \ll d_{i+1}$. Now the tangent vector magnitude of the spline is continuous at the intervening knot $u = i$. Then the behaviour of the two curve segments will be governed to some extent by the ratios $|\dot{\mathbf{r}}(i)|/d_i$ and $|\dot{\mathbf{r}}(i)|/d_{i+1}$, as we saw in the last paragraph but one. If the first ratio is large and the second small, the results may well be a loop followed by a close approximation to a straight line. Such behaviour is likely to be unacceptable in a practical context, and it stems from the use of a uniform knot parameter spacing with a highly non-uniform physical spacing of the knots. One way of overcoming the difficulty is to dispense with tangent magnitude continuity, making $|\dot{\mathbf{r}}(u)|$ small to the immediate left of the knot $u = i$ and large to its immediate right. In fact we may then reparametrise to regain tangent magnitude continuity; the resulting parameter spacing at the knots will then be found to reflect the actual physical spacing of the knots. For this reason, parametrisation according to chord-length (or some similar measure of the physical spacing of the knots) is to be recommended in cases of uneven knot spacing. On the other hand, composite Ferguson or Bézier curves, or uniformly parametrised splines, will generally give acceptable results when the knots are fairly evenly spaced.

Finally, we consider the end-conditions needed for the complete specification of a parametric spline on a given set of knots. It was explained in Section 6.3.4 that such a curve is equivalent to three planar splines. We therefore need three scalar conditions, which are often combined as a single vector condition, at each end. Of four methods given in Section 6.2.2 for dealing with end conditions of plane splines the first, third and fourth extend immediately to the parametric case. They require respectively that $\ddot{\mathbf{r}}(u) = 0$ at the end points, that $\mathbf{r}(u)$ is quadratic over the end spans, or that $\ddot{\mathbf{r}}(u)$ varies linearly over the first and last pairs of spans. But the remaining suggestion, involving the specification of end tangents, presents a problem in the parametric case. It is often easy to determine the direction desired for the end tangent, by measurement from a drawing or model. However, because we require a vector quantity, we also have to assign it a magnitude. From what has gone before we know that too small a choice for the magnitude may lead to excessive curvature in a small region near $u = 0$, while too large a choice may give an unwanted bulge or even a loop. A safe choice, if the angle between the end tangent and the chord of the first span is small, is to take $|\dot{\mathbf{r}}(0)|$ equal to, or only slightly greater than, the parametric length of the span. For larger angles it might be better to assign this value to the projection of $\dot{\mathbf{r}}(0)$ onto the chord direction rather than to $|\dot{\mathbf{r}}(0)|$ itself.

6.4 TWO FURTHER PRACTICAL SYSTEMS

In this section we outline two widely used practical curve-defining systems which do not fall exactly under any of our earlier section headings but which have interesting mathematical features.

First we examine the APT concept TABCYL. A TABCYL (TABulated CYLinder) is a surface generated by moving a straight line (the **generatrix**) along a plane curve (the **directrix**) so that it is always parallel to another fixed straight line. The curve is defined in terms of point data, and is generated as a sequence of plane cubic segments, each of which spans the interval between two successive data points. The equation of each segment is expressed in a local coordinate system having its origin at the first of the two data points (in input order) and with the positive axis of the of the independent variable lying along the chord line of the segment, as in Figure 6.14.

The equation of each segment therefore has the form

$$\eta(\xi) = \alpha\xi^3 + \beta\xi^2 + \gamma\xi \ ,$$

which contains no constant term since $\eta(0) = 0$. If d is the length of the chord-line of the segment, we must require $\eta(d) = 0$ also, which implies that

$$\gamma = -(\alpha d^2 + \beta d) .$$

Then the two coefficients α and β remain to be found for the segment to be completely specified.

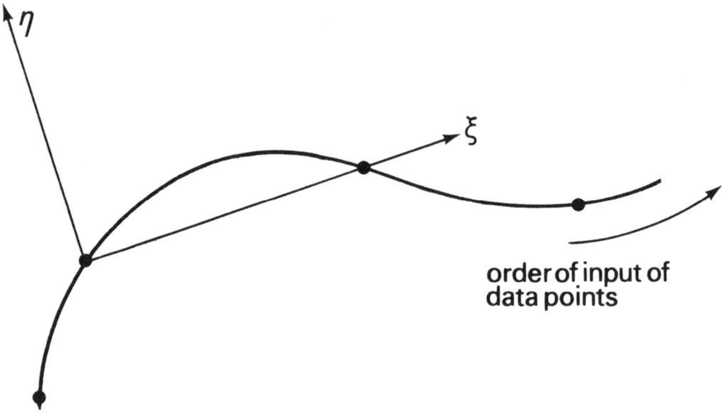

Figure 6.14 – A Local Axis System in TABCYL.

In the version of TABCYL in current general use, approximate values are calculated for the end tangents $\eta'(0)$ and $\eta'(d)$ in terms of neighbouring point data. Then $\eta(\xi)$ is required to have these slopes at its end-points, which fixes the values of α and β. Finally, the slopes at the interior points are adjusted in an iterative manner until the resulting TABCYL curve has curvature continuity.

TABCYL, then, is an example of a splining method which uses local axis redefinition to overcome the problems which can occur with nonparametric splines when large or infinite gradients are encountered. The two additional items of data needed to complete the definition of the spline may be supplied by the user in the form of end tangents. Failing this, the program will simply fit quadratics rather than cubics to the initial and terminal spans. This removes two degrees of freedom (the choice of the coefficient α for these spans) and hence leaves the spline completely determined.

While TABCYL will only construct plane curves, Thomas (1976) outlines a method which extends this general approach to the fitting of data in three dimensions.

We turn now to the curve-fitting procedure KURGLA. This was developed by Mehlum as part of his AUTOKON ship-design system (Mehlum, 1969; Mehlum and Sørenson, 1971), which has since found application in other branches of industry (Bates, 1972; Emmerson, 1976). KURGLA, like TABCYL, generates only plane curves, although in a later paper (Mehlum, 1974) a possible approach to three-dimensional curve fitting is outlined.

As mentioned in Section 6.2.2, a physical spline bent subject to constraints adopts the shape which minimises its mean squared curvature. Mehlum uses a variational analysis to show that between points of constraint the curvature κ

of such a curve varies linearly when referred to a certain fixed direction in its plane. It can be arranged that this fixed direction is that of the chord-line joining successive knots or constraint points, as shown in Figure 6.15.

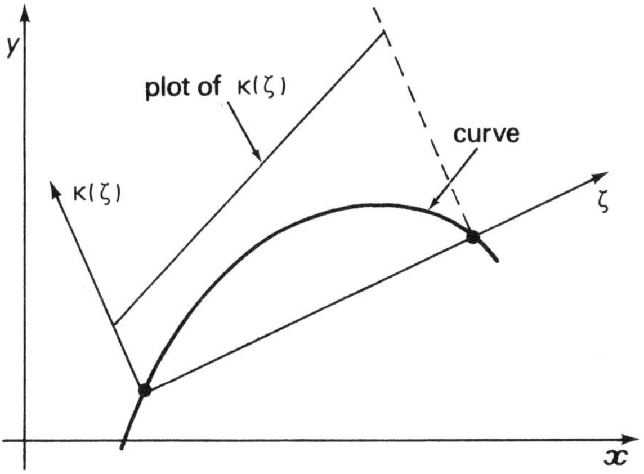

Figure 6.15 – A Segment with Linear Curvature Variation.

Thus far the analysis is exact, but in order to use curve segments of this type in a fitting procedure it is now necessary to make some approximations. Mehlum (1974) describes two versions of KURGLA, as follows:

KURGLA I approximates the linear curvature variation by a stepwise distribution (Figure 6.16). Each horizontal constant curvature step corresponds to an arc of a circle. The procedure incorporates a subroutine which calculates the number of arcs which must be taken for the discrepancy between the theoretical curve and its multi-arc approximation to lie within some specified tolerance. The centres and radii of the circular arcs are computed so that the approximation is continuous in position and gradient, having prescribed curvature at the two knots which are its end-points.

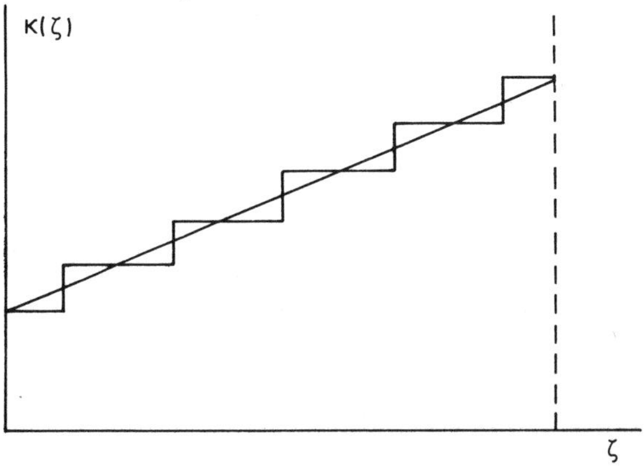

Figure 6.16 - Stepwise Approximation to Linear Curvature Variation.

KURGLA I uses curve segments of this kind to construct a composite curve which is continuous and has specified curvature at all the knots. These curvatures are estimated beforehand. The composite curve will in general be found to have discontinuities of slope at the knots, however, and the trial curvature values are now modified in an iterative manner until the kinks at the knots disappear. The final curve then has positional and gradient continuity everywhere, together with its stepwise curvature variation. The curvature discontinuities may be made arbitrarily small, of course, by taking a large enough number of circular arcs in the approximation of each span.

KURGLA II in effect approximates chord length by arc length on each span. Then each curve segment has curvature which varies linearly with arc length. The only curve having this property is the **Cornu spiral**, which is shown in Figure 6.17.

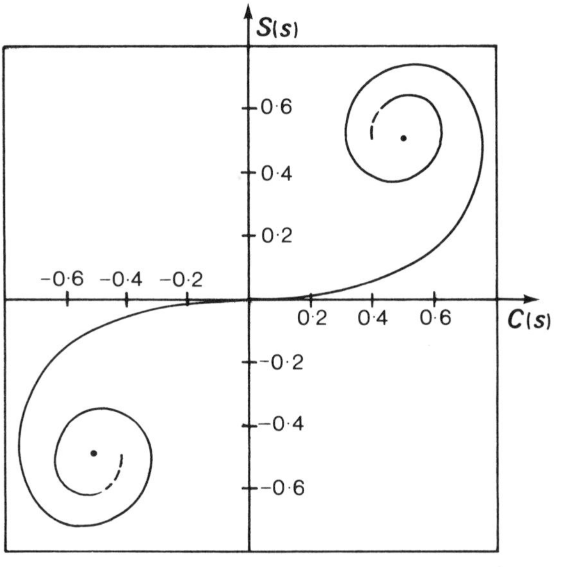

Figure 6.17 – The Cornu Spiral

In this diagram the coordinates of a point on the curve are given parametrically by

$$C(s) = \int_0^s \cos(\tfrac{1}{2}\pi\sigma^2)d\sigma\ , \qquad S(s) = \int_0^s \sin(\tfrac{1}{2}\pi\sigma^2)d\sigma\ ,$$

where s is arc length. $C(s)$ and $S(s)$ are the **Fresnel integrals**, whose main applications are in theoretical optics. KURGLA II selects the appropriate part of the spiral for each span of a composite curve, and scales and rotates it so that the eventual curve has positional, slope and curvature continuity. The actual computation of the Cornu spiral segments is based on well-known approximations for the Fresnel integrals (see, for example, Abramowitz and Stegun, 1964). Pal and Nutbourne (1977) have also described an algorithm for two-dimensional curve synthesis using segments of the same type.

The occurrence of the Cornu spiral in this context is easily explained on the basis of Section 1.2.4, where it was shown that any curve may be described parametrically in terms of arc-length s and curvature $\kappa(s)$ by the coupled differential equations

$$\left.\begin{aligned} \ddot{x} + \kappa(s)\dot{y} &= 0 \\ \ddot{y} - \kappa(s)\dot{x} &= 0 \end{aligned}\right\} .$$

For simplicity we assume that $x(0) = y(0) = \dot{y}(0) = \kappa(0) = 0,\ \dot{x}(0) = 1$. Other conditions merely result in translation and/or scaling of the basic curve. On multiplying the second equation by i and adding it to the first we obtain

$$\ddot{z} - i\kappa(s)\dot{z} = 0; \quad z(0) = 0,\ \dot{z}(0) = 1,$$

where $$z = x + iy.$$

This is a linear first-order equation in $\dot{z}$, with solution

$$\dot{z} = e^{i\int\kappa(s)ds} .$$

For the type of curve we are considering, we can write $\kappa(s) = as$, where a is a real constant. In this case,

$$\dot{z} = e^{\frac{1}{2}ias^2} .$$

Integration from $s = 0$ then leads to

$$z = \int_0^s e^{\frac{1}{2}ia\sigma^2} d\sigma = \int_0^s [\cos(\tfrac{1}{2}a\sigma^2) + i\sin(\tfrac{1}{2}a\sigma^2)]\, d\sigma ,$$

whence $$x + iy = \left(\frac{\pi}{a}\right)^{\frac{1}{2}} \left\{ C\left[\left(\frac{a}{\pi}\right)^{\frac{1}{2}} s\right] + iS\left[\left(\frac{a}{\pi}\right)^{\frac{1}{2}} s\right] \right\} .$$

This gives x and y parametrically in terms of s; the choice $a = \pi$ gives the basic curve portrayed in Figure 6.17.

There is an important distinction between the type of plane curve segment on which the KURGLA algorithms are based and the plane polynomial curve segments discussed earlier. A KURGLA segment $y = y(x)$ in principle has curvature

$$\kappa(x) = \frac{y''}{(1 + y'^2)^{3/2}} ,$$

which varies linearly with x. Suppose we have two such segments, $y_1(x)$ and $y_2(x)$, both spanning the same interval in x. Then if α, β are any constants, it is not difficult to see that in forming the linear combination $\alpha y_1(x) + \beta y_2(x)$ we destroy the property of linear curvature variation. Because a linear combination of KURGLA segments lacks the defining property of a KURGLA segment, second-order continuous curves constructed from elements of this type are described as **non-linear splines**.

By contrast, plane polynomial splines are linear splines, because any linear combination of polynomial splines of equal degree, defined on the same set of knots, is a spline of the same nature. As we saw in Section 6.2.2, such curves are based on the assumption that $y'(x)$ is everywhere small, in which case $\kappa(x) \simeq y''(x)$.

An obvious disadvantage of KURGLA is that its curve segments cannot be expressed exactly in terms of elementary mathematical functions which are easy to evaluate. For this reason, the computer can only produce an approximate representation of the desired curve. Some criterion of acceptability of the approximation has to be provided by the user, and a significant amount of computation may well be necessary for this criterion to be satisfied. A polynomial curve segment, on the other hand, can be held exactly by the computer, and is readily evaluated at any desired point. On the credit side, KURGLA can deal with large or infinite gradients. But then so can a plane polynomial spline, if local axis redefinition is used (as in TABCYL), or if we define it parametrically. Whether there is any virtue in modelling fairly precisely the behaviour of the physical spline traditionally used by draughtsmen is debatable. However, the fact remains that AUTOKON, with its KURGLA curves, has been for some years a successful system, and has now achieved widespread use.

We note in passing that splines in tension (Schweikert, 1966, Cline, 1974) are also non-linear, but the curve segments are expressible in terms of hyperbolic functions, which may be computed accurately using standard algorithms. However, they must be used with local axis redefinition, or parametrically, in the synthesis of curves having large or infinite gradients.

6.5 LOCAL ADJUSTMENT OF COMPOSITE CURVES

The need for local modification of a composite curve often arises in curve design. Some of the methods of this chapter permit this to be done easily, in particular those based on B-splines, as we have seen. But other standard methods may not allow a composite curve to assume a desired shape along part of its length because of the lack of flexibility remaining to individual segments once all positional and gradient continuity criteria have been satisfied. There are two basic ways of overcoming this difficulty. Either an offending segment may be split into two or more smaller segments, or the degree of the polynomials on which it is based may be raised. Both these expedients increase the scope for local modification of the curve. We give some simple examples below to illustrate the types of procedure involved.

We will demonstrate curve splitting for the Ferguson curve segment

$$\mathbf{r}(u) = \mathbf{a}_0 + \mathbf{a}_1 u + \mathbf{a}_2 u^2 + \mathbf{a}_3 u^3 \ .$$

If we make a linear parameter change in the form $u = (1 - u')u_0 + u'u_1$, as suggested in Section 5.3, we obtain a representation of the original segment in terms of the parameter u' with values $-u_0/(u_1 - u_0)$ and $(1 - u_0)/(u_1 - u_0)$ at the end points corresponding to $u = 0$, $u = 1$. Conversely, the values of u corresponding to $u' = 0,1$ are $u = u_0, u_1$. Then if $0 \leqslant u_0, u_1 \leqslant 1$, the portion of the reparametrised curve for which $0 \leqslant u' \leqslant 1$ is a standard Ferguson segment which exactly reproduces the part of the original curve for which $u_0 \leqslant u \leqslant u_1$.

Now let us choose $u_0 = 0$, $u_1 = \alpha$. Then u and u' are related by $u = \alpha u'$, or $u' = u/\alpha$. The Ferguson segment $\mathbf{r}_1(u')$ defined by $0 \leqslant u' \leqslant 1$ will correspond to that part of the original segment $\mathbf{r}(u)$ for which $0 \leqslant u \leqslant \alpha$.

Similarly, we may reproduce the remaining part of $\mathbf{r}(u)$, for which $\alpha \leqslant u \leqslant 1$, by choosing $u_0 = \alpha$, $u_1 = 1$. Then u and u' are related by $u = (1 - u')\alpha + u'$, or $u' = (u - \alpha)/(1 - \alpha)$, and $0 \leqslant u' \leqslant 1$ again defines a standard Ferguson segment, which we will call $\mathbf{r}_2(u')$. This completes the exact replacement of $\mathbf{r}(u)$ by $\mathbf{r}_1(u')$ and $\mathbf{r}_2(u')$, two segments of the same type which meet at the point where $u = \alpha$.

As we saw in Section 6.3.6, reparametrisation has the effect of changing tangent vector magnitudes. In terms of tangent vectors of $\mathbf{r}(u)$, we find that

$$\dot{\mathbf{r}}_1(0) = \alpha\dot{\mathbf{r}}(0) \ , \qquad \dot{\mathbf{r}}_1(1) = \alpha\dot{\mathbf{r}}(\alpha) \ ,$$

$$\dot{\mathbf{r}}_2(0) = (1 - \alpha)\dot{\mathbf{r}}(\alpha) \ , \qquad \dot{\mathbf{r}}_2(1) = (1 - \alpha)\dot{\mathbf{r}}(1) \ .$$

It follows that curve splitting destroys any continuity of tangent magnitude with the preceding and following segments of a composite curve which may previously have existed. Further, tangent vector magnitude continuity between $\mathbf{r}_1(u')$ and $\mathbf{r}_2(u')$ will only result if $\alpha = (1 - \alpha)$, or $\alpha = \frac{1}{2}$. On the other hand, this example reinforces our earlier conclusion that continuity of tangent magnitude is a restrictive and geometrically meaningless requirement. We have made an exact replacement of the original segment so that the geometry of the composite curve is unchanged, despite the different mathematical representation.

A similar method may be used to split a Bézier segment, and some of the details were covered in Section 5.3. The extension to rational cubic segments is then straightforward, because these can be expressed in Bézier form by the use of homogeneous coordinates (Section 5.2.2). Forrest (1968) also discusses the rational cubic case.

Curve splitting, as described above, must be used in the dissection of surface patches discussed in Section 7.7. However, if there is no reason why the original segment must be reproduced exactly, more flexibility is available. Suppose, for instance, that a composite Ferguson curve contains a segment with end conditions $\mathbf{r}(0)$, $\mathbf{r}(1)$, $\dot{\mathbf{r}}(0)$ and $\dot{\mathbf{r}}(1)$ which cannot be made to pass through some desired point $\mathbf{p}$. Then this segment can be removed and replaced by two segments with end conditions $\mathbf{r}(0)$, $\mathbf{p}$, $\dot{\mathbf{r}}(0)$, $\alpha_1\mathbf{T}$ and $\mathbf{p}$, $\mathbf{r}(1)$, $\alpha_2\mathbf{T}$, $\dot{\mathbf{r}}(1)$ respectively. The unit tangent vector $\mathbf{T}$, together with the scalars α_1 and α_2, are then at our disposal for shape modification.

To illustrate how flexibility may be enhanced by raising the degree of a curve segment, we again start with a cubic Ferguson segment,

$$\mathbf{r}^{[3]}(u) = \sum_{j=0}^{3} \mathbf{a}_j u^j \ ,$$

where the superscript in square brackets denotes degree. This is completely defined by its end points and tangents, which are related to the coefficients $\mathbf{a}_j$ by

$$\mathbf{r}^{[3]}(0) = \mathbf{a}_0 \ , \qquad \mathbf{r}^{[3]}(1) = \mathbf{a}_0 + \mathbf{a}_1 + \mathbf{a}_2 + \mathbf{a}_3 \ ,$$

$$\dot{\mathbf{r}}^{[3]}(0) = \mathbf{a}_1 \ , \qquad \dot{\mathbf{r}}^{[3]}(1) = \mathbf{a}_1 + 2\mathbf{a}_2 + 3\mathbf{a}_3 \quad .$$

Suppose we re-express this as a quartic segment

$$\mathbf{r}^{[4]}(u) = \sum_{j=0}^{3} \mathbf{b}_j u^j \ ,$$

whose end points and tangents are easily seen to be

$$\mathbf{r}^{[4]}(0) = \mathbf{b}_0 \ , \qquad \mathbf{r}^{[4]}(1) = \mathbf{b}_0 + \mathbf{b}_1 + \mathbf{b}_2 + \mathbf{b}_3 + \mathbf{b}_4 \ ,$$

$$\dot{\mathbf{r}}^{[4]}(0) = \mathbf{b}_1 \ , \qquad \dot{\mathbf{r}}^{[4]}(1) = \mathbf{b}_1 + 2\mathbf{b}_2 + 3\mathbf{b}_3 + 4\mathbf{b}_4 \ .$$

If we wish $\mathbf{r}^{[3]}(u)$ and $\mathbf{r}^{[4]}(u)$ to have the same end conditions, then clearly $\mathbf{b}_0 = \mathbf{a}_0$, $\mathbf{b}_1 = \mathbf{a}_1$ and consequently

$$\mathbf{b}_2 + \mathbf{b}_3 + \mathbf{b}_4 = \mathbf{a}_2 + \mathbf{a}_3$$

and

$$2\mathbf{b}_2 + 3\mathbf{b}_3 + 4\mathbf{b}_4 = 2\mathbf{a}_2 + 3\mathbf{a}_3 \ .$$

The most general solution of this last pair of equations is

$$\mathbf{b}_2 = \mathbf{a}_2 + \mathbf{k}\,, \qquad \mathbf{b}_3 = \mathbf{a}_3 - 2\mathbf{k}\,, \qquad \mathbf{b}_4 = \mathbf{k}\,,$$

where $\mathbf{k}$ is an arbitrary vector. Then it is possible to choose $\mathbf{k}$ so that $\mathbf{r}^{[4]}(u)$ passes through some specified point for a specified value of u, or so that the segment has a specified second derivative at one end.

Note that a further increase in degree would permit us to attain specified second derivatives at both ends of the segment, a more symmetrical situation. This is one reason why curve fitting and design procedures are usually based on polynomials of odd degree, which demand an even number of conditions to specify them completely. Apart from this, curves based on even-order segments can present certain mathematical problems. For example, a polynomial spline of even degree which interpolates values of a given single-valued function at prescribed points does not necessarily exist (Ahlberg, Nilson and Walsh, 1967).

Raising the degree of a Bézier curve segment involves finding a new characteristic polygon, with an increased number of sides, for the original curve segment. If the degrees of the equivalent segments are m and n, where $n > m$, we can write

$$\mathbf{r}(u) = [1 \ u \ u^2 \ \ldots \ u^m]\mathbf{M}_m \begin{bmatrix} \mathbf{r}_0^{[m]} \\ \mathbf{r}_1^{[m]} \\ \cdot \\ \cdot \\ \cdot \\ \cdot \\ \mathbf{r}_m^{[m]} \end{bmatrix} = [1 \ u \ u^2 \ \ldots . \ u^n]\mathbf{M}_n \begin{bmatrix} \mathbf{r}_0^{[n]} \\ \mathbf{r}_1^{[n]} \\ \cdot \\ \cdot \\ \cdot \\ \cdot \\ \mathbf{r}_n^{[n]} \end{bmatrix},$$

where the $\mathbf{r}_i^{[m]}$ and $\mathbf{r}_i^{[n]}$ are the sets of polygon vertices and $\mathbf{M}_k$ is the Bézier coefficient matrix for a curve of degree k. This is the generalisation of the matrix $\mathbf{M}$ defined by (5.6), and it is a $(k+1)\times(k+1)$ lower triangular matrix (all elements above the principal diagonal being zero) with elements

$$m_{k,ij} = \begin{cases} \dfrac{(-1)^{i-j}k!}{(i-j)!j!(k-i)!}\,, & i \geqslant j \\ 0 & i < j\,, \end{cases}$$

where i, j take the values $0, 1, \ldots, k$. We may write the equivalent curve segments more concisely as

$$\mathbf{r}(u) = \mathbf{U}_m \mathbf{M}_m \mathbf{R}_m = \mathbf{U}_n \mathbf{M}_n \mathbf{R}_n ,$$

and by stacking zeros below the matrix $\mathbf{M}_m$ we may write $\mathbf{U}_m \mathbf{M}_m$ as $\mathbf{U}_n \begin{bmatrix} \mathbf{M}_m \\ \hline \mathbf{0} \end{bmatrix}$.

Substituting the latter and equating coefficients of powers of u in the previous relation, we then obtain

$$\begin{bmatrix} \mathbf{M}_m \\ \hline \mathbf{0} \end{bmatrix} \mathbf{R}_m = \mathbf{M}_n \mathbf{R}_n ,$$

so that the new polygon vertices are given by

$$\mathbf{R}_n = \mathbf{M}_n^{-1} \begin{bmatrix} \mathbf{M}_m \\ \hline \mathbf{0} \end{bmatrix} \mathbf{R}_m , \qquad n > m .$$

It may be shown from this result that when the degree of a Bézier segment is increased by one, from m to $m+1$, the old and the new polygon vertices are related by

$$\mathbf{r}_i^{[m+1]} = \frac{1}{m+1} \left[i\mathbf{r}_{i-1}^{[m]} + (m+1-i)\mathbf{r}_i^{[m]} \right] , \quad i = 0, 1, \ldots, m+1.$$

The extra vertex which results may be manipulated so that a further condition is satisfied in addition to those imposed on the original segment. Again, the use of homogeneous coordinates enables us to extend this result to cover the rational polynomial curve segment.

Finally, we briefly mention two other methods which have been proposed for modifying individual curve segments. The first was suggested by Forrest (1968) who points out that the rational cubic curve segment defined by (6.20) reduces to an ordinary parametric cubic segment when $w_0 = w_1 = w_2 = w_3 = 1$. Then any segment of a composite parametric cubic curve can be regarded as a special case of a rational cubic curve segment, and modification is effected by letting the weights depart from unity. It is possible to do this in such a way that the original end points and end tangents are preserved.

Our last method is due to Coons (1977), who observes that the quintic function

$$\mathbf{q}(u) = (6u^5 - 15u^4 + 10u^3) [\mathbf{r}(1) - \mathbf{r}(0)] + \mathbf{r}(0) , \qquad 0 \leqslant u \leqslant 1,$$

has the properties

$$\mathbf{q}(0) = \mathbf{r}(0)\ , \qquad \dot{\mathbf{q}}(0) = 0\ , \qquad \ddot{\mathbf{q}}(0) = 0\ ,$$

$$\mathbf{q}(1) = \mathbf{r}(1)\ , \qquad \dot{\mathbf{q}}(1) = 0\ , \qquad \ddot{\mathbf{q}}(1) = 0\ .$$

Clearly this expression represents part of the straight line passing through $\mathbf{r}(0)$ and $\mathbf{r}(1)$. In fact, since the polynomial $(6u^5 - 15u^4 + 10u^3)$ increases monotonely from 0 to 1 over the interval $0 \leqslant u \leqslant 1$, $\mathbf{q}(u)$ gives precisely the straight line joining the points $\mathbf{r}(0)$ and $\mathbf{r}(1)$.

Now suppose that $\mathbf{r}(u)$, $0 \leqslant u \leqslant 1$ is any curve segment and α is any scalar. Then if $\mathbf{q}(u)$ is defined as above, we see that

$$\mathbf{r}^*(u) = (1 - \alpha)\mathbf{r}(u) + \alpha\mathbf{q}(u)$$

has the same end points as $\mathbf{r}(u)$, and that for $0 < \alpha < 1$ the curve $\mathbf{r}^*(u)$ will lie intermediate between $\mathbf{r}(u)$ and the linear interpolant between its end-points, as shown in Figure 6.18.

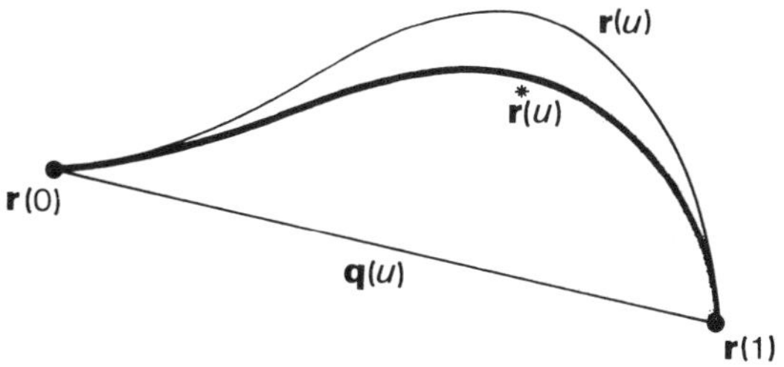

Figure 6.18 – Modification of a Curve Segment by Coons' Method.

By virtue of the properties of $\mathbf{q}(u)$, the tangent and second derivative vectors at the end-points of $\mathbf{r}^*(u)$ agree with those of $\mathbf{r}(u)$ in direction, though their magnitudes are multiplied by $(1 - \alpha)$. It follows that, if $\mathbf{r}(u)$ was originally part of a composite curve with continuity of tangent and second derivative in both direction and magnitude, these properties are maintained if each segment is modified using the same value of α. More generally, as is shown by Coons, we may let α be a function of the piecewise variable u. Continuity of $\alpha(u)$ and its first two derivatives over the whole length of the curve then ensures full second-order continuity of derivative vectors at joins between segments.

Coons points out that this simple method gives results qualitatively similar to those obtained using splines in tension (Schweikert, 1966; Cline, 1974; Nielsen, 1974). These are less prone to unwanted oscillation than ordinary splines, but the mathematics involved in their use is considerably more complicated.

Chapter 7

Composite surfaces

7.1 INTRODUCTION: COONS PATCHES

In this chapter we discuss a number of methods for defining surfaces. It was explained in Section 6.1 that some of these methods are based on a framework built up of two intersecting families of curves, and we start by considering procedures of this type. It will be assumed that the necessary mesh of curves has already been constructed using an appropriate method from Chapter 6. For greater generality we will work initially in terms of parametric curves, distinguishing fore-and-aft curves from transverse curves by defining them in terms of the parameters u and v respectively. We will later specialise our results to the nonparametric case.

The network of curves divides the surface into an assembly of topologically rectangular patches, each of which has as its boundaries two u-curves and two v-curves, as shown in Figure 7.1. Here it is assumed that u and v run from 0 to 1 along the relevant boundaries; then $\mathbf{r}(u,v)$, $0 < u,\ v < 1$, represents the interior of the surface patch, while $\mathbf{r}(u,0)$, $\mathbf{r}(1,v)$, $\mathbf{r}(u,1)$ and $\mathbf{r}(0,v)$ represent the four known boundary curves. The problem of defining a surface patch, then, is that of finding a suitably well-behaved function $\mathbf{r}(u,v)$ which reduces to the correct boundary curve when $u = 0, u = 1, v = 0$ or $v = 1$.

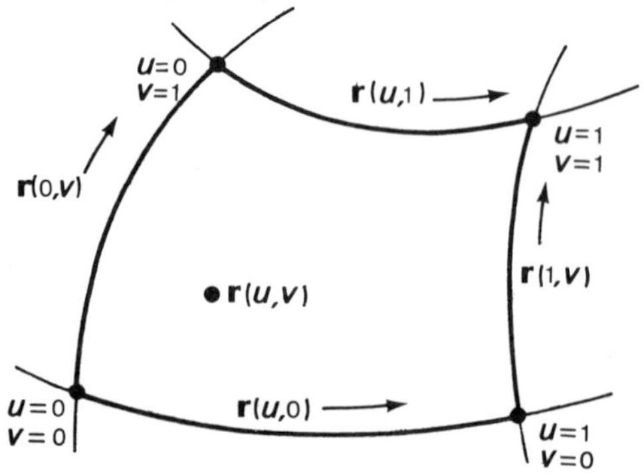

Figure 7.1. – A Parametric Surface Patch

Consider first the simpler problem of constructing a surface patch given only two of its boundaries, $\mathbf{r}(0,v)$ and $\mathbf{r}(1,v)$. If we use linear interpolation in the u-direction, we obtain the ruled surface

$$\mathbf{r}_1(u,v) = (1 - u)\mathbf{r}(0,v) + u\mathbf{r}(1,v) \ . \tag{7.1}$$

Alternatively, linear interpolation in the v-direction gives a surface which fits the other two boundaries:

$$\mathbf{r}_2(u,v) = (1 - v)\mathbf{r}(u,0) + v\mathbf{r}(u,1) \ . \tag{7.2}$$

These two surfaces are sketched in Figure 7.2 below.

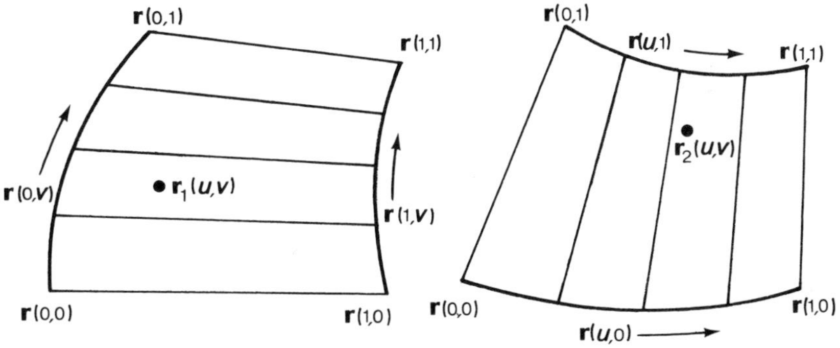

Figure 7.2 – Surface Patches Resulting from Linear Interpolation between Pairs of Boundaries

Now the sum $\mathbf{r}_1 + \mathbf{r}_2$ represents a surface patch each of whose boundaries is the sum of the desired boundary curve with the linear interpolant between the end-points of that curve. For instance, it is easily verified from (7.1) and (7.2) that the boundary corresponding to $v = 0$ is not $\mathbf{r}(u,0)$ but

$$\mathbf{r}(u,0) + [(1 - u)\mathbf{r}(0,0) + u\mathbf{r}(1,0)] \ .$$

Then if we can find a surface patch $\mathbf{r}_3(u,v)$ whose boundaries are the unwanted linear interpolants, we can recover the original boundary curves by forming $\mathbf{r}_1 + \mathbf{r}_2 - \mathbf{r}_3$. It is easy to construct $\mathbf{r}_3$; its $v = 0$ and $v = 1$ boundaries must be

$$(1 - u)\mathbf{r}(0,0) + ur(1,0), \qquad (1 - u)\mathbf{r}(0,1) + ur(1,1)$$

respectively, and a further linear interpolation in the v-direction then gives

$$\mathbf{r}_3(u,v) = (1 - u)(1 - v)\mathbf{r}(0,0) + u(1 - v)\mathbf{r}(1,0) + (1 - u)v\mathbf{r}(0,1) + uv\mathbf{r}(1,1). \tag{7.3}$$

The surface $\mathbf{r} = \mathbf{r}_1 + \mathbf{r}_2 - \mathbf{r}_3$ obtained from (7.1), (7.2) and (7.3) is conveniently expressed in the matrix form

$$\mathbf{r}(u,v) = [(1-u)\ u]\begin{bmatrix}\mathbf{r}(0,v)\\ \mathbf{r}(1,v)\end{bmatrix} + [\mathbf{r}(u,0)\ \ \mathbf{r}(u,1)]\begin{bmatrix}1-v\\ v\end{bmatrix}$$

$$-\ [(1-u)\ u]\begin{bmatrix}\mathbf{r}(0,0) & \mathbf{r}(0,1)\\ \mathbf{r}(1,0) & \mathbf{r}(1,1)\end{bmatrix}\begin{bmatrix}1-v\\ v\end{bmatrix}. \qquad (7.4)$$

Successive substitution of $u = 0, u = 1, v = 0$ and $v = 1$ quickly confirms that the patch defined by (7.4) has the four original curves as its boundaries.

This patch, constructed solely in terms of information given on its boundary and certain auxiliary scalar functions of u and v, is the most elementary of a class of surfaces originally studied by Coons (1967), and which have since become known as **Coons patches.** The treatment given here is essentially that of Forrest (1972a). The auxiliary functions u, $(1-u)$, v and $(1-v)$ are called **blending functions,** because their effect is to blend together four separate boundary curves to give a single well-defined surface.

An important generalisation of (7.4) may be made immediately. The linear blending functions in the patch equation result from the use of *uniform* linear interpolation. In fact, we may replace $(1-u)$, u by $\alpha_0(u)$, $\alpha_1(u)$ where the blending functions are now any functions such that $1 - \alpha_1 = \alpha_0$ and

$$\alpha_0(0) = 1, \qquad \alpha_0(1) = 0$$

$$\alpha_1(0) = 0, \qquad \alpha_1(1) = 1\ .$$

With this change, the u-interpolations performed in the derivation of (7.4) remain linear, but the rate of motion of $\mathbf{r}$ along the line of interpolation is now no longer constant as u increases uniformly from 0 to 1. The corresponding replacement of $(1-v)$, v by $\alpha_0(v)$, $\alpha_1(v)$ gives

$$\mathbf{r}(u,v) = [\alpha_0(u)\ \alpha_1(u)]\begin{bmatrix}\mathbf{r}(0,v)\\ \mathbf{r}(1,v)\end{bmatrix} + [\mathbf{r}(u,0)\ \mathbf{r}(u,1)]\begin{bmatrix}\alpha_0(v)\\ \alpha_1(v)\end{bmatrix}$$

$$-\ [\alpha_0(u)\ \alpha_1(u)]\begin{bmatrix}\mathbf{r}(0,0) & \mathbf{r}(0,1)\\ \mathbf{r}(1,0) & \mathbf{r}(1,1)\end{bmatrix}\begin{bmatrix}\alpha_0(v)\\ \alpha_1(v)\end{bmatrix}. \qquad (7.5)$$

The blending functions α_0 and α_1 are customarily chosen to be continuous and monotone (in other words continuously increasing or continuously decreasing) over the interval $0 \leqslant u \leqslant 1$. In practice polynomial blending functions are generally used, for both analytical and computational convenience. The condition that $1 - \alpha_1 = \alpha_0$ may be dispensed with, provided α_0 and α_1 satisfy the other stated requirements. In this case, however, the interpolations between opposite boundaries corresponding to equations (7.1) and (7.2) are nonlinear. In consequence the patch represented by (7.5) does not then reduce to a plane when all its boundaries are coplanar.

Given a network of curves, then, we can construct a composite surface made up of patches of the type described. This surface will have only positional continuity across patch boundaries, however. The gradient continuity essential for most practical applications may be achieved by using a less elementary kind of patch defined not only in terms of its boundary curves but also in terms of its cross-boundary slopes $\mathbf{r}_v(u,0)$, $\mathbf{r}_u(1,v)$, $\mathbf{r}_v(u,1)$ and $\mathbf{r}_u(0,v)$.†

The equation for such a patch may be derived in a similar manner to (7.5), except that we now use generalised Hermite interpolation rather than generalised linear interpolation. The surface

$$\mathbf{r}_1(u,v) = \alpha_0(u)\mathbf{r}(0,v) + \alpha_1(u)\mathbf{r}(1,v) + \beta_0(u)\mathbf{r}_u(0,v) + \beta_1(u)\mathbf{r}_u(1,v) \tag{7.6}$$

interpolates the two boundaries $\mathbf{r}(0,v)$ and $\mathbf{r}(1,v)$ and gives the specified slopes across these boundaries if the four blending functions satisfy

$$\begin{aligned} \alpha_0(0) &= 1\ , \qquad & \alpha_0(1) &= 0\ , \\ \alpha_1(0) &= 0\ , \qquad & \alpha_1(1) &= 1\ , \end{aligned} \tag{7.7}$$

$$\alpha_0'(0) = \alpha_0'(1) = \alpha_1'(0) = \alpha_1'(1) = 0\ , \tag{7.8}$$

$$\beta_0(0) = \beta_0(1) = \beta_1(0) = \beta_1(1) = 0 \tag{7.9}$$

and

$$\begin{aligned} \beta_0'(0) &= 1\ , \qquad & \beta_0'(1) &= 0\ , \\ \beta_1'(0) &= 0\ , \qquad & \beta_1'(1) &= 1\ . \end{aligned} \tag{7.10}$$

The reader may easily verify this for himself by setting $u = 0$ and $u = 1$ in (7.6) and in (7.6) differentiated partially with respect to u. Similarly the two curves $\mathbf{r}(u,0)$ and $\mathbf{r}(u,1)$, with their associated gradient data, are interpolated by

$$\mathbf{r}_2(u,v) = \alpha_0(v)\mathbf{r}(u,0) + \alpha_1(v)\mathbf{r}(u,1) + \beta_0(v)\mathbf{r}_v(u,0) + \beta_1(v)\mathbf{r}_v(u,1)\ . \tag{7.11}$$

†It is possible to ensure automatic continuity of cross-boundary gradient between elementary patches by suitably choosing the blending functions. But this allows less flexibility, and the resulting surfaces often exhibit unwanted flat regions (see Forrest's Appendix 2 in Bézier, 1972).

As in the derivation of (7.5), we find that $\mathbf{r}_1 + \mathbf{r}_2$ does not give us the surface we require. To get this, we have to subtract the further surface $\mathbf{r}_3(u,v)$ which results when we apply the same interpolation technique in both directions, using corner data alone. The resulting equation is:

$$\mathbf{r}(u,v) = [\alpha_0(u)\ \alpha_1(u)\ \beta_0(u)\ \beta_1(u)] \begin{bmatrix} \mathbf{r}(0,v) \\ \mathbf{r}(1,v) \\ \mathbf{r}_u(0,v) \\ \mathbf{r}_u(1,v) \end{bmatrix}$$

$$+ [\mathbf{r}(u,0)\ \mathbf{r}(u,1)\ \mathbf{r}_v(u,0)\ \mathbf{r}_v(u,1)] \begin{bmatrix} \alpha_0(v) \\ \alpha_1(v) \\ \beta_0(v) \\ \beta_1(v) \end{bmatrix}$$

$$- [\alpha_0(u)\,\alpha_1(u)\,\beta_0(u)\,\beta_1(u)] \begin{bmatrix} \mathbf{r}(0,0) & \mathbf{r}(0,1) & \mathbf{r}_v(0,0) & \mathbf{r}_v(0,1) \\ \mathbf{r}(1,0) & \mathbf{r}(1,1) & \mathbf{r}_v(1,0) & \mathbf{r}_v(1,1) \\ \mathbf{r}_u(0,0) & \mathbf{r}_u(0,1) & \mathbf{r}_{uv}(0,0) & \mathbf{r}_{uv}(0,1) \\ \mathbf{r}_u(1,0) & \mathbf{r}_u(1,1) & \mathbf{r}_{uv}(1,0) & \mathbf{r}_{uv}(1,1) \end{bmatrix} \begin{bmatrix} \alpha_0(v) \\ \alpha_1(v) \\ \beta_0(v) \\ \beta_1(v) \end{bmatrix} . \quad (7.12)$$

Here the three terms on the right are respectively $\mathbf{r}_1$, $\mathbf{r}_2$ and $\mathbf{r}_3$ in the foregoing argument. Note that corner values of the cross-derivative $\mathbf{r}_{uv}$ are necessary for the construction of $\mathbf{r}_3$. The reader may find it helpful to check in detail that, provided the blending functions satisfy (7.7) to (7.10), the surface represented by (7.12) has the specified boundary curves and cross-boundary gradients. Using patches of this kind we can match surface normal direction across boundaries and construct a composite surface which is everywhere smooth.

Working along similar lines, we may derive the equation of a patch with specified cross-boundary second derivatives. Two more blending functions are needed, and the square matrix involved will be of order 6X6. Most practical surface-defining systems avoid this degree of complexity, however.

The cross-derivative $\mathbf{r}_{uv}$ which occurs in the patch equation (7.12) appears frequently in this chapter, and a brief discussion of its physical significance may be useful. Let us first consider a surface $z = z(x,y)$ in Cartesian coordinates. Since $z_{xy} = \dfrac{\partial}{\partial x}\left(\dfrac{\partial z}{\partial y}\right)$, we see that the mixed or cross-derivative measures the rate of change in the x-direction of the slope of the surface in the y-direction. Then z_{xy} is a measure of *twist* in the surface. If z_{xy} is a continuous function of x and y, it may be shown analytically that $z_{yx} = z_{xy}$.

By analogy with the Cartesian case, the parametric cross-derivative $\mathbf{r}_{uv}$ is often called a **twist vector**. Although we follow custom and use this term ourselves in what follows, we must point out that it can be very misleading. This is because the value of $\mathbf{r}_{uv}$ at a point on a parametric surface depends not so much on the geometrical properties of the surface as on the way in which it is parametrised. A simple example serves to show this. Suppose that **a**, **b** and **c** are constant vectors such that $\mathbf{b} \times \mathbf{c} \neq \mathbf{0}$, and let u, v be any scalars. Then

$$\mathbf{r}(u,v) = \mathbf{a} + u\mathbf{b} + v\mathbf{c}$$

defines a flat plane containing the point whose position vector is **a**, and parallel to the vectors **b** and **c**. A flat plane is not a twisted surface, and it is therefore reassuring to find that $\mathbf{r}_{uv} = \mathbf{0}$. But now consider

$$\mathbf{r}(u,v) = \mathbf{a} + u\mathbf{b} + uv\mathbf{c} .$$

This reparametrisation of the previous equation defines the same flat plane, but now $\mathbf{r}_{uv} = \mathbf{c} \neq \mathbf{0}$. We conclude that caution must be exercised in interpreting the 'twist vector' in geometrical terms, since $\mathbf{r}_{uv} \neq \mathbf{0}$ does *not* necessarily imply a twist in a surface.

7.2 TENSOR-PRODUCT SURFACES

It is possible to simplify the Coons patch equation (7.12) considerably by suitably defining the boundary curves and cross-boundary gradients. Using a set of blending functions which satisfy the conditions (7.7) to (7.10), we may express a curve segment in terms of its end points and end tangents by

$$\mathbf{r}(u) = \alpha_0(u)\mathbf{r}(0) + \alpha_1(u)\mathbf{r}(1) + \beta_0(u)\dot{\mathbf{r}}(0) + \beta_1(u)\dot{\mathbf{r}}(1) . \quad (7.13)$$

This formulation permits our boundary curves and cross-boundary gradients to be written as

$$\mathbf{r}(i,v) = \alpha_0(v)\mathbf{r}(i,0) + \alpha_1(v)\mathbf{r}(i,1) + \beta_0(v)\mathbf{r}_v(i,0) + \beta_1(v)\mathbf{r}_v(i,1) ,$$

$$\mathbf{r}_u(i,v) = \alpha_0(v)\mathbf{r}_u(i,0) + \alpha_1(v)\mathbf{r}_u(i,1) + \beta_0(v)\mathbf{r}_{uv}(i,0) + \beta_1(v)\mathbf{r}_{uv}(i,1) , \quad (7.14)$$

where $i = 0, 1$, and

$$\mathbf{r}(u,j) = \alpha_0(u)\mathbf{r}(0,j) + \alpha_1(u)\mathbf{r}(1,j) + \beta_0(u)\mathbf{r}_u(0,j) + \beta_1(u)\mathbf{r}_u(1,j) ,$$

$$\mathbf{r}_v(u,j) = \alpha_0(u)\mathbf{r}_v(0,j) + \alpha_1(u)\mathbf{r}_v(1,j) + \beta_0(u)\mathbf{r}_{uv}(0,j) + \beta_1(u)\mathbf{r}_{uv}(1,j), \quad (7.15)$$

where $j = 0, 1$. When these forms are substituted into (7.12) it is found that all three terms are now identical but for the negative sign of the third. The equation therefore reduces to

$$\mathbf{r}(u,v) = [\alpha_0(u)\,\alpha_1(u)\,\beta_0(u)\,\beta_1(u)] \begin{bmatrix} \mathbf{r}(0,0) & \mathbf{r}(0,1) & \mathbf{r}_v(0,0) & \mathbf{r}_v(0,1) \\ \mathbf{r}(1,0) & \mathbf{r}(1,1) & \mathbf{r}_v(1,0) & \mathbf{r}_v(1,1) \\ \mathbf{r}_u(0,0) & \mathbf{r}_u(0,1) & \mathbf{r}_{uv}(0,0) & \mathbf{r}_{uv}(0,1) \\ \mathbf{r}_u(1,0) & \mathbf{r}_u(1,1) & \mathbf{r}_{uv}(1,0) & \mathbf{r}_{uv}(1,1) \end{bmatrix} \begin{bmatrix} \alpha_0(v) \\ \alpha_1(v) \\ \beta_0(v) \\ \beta_1(v) \end{bmatrix}$$

or, more concisely, $$\mathbf{r}(u,v) = \mathbf{F}(u)\mathbf{Q}\,\mathbf{F}^T(v). \tag{7.16}$$

This simplification results because we have defined our boundary data in terms of the same blending functions as are used in the construction of the original patch equation, (7.12). Note that the patch is now completely defined in terms of the vectors $\mathbf{r}$, $\mathbf{r}_u$, $\mathbf{r}_v$ and $\mathbf{r}_{uv}$ at its four corners. A patch of this type is often called a **tensor-product** (or **Cartesian product**) patch.

The fact that cross-boundary gradients are specified by (7.14) and (7.15) in terms of $\mathbf{r}_u$, $\mathbf{r}_v$ and $\mathbf{r}_{uv}$ *at the patch corners* has an important consequence when we come to construct composite surfaces from this type of patch. It is only necessary to match these three vector quantities at contiguous corners of adjacent patches to achieve gradient continuity across all boundaries.

7.2.1 Ferguson Surfaces

The simplest set of polynomial blending functions satisfying conditions (7.7) to (7.10) are the following cubics:

$$\alpha_0(u) = 1 - 3u^2 + 2u^3 \qquad \alpha_1(u) = 3u^2 - 2u^3$$

$$\beta_0(u) = u - 2u^2 + u^3 \qquad \beta_1(u) = -u^2 + u^3 .$$

On putting these into (7.13), we find that we obtain precisely the Ferguson cubic curve segment of Section 5.1.2 (see also Section 6.3.1). When using this set of blending functions, we are therefore constructing a composite surface on a curve network made up of Ferguson curves. We may write the blending function vectors in (7.16) as

$$\mathbf{F}(u) = [\alpha_0(u)\ \alpha_1(u)\ \beta_0(u)\ \beta_1(u)]$$

$$= [1\ u\ u^2\ u^3] \begin{bmatrix} 1 & 0 & 0 & 0 \\ 0 & 0 & 1 & 0 \\ -3 & 3 & -2 & -1 \\ 2 & -2 & 1 & 1 \end{bmatrix} = \mathbf{UC} \tag{7.17a}$$

and
$$\mathbf{F}^T(v) = \begin{bmatrix} \alpha_0(v) \\ \alpha_1(v) \\ \beta_0(v) \\ \beta_1(v) \end{bmatrix} = \begin{bmatrix} 1 & 0 & -3 & 2 \\ 0 & 0 & 3 & -2 \\ 0 & 1 & -2 & 1 \\ 0 & 0 & -1 & 1 \end{bmatrix} \begin{bmatrix} 1 \\ v \\ v^2 \\ v^3 \end{bmatrix} = \mathbf{C}^T \mathbf{V} \ . \tag{7.17b}$$

Then the Ferguson surface patch becomes

$$\mathbf{r}(u,v) = \mathbf{U\,C\,Q\,C}^T \mathbf{V} \ , \tag{7.18}$$

with the matrix **Q** as defined by (7.16). Earlier, in Section 5.1.6, we gave the formulation

$$\mathbf{r}(u,v) = \sum_{i=0}^{3} \sum_{j=0}^{3} \mathbf{a}_{ij} u^i v^j = \mathbf{U\,A\,V} \ ,$$

and comparison shows that $\mathbf{A} = \mathbf{CQC}^T$. The coefficients $\mathbf{a}_{ij}$, which are the elements of **A** and hence also of $\mathbf{CQC}^T$, are therefore linear combinations of position and derivative vectors at the patch corners.

In Ferguson's original papers (1963, 1964) the patch boundaries are blended in such a way that the cross-derivatives $\mathbf{r}_{uv}$ at patch corners are implicitly assumed to be zero, so that the lower right-hand 2X2 submatrix of **Q** is a null matrix. A notable application of this restricted type of patch is in the APT surface-fitting routine FMILL. Here the gradients $\mathbf{r}_u$ and $\mathbf{r}_v$ at mesh intersections are estimated in terms of local positional data.† Their directions are taken as parallel to the chord-line joining the preceding and the following data points on the appropriate mesh curve, while each magnitude is taken as the length of the chord joining the point concerned to either the preceding or the following point. The shorter chord is chosen to minimise the possibility of bulges or loops in the curve (see Section 6.3.6). The dangers of using estimated rather than measured gradients were touched upon in Section 6.2.1. However, the originators of FMILL considered them to be outweighed by the extra computation involved in calculating appropriate gradients using Ferguson's own proposed method (see Section 6.3.1). In practice FMILL usually works tolerably well because the effect of the simplifying assumptions are small enough to be masked by the scale of roughness of the unfinished machined FMILL surface. Some further details of this system are discussed in Section 7.7.

†Alternatively the user may specify all gradient vectors at the outset, but this option is rarely taken up.

Forrest, in his Appendix 2 to Bézier (1972), states that the assumption of $\mathbf{r}_{uv} = \mathbf{0}$ at all mesh intersections can lead to local flattening of the generated surface near patch corners. However, this effect can be minimised by arranging for the mesh lines to lie roughly along the lines of curvature of the surface (see Section 4.2.9), since there is no twist in a surface along all such lines.

If we permit non-zero twist vectors at patch corners, we are immediately faced with the problem of how to determine them. One way of doing this is to extend Ferguson's originally proposed method for calculating $\mathbf{r}_u$ and $\mathbf{r}_v$. Suppose the intersections of our curve network have position vectors $\mathbf{r}_{mn}$, where $m = 0, 1, 2, \ldots, M$ and $n = 0, 1, 2, \ldots, N$, as shown in Figure 7.3.

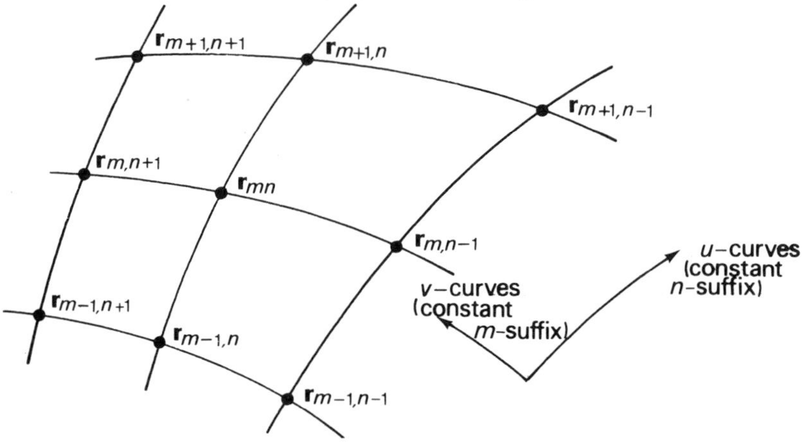

Figure 7.3

We recall from Section 6.3.1 that when first derivatives are matched in both direction and magnitude at the data points, the u-curves will have second derivative continuity if the tangent vectors $\mathbf{r}_{u,mn}$ are calculated from the system of equations

$$\mathbf{r}_{u,m-1,n} + 4\mathbf{r}_{u,mn} + \mathbf{r}_{u,m+1,n} = 3(\mathbf{r}_{m+1,n} - \mathbf{r}_{m-1,n}), \quad (7.19a)$$

$$m = 1, 2, \ldots, M-1.$$

Similarly, the v-curves will have second-order continuity if

$$\mathbf{r}_{v,m,n-1} + 4\mathbf{r}_{v,mn} + \mathbf{r}_{v,m,n+1} = 3(\mathbf{r}_{m,n+1} - \mathbf{r}_{m,n-1}), \quad (7.19b)$$

$$n = 1, 2, \ldots, N-1.$$

Now let us consider the conditions for continuity of $\mathbf{r}_{uu}$ between two patches which may be expressed using (7.18) as

$$\mathbf{r}^{(1)}(u,v) = \mathbf{UCQ}^{(1)}\mathbf{C}^T\mathbf{V}, \qquad \mathbf{r}^{(2)}(u,v) = \mathbf{UCQ}^{(2)}\mathbf{C}^T\mathbf{V},$$

and which have the common boundary $\mathbf{r}^{(1)}(1,v) = \mathbf{r}^{(2)}(0,v)$.

On differentiating the patch equations twice with respect to u, we find that

$$\mathbf{r}_{uu}^{(1)}(1,v) = [0 \quad 0 \quad 2 \quad 6]\,\mathbf{CQ}^{(1)}\mathbf{C}^T\mathbf{V}$$

and

$$\mathbf{r}_{uu}^{(2)}(0,v) = [0 \quad 0 \quad 2 \quad 0]\,\mathbf{CQ}^{(2)}\mathbf{C}^T\mathbf{V} .$$

The right hand sides are both cubic polynomials in v. For continuity of $\mathbf{r}_{uu}$ across the boundary for $0 \leqslant v \leqslant 1$ they must be equal, and so we may equate coefficients to obtain

$$[0 \quad 0 \quad 2 \quad 6]\,\mathbf{CQ}^{(1)}\mathbf{C}^T = [0 \quad 0 \quad 2 \quad 0]\,\mathbf{CQ}^{(2)}\mathbf{C}^T .$$

Since $\mathbf{C}^T$ is square and nonsingular we may postmultiply by $(\mathbf{C}^T)^{-1}$, which leads to

$$[6 \quad -6 \quad 2 \quad 4]\,\mathbf{Q}^{(1)} = [-6 \quad 6 \quad -4 \quad -2]\,\mathbf{Q}^{(2)} .$$

When written in scalar form, this is a set of four equations which involve values of $\mathbf{r}$, $\mathbf{r}_u$, $\mathbf{r}_v$ and $\mathbf{r}_{uv}$ at the corners of the two patches. These simplify if we take into account the conditions $\mathbf{r}^{(1)}(1,0) = \mathbf{r}^{(2)}(0,0)$, $\mathbf{r}^{(1)}(1,1) = \mathbf{r}^{(2)}(0,1)$, $\mathbf{r}_u^{(1)}(1,0) = \mathbf{r}_u^{(2)}(0,0)$ and $\mathbf{r}_u^{(1)}(1,1) = \mathbf{r}_u^{(2)}(0,1)$ which must be imposed for positional and gradient continuity across the boundary. We then find that two of the equations boil down to the conditions required by (7.19a) for second-order continuity of the composite boundary curves of the two-patch surface. The remaining two are the extra conditions which must be met for second-order continuity of the surface as a whole in the u-direction. If these are applied across all the v-boundaries of a larger composite surface, they may be summarised as

$$\mathbf{r}_{uv,m-1,n} + 4r_{uv,mn} + \mathbf{r}_{uv,m+1,n} = 3(\mathbf{r}_{v,m+1,n} - \mathbf{r}_{v,m-1,n}) , \quad (7.20a)$$

$$m = 1, 2, \ldots, M-1.$$

Here the gradients $\mathbf{r}_v$ on the right-hand side may be considered known from (7.19b).

Alternatively, we may achieve continuity of $\mathbf{r}_{vv}$ in the v-direction over the entire composite surface by satisfying the analogous equation

$$\mathbf{r}_{vu,m,n-1} + 4\mathbf{r}_{vu,mn} + \mathbf{r}_{vu,m,n+1} = 3(\mathbf{r}_{u,m,n+1} - \mathbf{r}_{u,m,n-1}), \quad (7.20b)$$

$$n = 1, 2, \ldots, N-1.$$

Now the formulation of the tensor-product patch assumes that $\mathbf{r}_{uv} = \mathbf{r}_{vu}$ at patch corners, and it is by no means obvious that the solutions $\mathbf{r}_{uv}$ of (7.20a) are going to be the same as the solutions $\mathbf{r}_{vu}$ of (7.20b). If they differ, we will not be able to satisfy both equations simultaneously, so that second-order continuity can be achieved only in one direction. Fortunately however, the two equations both give the same set of twist vectors, and overall second-order continuity is therefore attainable. An outline proof of this statement follows. The reader may not wish to reconstruct all the details for himself, but will find it instructive to follow the general lines of the proof, which embodies a practical method for computing a class of curvature-continuous surfaces based on splines.

For brevity, we will write $\mathbf{r}_u = \mathbf{s}$, $\mathbf{r}_v = \mathbf{t}$, $\mathbf{r}_{uv} = \mathbf{x}_1$ and $\mathbf{r}_{vu} = \mathbf{x}_2$. The equation for the $\mathbf{r}_u$-values may then be written in matrix form as

$$\begin{bmatrix} 1 & 4 & 1 & & & \\ & 1 & 4 & 1 & & \\ & & \cdot & & & \\ & & & \cdot & & \\ & & & & \cdot & \\ & & & 1 & 4 & 1 \end{bmatrix} \begin{bmatrix} \mathbf{s}_{00} & \mathbf{s}_{01} & \cdots & \mathbf{s}_{0N} \\ \mathbf{s}_{10} & \mathbf{s}_{11} & \cdots & \mathbf{s}_{1N} \\ \vdots & \vdots & & \vdots \\ \mathbf{s}_{M0} & \mathbf{s}_{M1} & \cdots & \mathbf{s}_{MN} \end{bmatrix} =$$

$$\begin{bmatrix} -3 & 0 & 3 & & & \\ & -3 & 0 & 3 & & \\ & & \cdot & & & \\ & & & \cdot & & \\ & & & & \cdot & \\ & & & -3 & 0 & 3 \end{bmatrix} \begin{bmatrix} \mathbf{r}_{00} & \mathbf{r}_{01} & \cdots & \mathbf{r}_{0N} \\ \mathbf{r}_{10} & \mathbf{r}_{11} & \cdots & \mathbf{r}_{1N} \\ \vdots & \vdots & & \vdots \\ \mathbf{r}_{M0} & \mathbf{r}_{M1} & \cdots & \mathbf{r}_{MN} \end{bmatrix} ,$$

or

$$\mathbf{A}_1\mathbf{S} = \mathbf{B}_1\mathbf{R} ,$$

where $\mathbf{A}_1$ and $\mathbf{B}_1$ are of order $(M-1)\times(M+1)$ while $\mathbf{S}$ and $\mathbf{R}$ are of order $(M+1)\times(N+1)$. We will presume the elements in the first and last rows of $\mathbf{S}$ to be known, in which case the remaining s-values may be uniquely determined.

Similarly, the equation for the $\mathbf{r}_v$-values has the matrix form

$$\mathbf{A}_2\mathbf{T}^T = \mathbf{B}_2\mathbf{R}^T ,$$

where $\mathbf{A}_2$ and $\mathbf{B}_2$ differ from $\mathbf{A}_1$ and $\mathbf{B}_1$ respectively only in that they are of order $(N-1)\times(N+1)$. The matrix $T = [\mathbf{t}_{mn}]$ has order $(M+1)\times(N+1)$, and the elements of its first and last columns are presumed known in order that its remaining elements may be uniquely determined.

With $\mathbf{S}$ and $\mathbf{T}$ known, we have two alternative ways of calculating the twist vectors, using either (7.20a) or (7.20b). In matrix form these are

$$\mathbf{A}_1\mathbf{X}_1 = \mathbf{B}_1\mathbf{T}\ , \qquad \mathbf{A}_2\mathbf{X}_2{}^T = \mathbf{B}_2\mathbf{S}^T\ ,$$

where $\mathbf{X}_1 = [\mathbf{x}_{1,mn}]$ and $\mathbf{X}_2 = [\mathbf{x}_{2,mn}]$. The first and last rows of $\mathbf{X}_1$ and the first and last columns of $\mathbf{X}_2$ are assumed to be known. Elimination of $\mathbf{S}$ and $\mathbf{T}$ from our four matrix equations shows that

$$\mathbf{A}_1\mathbf{X}_1\mathbf{A}_2{}^T = \mathbf{A}_1\mathbf{X}_2\mathbf{A}_2{}^T = \mathbf{B}_1\mathbf{R}\mathbf{B}_2{}^T,$$

whence
$$\mathbf{A}_1(\mathbf{X}_2 - \mathbf{X}_1)\mathbf{A}_2{}^T = \mathbf{0}\ .$$

Now it turns out (as we shall shortly show) that $\mathbf{X}_1$ and $\mathbf{X}_2$ are identical in their first and last rows and columns. Then all the elements on the periphery of $(\mathbf{X}_2 - \mathbf{X}_1)$ are zero, and the last equation may be rewritten as

$$\begin{bmatrix} 4 & 1 & & & \\ 1 & 4 & 1 & & \\ & \cdot & & & \\ & & \cdot & & \\ & & & \cdot & \\ & & & 1 & 4 \end{bmatrix} \begin{bmatrix} \Delta\mathbf{x}_{11} & \Delta\mathbf{x}_{12} & \cdots & \Delta\mathbf{x}_{1,N-1} \\ \Delta\mathbf{x}_{21} & \Delta\mathbf{x}_{22} & \cdots & \Delta\mathbf{x}_{2,N-1} \\ \vdots & \vdots & & \vdots \\ \Delta\mathbf{x}_{M-1,1} & \Delta\mathbf{x}_{M-1,2} & \cdots & \Delta\mathbf{x}_{M-1,N-1} \end{bmatrix} \begin{bmatrix} 4 & 1 & & & \\ 1 & 4 & 1 & & \\ & \cdot & & & \\ & & \cdot & & \\ & & & \cdot & \\ & & & 1 & 4 \end{bmatrix} = \mathbf{0}\ ,$$

in which $\Delta\mathbf{x}_{mn} = \mathbf{x}_{2,mn} - \mathbf{x}_{1,mn}$. The first and third factors are now square non-singular matrices, and it follows that the middle factor must be a null matrix. Then $\mathbf{X}_2 = \mathbf{X}_1$, so that (7.20a) and (7.20b) both yield the same set of values for the corner twist vectors $\mathbf{r}_{uv}$.

The truth of our assertion that $\mathbf{X}_1$ and $\mathbf{X}_2$ have identical peripheral elements may be established by considering the way in which these elements are determined. If we use (7.20a) to compute the $\mathbf{x}$'s, we need the values of $\mathbf{r}_{uv,0n}$ and $\mathbf{r}_{uv,Mn}$ to make the system fully determinate for each value of n. We may obtain these values in terms of the four twist vectors $\mathbf{r}_{uv,00}$, $\mathbf{r}_{uv,0N}$, $\mathbf{r}_{uv,M0}$ and $\mathbf{r}_{uv,MN}$ alone by using (7.20b) with $m = 0$ and $m = M$, because the values of $\mathbf{r}_u$ were previously assumed to be known along the $m = 0$ and $m = M$ boundaries of the composite surface. On the other hand, if we use (7.20b) to calculate the $\mathbf{x}$'s, we must first specify the values of $\mathbf{r}_{uv,m0}$ and $\mathbf{r}_{uv,mN}$ for each value of m. But these may also be determined in terms of the twist vectors at the four corners of the composite surface by using (7.20a) with $n = 0$ and $n = N$, because the values of $\mathbf{r}_v$ are assumed known along the $n = 0$ and $n = N$ boundaries. Thus in either case the peripheral elements of $\mathbf{X}_1$ and $\mathbf{X}_2$ are calculated in the same manner from the same information, and consequently have the same values. This completes the proof.

We note that the construction of the entire second-order continuous composite surface requires tangent vectors to be specified at all mesh points on the boundary, but twist vectors only at the four corner points. The total data needed may therefore be expressed diagrammatically as follows:

$\mathbf{x}_{00}$	$\mathbf{s}_{00}$	$\mathbf{s}_{01}$	....	$\mathbf{s}_{0N}$	$\mathbf{x}_{0N}$
$\mathbf{t}_{00}$	$\mathbf{r}_{00}$	$\mathbf{r}_{01}$	....	$\mathbf{r}_{0N}$	$\mathbf{t}_{0N}$
$\mathbf{t}_{10}$	$\mathbf{r}_{10}$	$\mathbf{r}_{11}$	....	$\mathbf{r}_{1N}$	$\mathbf{t}_{1N}$
.	.	.		.	.
.	.	.		.	.
.	.	.		.	.
$\mathbf{t}_{M0}$	$\mathbf{r}_{M0}$	$\mathbf{r}_{M1}$	....	$\mathbf{r}_{MN}$	$\mathbf{t}_{MN}$
$\mathbf{x}_{M0}$	$\mathbf{s}_{M0}$	$\mathbf{s}_{M1}$	....	$\mathbf{s}_{MN}$	$\mathbf{x}_{MN}$

Here, as previously, $\mathbf{s} = \mathbf{r}_u$, $\mathbf{t} = \mathbf{r}_v$ and $\mathbf{x} = \mathbf{r}_{uv}$.

The surface which results from this procedure is, in fact, a special case of the parametric spline surfaces discussed later in Section 7.2.4. We will there deal with a more general situation where u and v are not restricted to integer values at mesh intersections, and will also give a more detailed account of the computational procedure.

7.2.2 Bézier (UNISURF) Surfaces

As mentioned earlier in Section 5.1.3, the cubic Bézier curve is simply a Ferguson curve reformulated in a way which avoids any need for tangents to be specified. There is a correspondingly close relationship between Ferguson and Bézier surface patches. We saw in Section 5.1.7 that a Bézier patch is designed in terms of a 'characteristic polyhedron', which is specified in terms of the position vectors $\mathbf{r}_{ij}$ of its 16 vertices.

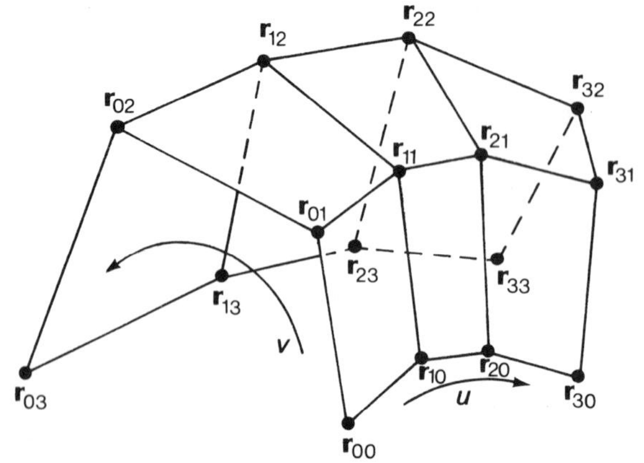

Figure 7.4 – The Characteristic Polyhedron of a cubic Bézier patch

The patch is in a sense an approximation to the polyhedron, though in general the only points they have in common are the corner points $\mathbf{r}_{00}$, $\mathbf{r}_{03}$, $\mathbf{r}_{30}$, and $\mathbf{r}_{33}$. Note that the $\mathbf{r}_{ij}$ notation differs from the $\mathbf{r}_{mn}$ notation of the last section in that each point $\mathbf{r}_{ij}$ lies on the characteristic polyhedron rather than on the surface itself. The configuration of the polyhedron gives the designer a good indication of the general shape of the corresponding patch, and modification of one or more of the vectors $\mathbf{r}_{ij}$ alters the patch in a predictable way. No gradients or twist vectors need be specified.

The equation defining each patch is

$$\mathbf{r}(u,v) = \sum_{i=0}^{3}\sum_{j=0}^{3} \mathbf{r}_{ij}\, g_i(u)\, g_j(v)\,, \qquad 0 \leqslant u,\, v \leqslant 1\,, \tag{7.21}$$

where $$g_k(t) = \frac{3!}{k!(3-k)!} t^k (1-t)^{3-k}\,, \qquad k = 0, 1, 2, 3.$$

The g_k are the cubic Bernstein basis functions defined in Appendix 3. It is easy to check that $\mathbf{r}(0,v)$ is a cubic Bézier curve whose characteristic polygon is defined by $\mathbf{r}_{00}$, $\mathbf{r}_{01}$, $\mathbf{r}_{02}$ and $\mathbf{r}_{03}$; reference to Figure 7.4 shows that this curve is one of the boundaries of the patch. Correspondingly, the other three patch boundaries $\mathbf{r}(1,v)$, $\mathbf{r}(u,0)$ and $\mathbf{r}(u,1)$ are curves of the same type.

Now let us examine the relation between the Bézier and Ferguson patch formulations. Equation (7.21) may be written in matrix form as

$$\mathbf{r}(u,v) = \mathbf{G}(u)\,\mathbf{B}\,\mathbf{G}^T(v)\,,$$

where $\mathbf{B} = [\mathbf{r}_{ij}]$ is the 4×4 matrix of polyhedron vertices and

$$\begin{aligned}
\mathbf{G}(u) &= [g_0(u)\ g_1(u)\ g_2(u)\ g_3(u)] \\
&= [(1-u)^3 \quad 3u(1-u)^2 \quad 3u^2(1-u) \quad u^3] \\
&= [1\ u\ u^2\ u^3] \begin{bmatrix} 1 & 0 & 0 & 0 \\ -3 & 3 & 0 & 0 \\ 3 & -6 & 3 & 0 \\ -1 & 3 & -3 & 1 \end{bmatrix} \\
&= \mathbf{UM}\,.
\end{aligned} \tag{7.22}$$

Similarly, $\mathbf{G}(v) = \mathbf{M}^T\mathbf{V}$, with $\mathbf{M}$ as defined by (7.22) and $\mathbf{V} = [1\ v\ v^2\ v^3]^T$. Using these results, we may express the patch equation as

$$\mathbf{r}(u,v) = \mathbf{UMBM}^T\mathbf{V}\,. \tag{7.23}$$

Now the Ferguson patch equation derived in the last section was

$$\mathbf{r}(u,v) = \mathbf{UCQC}^T\mathbf{V}\ ,$$

where $\mathbf{Q}$ is defined by (7.16) and $\mathbf{C}$ by (7.17). For the two to be equivalent we must have

$$\mathbf{CQC}^T = \mathbf{MBM}^T\ ,$$

or

$$\mathbf{Q} = (\mathbf{C}^{-1}\mathbf{M})\mathbf{B}(\mathbf{C}^{-1}\mathbf{M})^T\ .$$

On inverting $\mathbf{C}$ and multiplying out the right-hand side we obtain

$$\mathbf{Q} = \begin{bmatrix} \mathbf{r}(0,0) & \mathbf{r}(0,1) & \mathbf{r}_v(0,0) & \mathbf{r}_v(0,1) \\ \mathbf{r}(1,0) & \mathbf{r}(1,1) & \mathbf{r}_v(1,0) & \mathbf{r}_v(1,1) \\ \mathbf{r}_u(0,0) & \mathbf{r}_u(0,1) & \mathbf{r}_{uv}(0,0) & \mathbf{r}_{uv}(0,1) \\ \mathbf{r}_u(1,0) & \mathbf{r}_u(1,1) & \mathbf{r}_{uv}(1,0) & \mathbf{r}_{uv}(1,1) \end{bmatrix} =$$

$$\begin{bmatrix} \mathbf{r}_{00} & \mathbf{r}_{03} & 3(\mathbf{r}_{01}-\mathbf{r}_{00}) & 3(\mathbf{r}_{03}-\mathbf{r}_{02}) \\ \mathbf{r}_{30} & \mathbf{r}_{33} & 3(\mathbf{r}_{31}-\mathbf{r}_{30}) & 3(\mathbf{r}_{33}-\mathbf{r}_{32}) \\ 3(\mathbf{r}_{10}-\mathbf{r}_{00}) & 3(\mathbf{r}_{13}-\mathbf{r}_{03}) & 9(\mathbf{r}_{00}-\mathbf{r}_{10}-\mathbf{r}_{01}+\mathbf{r}_{11}) & 9(\mathbf{r}_{02}-\mathbf{r}_{12}-\mathbf{r}_{03}+\mathbf{r}_{13}) \\ 3(\mathbf{r}_{30}-\mathbf{r}_{20}) & 3(\mathbf{r}_{33}-\mathbf{r}_{23}) & 9(\mathbf{r}_{20}-\mathbf{r}_{30}-\mathbf{r}_{21}+\mathbf{r}_{31}) & 9(\mathbf{r}_{22}-\mathbf{r}_{32}-\mathbf{r}_{23}+\mathbf{r}_{33}) \end{bmatrix}. \tag{7.24}$$

From this last equation we can read off the gradient and twist vectors at the patch corners, expressed solely in terms of the vertices of the characteristic polyhedron. The gradient vectors depend only upon points lying on the boundary of the polyhedron, while the four interior points $\mathbf{r}_{11}$, $\mathbf{r}_{12}$, $\mathbf{r}_{21}$ and $\mathbf{r}_{22}$ influence the values of the cross-derivatives.

The foregoing approach is implemented in the UNISURF surface design system developed by Bézier (1972) for the Renault car company. It has the great virtue that the designer has to specify neither gradients nor twist vectors in constructing his patch. Thus the need for Ferguson's rather unsatisfactory simplifying assumption that $\mathbf{r}_{uv} = \mathbf{0}$ at patch corners is avoided. Only the position vectors $\mathbf{r}_{ij}$ of the 16 polyhedron vertices need be supplied. These have a fairly obvious geometrical interpretation, and the system is therefore suitable for use by an operator with no advanced mathematical training.

Let us now consider how continuity may be achieved for a composite

Bézier surface. Figure 7.5 depicts two adjacent cubic Bézier patches, whose equations are respectively

$$\mathbf{r}^{(1)}(u,v) = \mathbf{UMB}^{(1)}\mathbf{M}^T\mathbf{V} \ , \quad \mathbf{r}^{(2)}(u,v) = \mathbf{UMB}^{(2)}\mathbf{M}^T\mathbf{V} \ . \qquad (7.25)$$

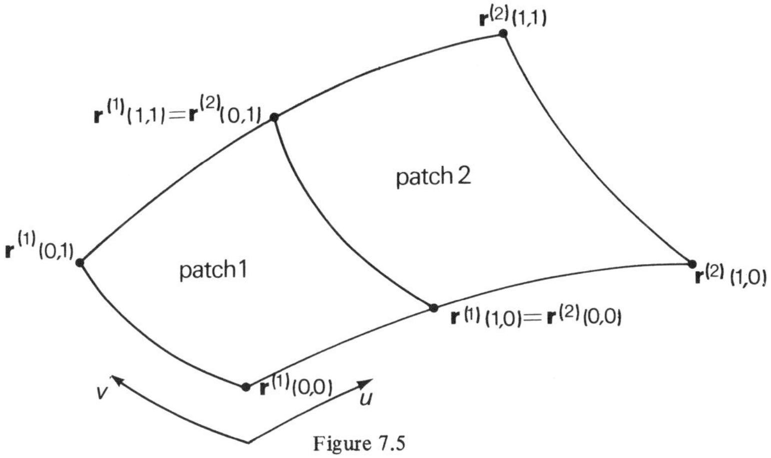

Figure 7.5

Positional continuity across the boundary will result if $\mathbf{r}^{(1)}(1,v) = \mathbf{r}^{(2)}(0,v)$ for all v such that $0 \leqslant v \leqslant 1$. We may use (7.25) to write this condition as

$$[1 \ \ 1 \ \ 1 \ \ 1]\mathbf{MB}^{(1)}\mathbf{M}^T\mathbf{V} = [1 \ \ 0 \ \ 0 \ \ 0]\mathbf{MB}^{(2)}\mathbf{M}^T\mathbf{V} \ .$$

Both sides are cubic polynomials in v. Equating coefficients, we obtain

$$[1 \ \ 1 \ \ 1 \ \ 1]\mathbf{MB}^{(1)}\mathbf{M}^T = [1 \ \ 0 \ \ 0 \ \ 0]\mathbf{MB}^{(2)}\mathbf{M}^T \ ,$$

which may be postmultiplied by $(\mathbf{M}^T)^{-1}$ and then multiplied out to give the set of four relations

$$\mathbf{r}_{3i}{}^{(1)} = \mathbf{r}_{0i}{}^{(2)} \ , \qquad i = 0, 1, 2, 3.$$

These imply, reasonably enough, that a common boundary curve between the two patches requires a common boundary polygon between the two characteristic polyhedra (see Figure 7.6).

For gradient continuity across the boundary, the tangent plane of Patch 1 on $u = 1$ must coincide with that of Patch 2 on $u = 0$, for all v such that $0 \leqslant v \leqslant 1$. Then the direction of the surface normal will be continuous across the boundary, and the condition

$$\mathbf{r}_u{}^{(2)}(0,v) \times \mathbf{r}_v{}^{(2)}(0,v) = \lambda(v)\mathbf{r}_u{}^{(1)}(1,v) \times \mathbf{r}_v{}^{(1)}(1,v) \qquad (7.26)$$

must apply. The presence of the positive-valued scalar function $\lambda(v)$ takes account of any discontinuity in the *magnitude* of the surface normal vector. We will examine two ways in which this equation can be satisfied.

Case 1:
Since $\mathbf{r}_v^{(2)}(0,v) = \mathbf{r}_v^{(1)}(1,v)$, the simplest solution is to take

$$\mathbf{r}_u^{(2)}(0,v) = \lambda(v)\mathbf{r}_u^{(1)}(1,v) \tag{7.27}$$

in (7.26). This implies that all lines of constant v in the composite surface will have continuity of gradient direction, which seems reasonable if this condition is desired to hold for the composite curves forming the patch boundaries. We can use (7.25) to express (7.27) in matrix form as

$$[0 \quad 1 \quad 0 \quad 0]\mathbf{MB}^{(2)}\mathbf{M}^T\mathbf{V} = \lambda(v)\,[0 \quad 1 \quad 2 \quad 3]\mathbf{MB}^{(1)}\mathbf{M}^T\mathbf{V}\ . \tag{7.28}$$

Since the left-hand side is a cubic polynomial in v, we must clearly take $\lambda(v) = \lambda$, a (positive) constant; otherwise the right-hand side will not be cubic. The equation must hold for all relevant v, and therefore we equate coefficients and post-multiply by $(\mathbf{M}^T)^{-1}$, as previously, to obtain the four equations

$$(\mathbf{r}_{1i}^{(2)} - \mathbf{r}_{0i}^{(2)}) = \lambda(\mathbf{r}_{3i}^{(1)} - \mathbf{r}_{2i}^{(1)})\ , \quad i = 0, 1, 2, 3. \tag{7.29}$$

In terms of the characteristic polyhedra of the two patches, these require that the four pairs of polyhedron edges which meet at the boundary must be collinear, as shown in Figure 7.6.

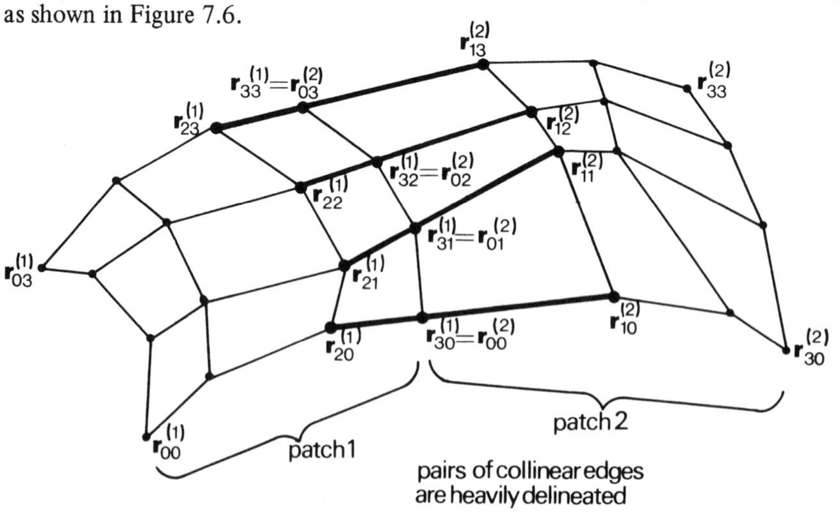

Figure 7.6 – Configuration of Characteristic Polyhedra giving Positional and Gradient Continuity between Cubic Bézier Patches (Case (i)).

The foregoing analysis has shown that the cross-boundary tangent magnitude ratio λ must be constant along the common boundary. Because of the way in which conditions are matched at patch corners in a composite surface, this implies that the ratio of u-derivative magnitudes must be constant across any composite v-curve in the patch boundary network, and *vice versa,* as illustrated in Figure 7.7. By virtue of (7.27), all the composite u-curves and v-curves will be smooth as well as continuous.

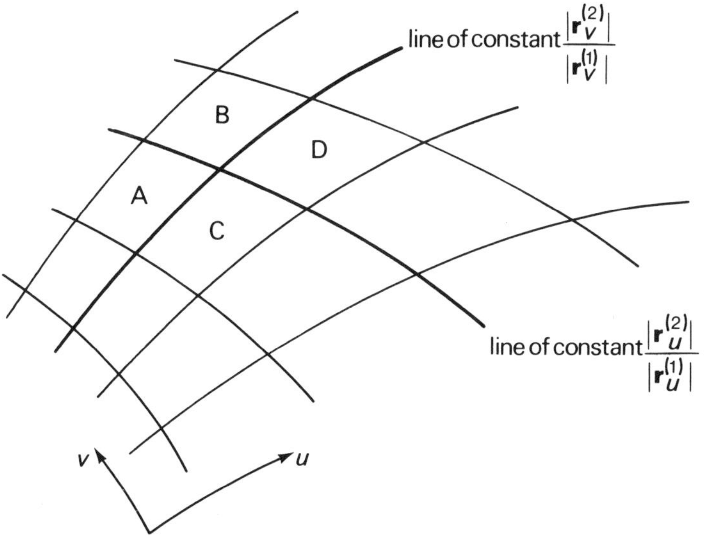

Figure 7.7 – Constant Tangent Magnitude Ratios across Composite Patch Boundary Curves.

In practice, the restrictions imposed by this constancy of tangent ratio are severe. Consider, for example, the construction of the smooth composite surface shown in Figure 7.7, starting with patch A and continuing with patches B, C and D, in that order. We can choose the sixteen vertices of the characteristic polyhedron of patch A as we wish. We must then specify a tangent ratio across the boundary between patches A and B. Once this has been done, we find that no less than eight vertices of the characteristic polyhedron of B are fixed by the conditions for positional and gradient continuity, leaving only eight free to be chosen. Similarly, once the tangent ratio between patches A and C is specified, we are only left with eight vertices to choose in defining patch C. With patch D we are even more constrained; the tangent ratios between B and D and between C and D are already fixed, and we find that we are only free to choose four of the sixteen vertices which define this patch.

If the composite surface is being constructed on a predetermined mesh of cubic Bézier curves, the polyhedron corresponding to each patch has all its twelve boundary vertices already fixed. The number of internal vertices which can be freely chosen in patches A, B, C and D are then respectively 4, 2, 2 and 1;

these correspond to the number of corners at which the twist vector $\mathbf{r}_{uv}$ has not yet been fixed.

Case 2:

To obtain more freedom in constructing composite surfaces, Bézier (1972) rejects (7.27) as the gradient matching condition, and satisfies (7.26) by taking

$$\mathbf{r}_u^{(2)}(0,v) = \lambda(v)\mathbf{r}_u^{(1)}(1,v) + \mu(v)\mathbf{r}_v^{(1)}(1,v) , \qquad (7.30)$$

in which $\mu(v)$ is another scalar function of v. This simply requires $\mathbf{r}_u^{(2)}(0,v)$ to lie in the same plane as $\mathbf{r}_u^{(1)}(1,v)$ and $\mathbf{r}_v^{(1)}(1,v)$, that is in the tangent plane of Patch 1 at the boundary point concerned. Much more scope is now available. In matrix terms, (7.30) is

$$[0 \quad 1 \quad 0 \quad 0]\mathbf{M}\mathbf{B}^{(2)}\mathbf{M}^T\mathbf{V} = \lambda(v)[0 \quad 1 \quad 2 \quad 3]\mathbf{M}\mathbf{B}^{(1)}\mathbf{M}^T\mathbf{V}$$

$$+ \mu(v)\ [1 \quad 1 \quad 1 \quad 1]\mathbf{M}\mathbf{B}^{(1)}\mathbf{M}^T\begin{bmatrix}0\\1\\2v\\3v^2\end{bmatrix} . \qquad (7.31)$$

If we wish to work entirely in terms of cubic patches, we see that $\lambda(v)$ may be any positive constant and $\mu(v)$ any linear function of v.† It is important to realise that when (7.30) is used as the smoothness condition, cross-boundary gradient vectors are no longer continuous in direction across patch boundaries. The collinearity condition on polyhedron edges meeting at a boundary may therefore be discarded. Using our more general criterion for gradient continuity, we can build up a smooth continuous surface on which the composite patch boundaries, in both the u- and v-directions, will have positional continuity but gradient discontinuity at all patch corners. We can infer from (7.30), applied at a mesh intersection, that the tangent directions of all four patch boundaries meeting at an intersection must be coplanar, however.

The interpretation of gradient continuity between patches in terms of their characteristic polyhedra is less straightforward than previously. If we set $\lambda(v) = \lambda$ and $\mu(v) = \mu_0 + \mu_1 v$ in (7.31) and then equate coefficients of powers of v, as was done for (7.28) earlier, we obtain the four following relations:

$$(\mathbf{r}_{10}^{(2)} - \mathbf{r}_{00}^{(2)}) = \lambda(\mathbf{r}_{30}^{(1)} - \mathbf{r}_{20}^{(1)}) + \mu_0(\mathbf{r}_{31}^{(1)} - \mathbf{r}_{30}^{(1)}),$$

†Note, however, that these functions may be used to obtain smooth joins between patches of different degree. For instance, if $\lambda(v)$ is quadratic in v and $\mu(v)$ cubic, Patch 2 is a quintic (fifth-degree) patch.

$$(\mathbf{r}_{11}{}^{(2)} - \mathbf{r}_{01}{}^{(2)}) = \lambda(\mathbf{r}_{31}{}^{(1)} - \mathbf{r}_{21}{}^{(1)}) + \frac{1}{3}\mu_0(2\mathbf{r}_{32}{}^{(1)} - \mathbf{r}_{31}{}^{(1)} - \mathbf{r}_{30}{}^{(1)}) + \frac{1}{3}\mu_1(\mathbf{r}_{31}{}^{(1)} - \mathbf{r}_{30}{}^{(1)}),$$

$$(\mathbf{r}_{12}{}^{(2)} - \mathbf{r}_{02}{}^{(2)}) = \lambda(\mathbf{r}_{32}{}^{(1)} - \mathbf{r}_{22}{}^{(1)}) + \frac{1}{3}\mu_0(\mathbf{r}_{33}{}^{(1)} + \mathbf{r}_{32}{}^{(1)} - 2\mathbf{r}_{31}{}^{(1)}) + \frac{2}{3}\mu_1(\mathbf{r}_{32}{}^{(1)} - \mathbf{r}_{31}{}^{(1)}),$$

$$(\mathbf{r}_{13}{}^{(2)} - \mathbf{r}_{03}{}^{(2)}) = \lambda(\mathbf{r}_{33}{}^{(1)} - \mathbf{r}_{23}{}^{(1)}) + (\mu_0 + \mu_1)(\mathbf{r}_{33}{}^{(1)} - \mathbf{r}_{32}{}^{(1)}).$$

With $\mu_0 = \mu_1 = 0$ these reduce to (7.29). Since for positional continuity $\mathbf{r}_{00}{}^{(2)} = \mathbf{r}_{30}{}^{(1)}$, the first equation shows that the three vectors $(\mathbf{r}_{10}{}^{(2)} - \mathbf{r}_{00}{}^{(2)})$, $(\mathbf{r}_{30}{}^{(1)} - \mathbf{r}_{20}{}^{(1)})$ and $(\mathbf{r}_{31}{}^{(1)} - \mathbf{r}_{30}{}^{(1)})$ must be coplanar. But we know from the properties of Bézier patches (see Section 5.1.7) that these vectors are scalar multiples of $\mathbf{r}_u{}^{(2)}(0,0)$, $\mathbf{r}_u{}^{(1)}(1,0)$ and $\mathbf{r}_v{}^{(1)}(1,0)$ respectively. We thus confirm the conclusion of the last paragraph that tangent directions of patch boundaries at a mesh intersection must be coplanar. The fourth equation similarly shows that $\mathbf{r}_u{}^{(2)}(0,1)$, $\mathbf{r}_u{}^{(1)}(1,1)$ and $\mathbf{r}_v{}^{(1)}(1,1)$ must be coplanar. These two coplanarity conditions are illustrated in terms of characteristic polyhedra in Figure 7.8. The second and third of the gradient continuity equations have no such simple geometrical interpretation.

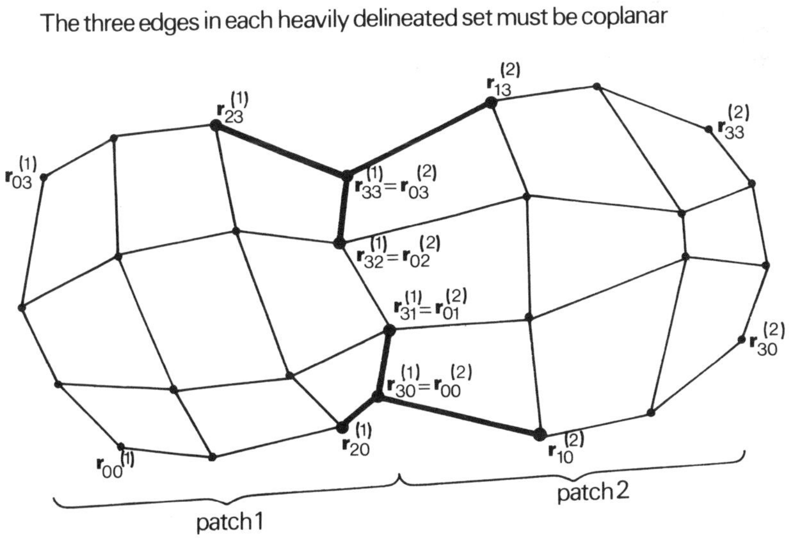

Figure 7.8 – Configuration of Characteristic Polyhedra giving Positional and Gradient Continuity between cubic Bézier Patches (Case (ii)).

Even with the greater flexibility permitted by the use of (7.30) rather than (7.27), there is insufficient freedom for curvature continuity to be achieved across all patch boundaries if a composite surface is to be constructed by first designing a single patch and then working outwards from it as suggested earlier. It would be necessary to use patches of higher than cubic degree if this were required. The definitive version of UNISURF (Bézier, 1974b) permits the use of such patches, and we therefore note here that a generalised Bézier surface patch is defined by

$$\mathbf{r}(u,v) = \sum_{i=0}^{p} \sum_{j=0}^{q} \mathbf{r}_{ij}\, g_i^{(p)}(u)\, g_j^{(q)}(v) \; .$$

Here $g_i^{(p)}$ and $g_j^{(q)}$ are Bernstein basis functions of degree p and q respectively, while the patch is now specified in terms of a characteristic polyhedron with $(p + 1) \times (q + 1)$ vertices. The conditions ensuring various orders of continuity may be derived by extending the methods of this section. Second and higher order continuity of generalised Bézier surfaces is the subject of a paper by Veron *et al.* (1976).

7.2.3 Rational Parametric Surfaces

The rational parametric surface is capable of great generality, but because no practical system based on its general form has yet achieved widespread currency, we will here give only a brief survey of some of its notable features. We examine first the rational bicubic patch which is based on the rational cubic curve segment

$$\mathbf{r}(u) = \frac{w_0\mathbf{r}_0(1-u)^3 + 3w_1\mathbf{r}_1u(1-u)^2 + 3w_2\mathbf{r}_2u^2(1-u) + w_3\mathbf{r}_3u^3}{w_0(1-u)^3 + 3w_1u(1-u)^2 + 3w_2u^2(1-u) + w_3u^3} \, ,$$

$$0 \leqslant u \leqslant 1 \, , \qquad (7.32)$$

discussed earlier in Sections 5.2.2 and 6.3.3. We recall that

$$\begin{aligned} \mathbf{r}(0) &= \mathbf{r}_0 \, , & \mathbf{r}(1) &= \mathbf{r}_3 \, , \\ \dot{\mathbf{r}}(0) &= 3\frac{w_1}{w_0}(\mathbf{r}_1 - \mathbf{r}_0) \, , & \dot{\mathbf{r}}(1) &= 3\frac{w_2}{w_3}(\mathbf{r}_3 - \mathbf{r}_2) \, . \end{aligned} \qquad (7.33)$$

We can express the curve segment in the form

$$\mathbf{r}(u) = \alpha_0(u)\mathbf{r}(0) + \alpha_1(u)\mathbf{r}(1) + \beta_0(u)\dot{\mathbf{r}}(0) + \beta_1(u)\dot{\mathbf{r}}(1) \, ,$$

in terms of its end points, end tangents and blending functions. Substitution of (7.33) and comparison of the result with (7.32) shows that the four blending functions must be

$$\alpha_0(u) = \frac{w_0(1-u)^3 + 3w_1u(1-u)^2}{w(u)}, \quad \alpha_1(u) = \frac{w_3u^3 + 3w_2u^2(1-u)}{w(u)},$$

$$\beta_0(u) = \frac{w_0u(1-u)^2}{w(u)}, \qquad \beta_1(u) = -\frac{w_3u^2(1-u)}{w(u)},$$

in which $w(u)$ denotes the denominator in (7.32). Reasonably enough, these scalar functions are rational cubics. A little algebra shows them to satisfy conditions (7.7) to (7.10) which we earlier required to hold for the blending functions of a patch with specified boundary curves and cross-boundary derivatives. We may therefore construct a rational bicubic tensor-product patch of this type by substituting the foregoing set of blending functions into (7.16). The result can be expressed in the form

$$\mathbf{r}(u,v) = w^{-1}(u)w^{-1}(v)\mathbf{UCWQW}^T\mathbf{C}^T\mathbf{V}, \tag{7.34}$$

where $\mathbf{Q}$ is defined by (7.16), $\mathbf{U}$, $\mathbf{V}$ and $\mathbf{C}$ by (7.17), and

$$\mathbf{W} = \begin{bmatrix} w_0 & 0 & 0 & 0 \\ 0 & w_3 & 0 & 0 \\ 3(w_1 - w_0) & 0 & w_0 & 0 \\ 0 & 3(w_3 - w_2) & 0 & w_3 \end{bmatrix}.$$

Alternatively we may start from the Bézier point of view, and define the surface patch in terms of the sixteen vertices of a characteristic polyhedron. Since a rational cubic curve segment can be expressed as

$$\mathbf{r}(u) = w^{-1}(u) \sum_{i=0}^{3} w_i\mathbf{r}_i g_i(u),$$

where the g_i are the Bernstein basis functions of (7.21), we can define our bicubic surface patch analogously by

$$\mathbf{r}(u,v) = w^{-1}(u)w^{-1}(v) \sum_{i=0}^{3}\sum_{j=0}^{3} w_iw_j\mathbf{r}_{ij}g_i(u)g_j(v). \tag{7.35}$$

This equation may be written in matrix form as

$$\mathbf{r}(u,v) = w^{-1}(u)w^{-1}(v)\mathbf{UMDBDM}^T\mathbf{V}, \tag{7.36}$$

in which $\mathbf{M}$ is defined by (7.22), $\mathbf{B} = [\mathbf{r}_{ij}]$ is the 4X4 matrix of polyhedron vertices and $\mathbf{D} = \mathrm{diag}(w_0, w_1, w_2, w_3)$. Comparison of (7.34) and (7.36) shows that

$$\mathbf{Q} = (\mathbf{W}^{-1}\mathbf{C}^{-1}\mathbf{MD})\mathbf{B}(\mathbf{W}^{-1}\mathbf{C}^{-1}\mathbf{MD})^T ,$$

or

$$\begin{bmatrix} \mathbf{r}(0,0) & \mathbf{r}(0,1) & \mathbf{r}_v(0,0) & \mathbf{r}_v(0,1) \\ \mathbf{r}(1,0) & \mathbf{r}(1,1) & \mathbf{r}_v(1,0) & \mathbf{r}_v(1,1) \\ \mathbf{r}_u(0,0) & \mathbf{r}_u(0,1) & \mathbf{r}_{uv}(0,0) & \mathbf{r}_{uv}(0,1) \\ \mathbf{r}_u(1,0) & \mathbf{r}_u(1,1) & \mathbf{r}_{uv}(1,0) & \mathbf{r}_{uv}(1,1) \end{bmatrix} = \tag{7.37}$$

$$\begin{bmatrix} \mathbf{r}_{00} & \mathbf{r}_{03} & 3\frac{w_1}{w_0}(\mathbf{r}_{01}-\mathbf{r}_{00}) & 3\frac{w_2}{w_3}(\mathbf{r}_{03}-\mathbf{r}_{02}) \\ \mathbf{r}_{30} & \mathbf{r}_{33} & 3\frac{w_1}{w_0}(\mathbf{r}_{31}-\mathbf{r}_{30}) & 3\frac{w_2}{w_3}(\mathbf{r}_{33}-\mathbf{r}_{32}) \\ 3\frac{w_1}{w_0}(\mathbf{r}_{10}-\mathbf{r}_{00}) & 3\frac{w_1}{w_0}(\mathbf{r}_{13}-\mathbf{r}_{03}) & 9\left(\frac{w_1}{w_0}\right)^2(\mathbf{r}_{00}-\mathbf{r}_{10}-\mathbf{r}_{01}+\mathbf{r}_{11}) & 9\frac{w_1 w_2}{w_0 w_3}(\mathbf{r}_{02}-\mathbf{r}_{12}-\mathbf{r}_{03}+\mathbf{r}_{13}) \\ 3\frac{w_2}{w_3}(\mathbf{r}_{30}-\mathbf{r}_{20}) & 3\frac{w_2}{w_3}(\mathbf{r}_{33}-\mathbf{r}_{23}) & 9\frac{w_1 w_2}{w_0 w_3}(\mathbf{r}_{20}-\mathbf{r}_{30}-\mathbf{r}_{21}+\mathbf{r}_{31}) & 9\left(\frac{w_2}{w_3}\right)^2(\mathbf{r}_{22}-\mathbf{r}_{32}-\mathbf{r}_{23}+\mathbf{r}_{33}) \end{bmatrix} .$$

On comparing this with (7.24) we can see that this rational bicubic patch is a generalisation of the cubic Bézier patch. In both cases the patch corners coincide with the polyhedron vertices $\mathbf{r}_{00}$, $\mathbf{r}_{03}$, $\mathbf{r}_{30}$ and $\mathbf{r}_{33}$, while $\mathbf{r}_u$ and $\mathbf{r}_v$ at the patch corners are in the directions of the corresponding polyhedron edges. However, the rational bicubic patch is more flexible because we can adjust the ratios w_1/w_0 and w_2/w_3 to modify the gradient and twist vector magnitudes.

The patch as defined by (7.36) has the property that each of its boundary curves is a rational cubic curve whose weights are w_0, w_1, w_2 and w_3. We can obtain yet more freedom by retaining these weights for the u-boundaries but choosing another set of weights for the v-boundaries, say w_0', w_1', w_2' and w_3'. In effect, we are then using one set of blending functions in the u-direction and another, based on the new set of weights, in the v-direction. The foregoing analysis still holds, provided that primes are added to all w's related to the v-direction. In consequence, the common factors w_1/w_0 and w_2/w_3 in columns 3 and 4 of the right-hand side of (7.37) become w_1'/w_0' and w_2'/w_3' respectively. These two additional ratios are then at our disposal as shape-modifying parameters.

The ultimate in generality for the parametric rational bicubic patch is obtained on replacing (7.35) by (5.37) from Section 5.2.3:

$$\mathbf{P}(u,v) = \mathbf{UMB^*M}^T\mathbf{V} . \tag{7.38}$$

Here $\mathbf{P} = (xw,yw,zw,w)^T$ and $\mathbf{B}^* = [\mathbf{P}_{ij}]$, where $\mathbf{P}_{ij} = (x_{ij}w_{ij},y_{ij}w_{ij},z_{ij}w_{ij},w_{ij})^T$. The $\mathbf{P}_{ij}$ are polyhedron vertices expressed in terms of homogeneous coordinates, and $\mathbf{P}(u,v)$ similarly represents a point on the surface patch. We can write this less concisely, but in terms of a more familiar notation, by noting that in partitioned form

$$\mathbf{P} = \begin{bmatrix} w\mathbf{r} \\ \hline w \end{bmatrix} \quad \text{and} \quad \mathbf{P}_{ij} = \begin{bmatrix} w_{ij}\mathbf{r}_{ij} \\ \hline w_{ij} \end{bmatrix} .$$

Consequently

$$\begin{bmatrix} w\mathbf{r} \\ \hline w \end{bmatrix} = \mathbf{UM}\left[\begin{bmatrix} w_{ij}\mathbf{r}_{ij} \\ \hline w_{ij} \end{bmatrix}\right]\mathbf{M}^T\mathbf{V} ,$$

where the double bracket denotes a matrix whose elements are the indicated partitioned vectors. Hence

$$w\mathbf{r} = \mathbf{UM}\left[w_{ij}\mathbf{r}_{ij}\right]\mathbf{M}^T\mathbf{V}$$

and

$$w = \mathbf{UM}\left[w_{ij}\right]\mathbf{M}^T\mathbf{V} .$$

A final division then gives

$$\mathbf{r}(u,v) = \frac{\mathbf{UM}\left[w_{ij}\mathbf{r}_{ij}\right]\mathbf{M}^T\mathbf{V}}{\mathbf{UM}\left[w_{ij}\right]\mathbf{M}^T\mathbf{V}} .$$

We note the following features of this very general rational bicubic patch:

(i) No less than sixteen weights w_{ij} are involved in the patch definition. As previously, however, it is *ratios* between weights which determine the shape of the patch surface, and just twelve such ratios are required to determine the tangents and twist vectors at the patch corners. For example, at the corner with $u = v = 0$ we have

$$\mathbf{r}(0,0) = \mathbf{r}_{00}$$

$$\mathbf{r}_u(0,0) = 3\frac{w_{10}}{w_{00}}(\mathbf{r}_{10} - \mathbf{r}_{00}) , \quad \mathbf{r}_v(0,0) = 3\frac{w_{01}}{w_{00}}(\mathbf{r}_{01} - \mathbf{r}_{00})$$

and

$$\mathbf{r}_{uv}(0,0) = 9\left\{\frac{w_{11}}{w_{00}}(\mathbf{r}_{11} - \mathbf{r}_{00}) + \frac{w_{01}w_{10}}{w_{00}^2}(2\mathbf{r}_{00} - \mathbf{r}_{01} - \mathbf{r}_{10})\right\} ,$$

which involve the three ratios w_{01}/w_{00}, w_{10}/w_{00} and w_{11}/w_{00}; analogously, three ratios are important at each of the remaining corners.

(ii) Each patch boundary is a rational cubic curve, but opposite boundaries are now defined in terms of different sets of weights. Application of the Coons patch theory of Section 7.1 is therefore not entirely straightforward, because this envisages that opposite boundary curves are similarly defined. However, the homogeneous coordinate formulation comes to our aid here. Putting $v = 0$ in (7.38) we obtain

$$\mathbf{P}(u,0) = (1-u)^3\mathbf{P}_{00} + 3u(1-u)^2\mathbf{P}_{10} + 3u^2(1-u)\mathbf{P}_{20} + u^3\mathbf{P}_{30},$$

while similar equations represent the other three boundaries. The differences in weights are now concealed in the **P**-vectors, and the blending functions depend only on u or v. Provided we work in terms of **P**'s rather than **r**'s, then, the Coons theory applies directly. On noting that (7.38) is identical with (7.23) but for the substitution of $\mathbf{B}^*$ for $\mathbf{B}$, we conclude that (7.24) holds for the rational bicubic patch if we replace all the **r**'s by **P**'s. Although expressions such as $\mathbf{P}_u(0,0)$ or $\mathbf{P}_{uv}(0,0)$ do not have the immediate physical significance of $\mathbf{r}_u(0,0)$ or $\mathbf{r}_{uv}(0,0)$, a little algebra will translate the relationships between **P**-derivatives and the $\mathbf{P}_{ij}$ into relationships between **r**-derivatives and the $\mathbf{r}_{ij}$. The results for the corner point (0,0) have been quoted under (i).

By analogy with (7.38), we may define a rational biquadratic patch, whose boundaries will be conic sections (see Section 5.2.1). We may write it as

$$\mathbf{P}(u,v) = [1 \quad u \quad u^2]\begin{bmatrix} 1 & 0 & 0 \\ -2 & 2 & 0 \\ 1 & -2 & 1 \end{bmatrix}\begin{bmatrix} \mathbf{P}_{00} & \mathbf{P}_{01} & \mathbf{P}_{02} \\ \mathbf{P}_{10} & \mathbf{P}_{11} & \mathbf{P}_{12} \\ \mathbf{P}_{20} & \mathbf{P}_{21} & \mathbf{P}_{22} \end{bmatrix}\begin{bmatrix} 1 & -2 & 1 \\ 0 & 2 & -2 \\ 0 & 0 & 1 \end{bmatrix}\begin{bmatrix} 1 \\ v \\ v^2 \end{bmatrix},$$

in terms of homogeneous coordinates, by analogy with a quadratic Bézier patch. This type of patch also has considerable generality, and includes all quadric surfaces as special cases. It cannot be expressed in the form of (7.16), however, because although the corner points and corner gradients can be chosen arbitrarily, the corner twist vectors cannot.

It is evident that great flexibility results from the use of rational polynomial patches. One problem which arises is that the very generality of these functions presents difficulties in the design of a practical system for specifying surfaces. It is desirable that any such system can be operated by a non-mathematician, but it is by no means obvious how all the available degrees of freedom can be incorporated into the system so that each has a fairly obvious geometrical significance. In a practical implementation positive values will normally be chosen for all the weights, so that the denominators of the rational functions will have no zeros for $0 \leqslant u, v \leqslant 1$.

We shall not enter here into a discussion of continuity conditions between rational parametric patches. Considerable scope is clearly available, and the necessary analysis follows the lines of that already given for the composite Bézier surface.

7.2.4 Parametric Spline Surfaces

The construction of parametric spline curves in three dimensions was dealt with in Section 6.3.4. Given a suitable array of points, we can compute a mesh of such curves in which each of the specified points is a knot of both a u-curve and a v-curve. If the splines are cubic, all these curves will have continuity of gradient and curvature at all internal knots, both in direction and magnitude. This is because

$$\dot{\mathbf{r}}(u) = \begin{bmatrix} \dot{x}(u) \\ \dot{y}(u) \\ \dot{z}(u) \end{bmatrix} , \qquad \ddot{\mathbf{r}}(u) = \begin{bmatrix} \ddot{x}(u) \\ \ddot{y}(u) \\ \ddot{z}(u) \end{bmatrix} ,$$

and the splines are computed so that the scalar components of each of these vectors are continuous at the knots. The spline curves form the boundaries of the topologically rectangular (or curvilinear quadrilateral) patches of a composite surface.

The parametrisation of spline surfaces can present some problems, and we will discuss these later. For the moment, however, we consider a patch such that $u_m \leqslant u \leqslant u_{m+1}$ along both its u-boundaries and $v_n \leqslant v \leqslant v_{n+1}$ along both its v-boundaries. Since each boundary is one span of a cubic spline, the patch surface may be represented by a bicubic, which we will write as

$$\mathbf{r}(u,v) = \sum_{i=0}^{3}\sum_{j=0}^{3} \mathbf{a}_{ij}\left(\frac{u - u_m}{h_m}\right)^i\left(\frac{v - v_n}{k_n}\right)^j , \tag{7.39}$$

where $h_m = u_{m+1} - u_m$ and $k_n = v_{n+1} - v_n$. If we set $u' = (u - u_m)/h_m$ and $v' = (v - v_n)/k_n$, so that $0 \leqslant u', v' \leqslant 1$, this reduces to a standard Ferguson bicubic patch formula,

$$\mathbf{r}(u',v') = \sum_{i=0}^{3}\sum_{j=0}^{3} \mathbf{a}_{ij}u'^i v'^j = \mathbf{U}'\mathbf{A}\mathbf{V}' ,$$

in which $\mathbf{A} = [\mathbf{a}_{ij}]$, $\mathbf{U}' = (1,u',u'^2,u'^3)$ and $\mathbf{V}' = (1,v',v'^2,v'^3)^T$. We can use (7.18) to express this last equation as

$$\mathbf{r}(u',v') = \mathbf{U}'\mathbf{C}\mathbf{Q}'\mathbf{C}^T\mathbf{V}', \tag{7.40}$$

where $\mathbf{Q}'$ is as defined by (7.16) but with all derivatives taken with respect to u' and v', while $\mathbf{C}$ is defined by (7.17). But because

$$\mathbf{r}_{u'}(u',v') = \frac{\partial \mathbf{r}(u',v')}{\partial u'} = \frac{\partial \mathbf{r}(u,v)}{\partial u}\frac{du}{du'} = h_m \mathbf{r}_u(u,v)$$

and so on, it is easy to show that

$$\mathbf{Q}' = \mathbf{H}_m \mathbf{Q} \mathbf{K}_n \,, \tag{7.41}$$

where $\mathbf{H}_m = \text{diag}\,[1,1,h_m,h_m]$, $\mathbf{K}_n = \text{diag}\,[1,1,k_n,k_n]$ and

$$\mathbf{Q} = \begin{bmatrix} \mathbf{r}(u_m,v_n) & \mathbf{r}(u_m,v_{n+1}) & \mathbf{r}_v(u_m,v_n) & \mathbf{r}_v(u_m,v_{n+1}) \\ \mathbf{r}(u_{m+1},v_n) & \mathbf{r}(u_{m+1},v_{n+1}) & \mathbf{r}_v(u_{m+1},v_n) & \mathbf{r}_v(u_{m+1},v_{n+1}) \\ \mathbf{r}_u(u_m,v_n) & \mathbf{r}_u(u_m,v_{n+1}) & \mathbf{r}_{uv}(u_m,v_n) & \mathbf{r}_{uv}(u_m,v_{n+1}) \\ \mathbf{r}_u(u_{m+1},v_n) & \mathbf{r}_u(u_{m+1},v_{n+1}) & \mathbf{r}_{uv}(u_{m+1},v_n) & \mathbf{r}_{uv}(u_{m+1},v_{n+1}) \end{bmatrix} .$$

This last matrix is the generalisation of the matrix **Q** of (7.16) to the case of parameter ranges other than [0,1]. Putting (7.41) into (7.40) and comparing the two equations for $\mathbf{r}(u',v')$ then, we find that

$$\mathbf{A} = \mathbf{C}\mathbf{H}_m \mathbf{Q} \mathbf{K}_n \mathbf{C}^T \,, \tag{7.42}$$

which gives the coefficients $\mathbf{a}_{ij}$ in (7.39) in terms of $\mathbf{r}$, $\mathbf{r}_u$, $\mathbf{r}_v$ and $\mathbf{r}_{uv}$ at the patch corners.

Now the positions of the patch corners are specified at the outset, while the corner tangent vectors are a by-product of the construction of the spline curve network. The complete specification of the surface patch fitting each mesh cell therefore only requires the twist vectors $\mathbf{r}_{uv}$ at the patch corners to be supplied. These may be determined in such a way that the entire composite surface has continuity of curvature in the sense that $\ddot{\mathbf{r}}(u)$ is continuous across all v-boundaries, and *vice versa.* The procedure follows that outlined in Section 7.2.1 for Ferguson surfaces with curvature continuity; in fact those surfaces are merely special cases of spline surfaces in which the parameters u and v are translated and scaled so as to run from 0 to 1 on each patch. The procedure for fitting a spline surface to a topologically rectangular array of points $\mathbf{r}_{mn}$, $m = 0, 1, \ldots, M$, $n = 0, 1, \ldots, N$, is as given below. Note that $[\mathbf{r}_m]$, $m = 0, 1, \ldots, M$, is used to denote the set of points $\mathbf{r}_0, \mathbf{r}_1, \ldots, \mathbf{r}_M$.

1) Spline curves are computed in the u-direction through the $(N + 1)$ sets of points $[\mathbf{r}_{m0}]$, $[\mathbf{r}_{m1}]$, ..., $[\mathbf{r}_{mN}]$, $m = 0, 1, \ldots, M$. Each curve will require two additional items of data, which will here assume to be values of $\mathbf{r}_u$ at the end points.
2) Spline curves are now computed in the v-direction through the $(M + 1)$ sets of points $[\mathbf{r}_{0n}]$, $[\mathbf{r}_{1n}]$, $[\mathbf{r}_{2n}]$, ..., $[\mathbf{r}_{Mn}]$, $n = 0, 1, \ldots, N$, using given values of $\mathbf{r}_v$ at the ends of each curve.

This establishes the curve network and the values of $\mathbf{r}_u$ and $\mathbf{r}_v$ at all mesh intersections may now be considered as known.

3) Splines are computed in the v-direction which fit the two sets of gradient vectors $[\mathbf{r}_{u,0n}]$, $[\mathbf{r}_{u,Mn}]$, $n = 0, 1, \ldots, N$. From these we obtain the twist vectors $\mathbf{r}_{uv}$ along the v-boundaries of the composite surface. The corner twist vectors $\mathbf{r}_{uv,00}$, $\mathbf{r}_{uv,0N}$, $\mathbf{r}_{uv,M0}$ and $\mathbf{r}_{uv,MN}$ are needed in constructing these two curves.
4) Finally, $(N + 1)$ splines are computed in the u-direction which fit the sets of gradient vectors $[\mathbf{r}_{v,m0}]$, $[\mathbf{r}_{v,m1}]$, $\ldots$, $[\mathbf{r}_{v,mN}]$, $m = 0, 1, \ldots, M$. The end-point values of $\mathbf{r}_{uv}$ needed for this are already available from Step (3). From this last set of curves we obtain the twist vectors $\mathbf{r}_{uv}$ at all the patch corners. We now have all the elements of all the **Q**-matrices, and (7.42) and (7.39) may be used to construct the equation for each patch of the composite surface.

The roles of u and v in Steps (3) and (4) may be interchanged, of course, but the same curvature-continuous surface will result whichever course is adopted. This may be proved by a generalisation of the argument given in Section 7.2.1. The entire surface is therefore uniquely determined in terms of (i) the points $\mathbf{r}_{mn}$, (ii) the tangent vectors $\mathbf{r}_u$ along the v-boundaries and $\mathbf{r}_v$ along the u-boundaries of the composite surface as a whole, and (iii) the twist vectors $\mathbf{r}_{uv}$ at its corners, which in practice are often taken to be zero for the sake of simplicity.

Each of the parametric curves established in the foregoing process is calculated in terms of plane splines, as explained in Section 6.3.4. These may be computed directly, by the method of Section 6.2.2, or in terms of B-splines as suggested in Section 6.2.3. The latter approach enables local modifications to be made to the surface; it is possible to make a change affecting no more than 16 patches (in the cubic case) out of what may be a much larger total number.

Parametric spline surfaces are the basis of the Numerical Master Geometry (NMG) system developed by Sabin at the British Aircraft Corporation (Weybridge), and the Cambridge Computer-Aided Design Centre's POLYSURF system. Details are given in Sabin (1971), Flutter (1974), CAD Centre (1972), Flutter and Rolph (1976). Coles (1977) describes the rôle played by NMG in an integrated system for computer-aided design and manufacture.

Finally, we turn briefly to the question of how best to parametrise a spline surface. It was suggested in Sections 6.3.4 and 6.3.6 that chord-length parametrisation is convenient for single curves, and this is indeed so. But there are many other possibilities, the simplest of which is simply to assign integer values $u = 0, 1, 2, 3, \ldots$ to successive knots on the curve. Experience has shown this to be satisfactory provided the physical spacing of the data points is fairly uniform. If the points are very unevenly spaced, an integer or other uniform parametrisation of the curve can give rise to unwanted flat regions on long spans and oscillations (or even loops) on short spans. This tendency is lessened if the parameter assignment bears some close relationship to the actual separation between knots, and chord-length parametrisation is the simplest way of achieving this object.

When we come to construct a spline *surface*, however, matters are less straightforward. It will generally be found that opposite sides of any patch have different chord lengths, whereas our patch equation, (7.39), requires u and v to vary over the same range along opposite patch boundaries. Simple chord-length parametrisation will not be satisfactory, then. A practical system must provide some way of overcoming this difficulty. Butterfield (1978) achieves this by using bivariate blending functions, each depending on both u and v. These are special cases of the **boundary dependent blending functions** introduced in Chapter 17 of Forrest (1968). Alternatively, as in certain versions of NMG, the choice of parametrisation may be left to the operator, who can choose either a uniform or a non-uniform assignment of parameter values to the knots in the u and v directions, depending on the set of data points he is working with. Corresponding knots on all u-curves must have the same parameter values, however, and similarly on all v-curves. For instance, a reasonable assignment of parameter values for the set of points shown in Figure 7.9 might be

u-curve knots $u = 0, 1, 2, 3, 4, 8, 12, 16, 17, 18,$
v-curve knots $v = 0, 1, 2, 3.$

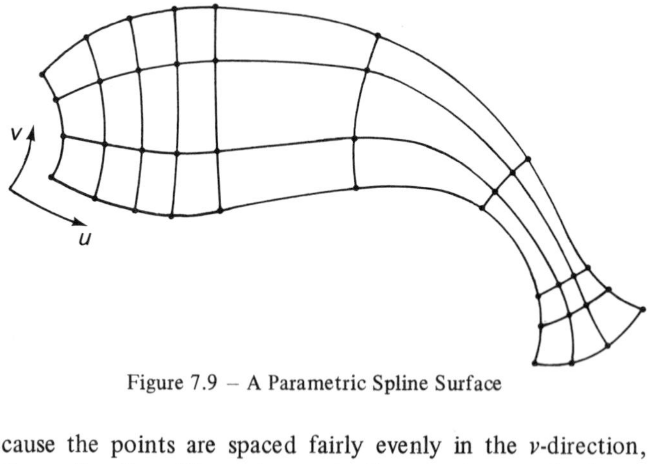

Figure 7.9 – A Parametric Spline Surface

This is because the points are spaced fairly evenly in the v-direction, but unevenly in the u-direction. The assignment of parameters to the knots reflects these facts. Obviously, it is not always possible to make such a clear-cut decision. We illustrate in Figure 7.10 a more difficult example. For this case we would probably cross our fingers and try a uniform knot parameter assignment in both directions. Fortunately such awkward instances can usually be avoided in practice.

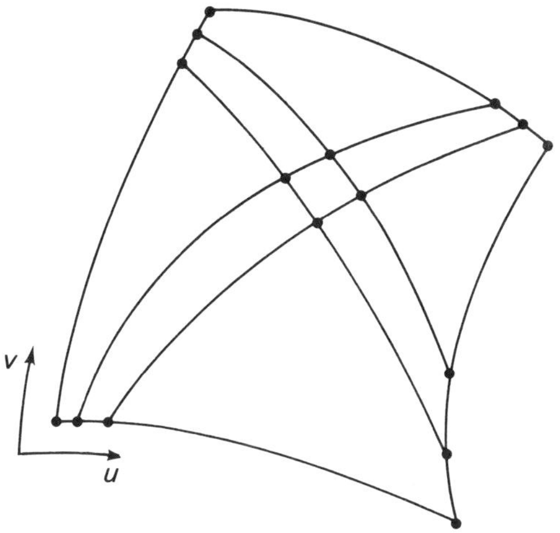

Figure 7.10 – A Surface with highly uneven knot spacing

7.2.5 B-spline Surfaces

B-spline curves were introduced in Section 6.3.5. They are related to B-spline surfaces in exactly the same way as Bézier curves are related to Bézier surfaces (see Sections 5.1.7, 7.2.2). Like the Bézier surface patch, the B-spline patch is defined in terms of a characteristic polyhedron. This mimics fairly closely the gross geometrical properties of the patch, which is defined for a topologically rectangular set of polyhedron vertices $\mathbf{r}_{ij}$, $i = 0, 1, 2, \ldots, p$, $j = 0, 1, 2, \ldots, q$, by

$$\mathbf{r}(u,v) = \sum_{i=0}^{p} \sum_{j=0}^{q} \mathbf{r}_{ij} N_{4,i+1}(u) N_{4,j+1}(v) \tag{7.43}$$

in analogy with (6.22). Here we are using cubic B-splines, as earlier, and $0 \leqslant u \leqslant p{-}2$, $0 \leqslant v \leqslant q{-}2$. The properties of the B-splines ensure that the surface defined by (7.43) will have overall continuity of gradient and curvature.

Surfaces of this type are a recent development, and not much has yet been published concerning them, but they seem to be a very promising innovation (Gordon and Riesenfeld, 1974a). One attraction arises because B-splines are only locally non-zero. Let us examine the effect of changing a single vector coefficient, $\mathbf{r}_{\alpha\beta}$, in (7.43). Now $N_{4,\alpha+1}(u)$ is only non-zero for $\alpha{-}3 < u < \alpha{+}1$ and $N_{4,\beta+1}(v)$ is only non-zero for $\beta{-}3 < v < \beta{+}1$. Thus the only part of the B-spline surface to be affected by the change is the region for which u and v lie in these ranges; the rest is unaltered. Note that the stated inequalities must be modified near the edges of the surface because of the restrictions on u and v given earlier.

A cubic Bézier patch, it will be recalled, is based on a polyhedron defined by just 16 vertices. A single patch of this kind can only represent a surface element having a fairly simple topography. As we have seen, moreover, Ferguson and cubic spline patches are mathematically equivalent to Bézier patches, and they consequently have the same limitation. By contrast, there is no limit on the number of vertices defining the characteristic polyhedron of a cubic B-spline patch. It is therefore possible, by choosing a sufficiently complex polyhedron, to represent a highly convoluted surface element by means of a single equation such as (7.43). We conclude that an individual B-spline patch, based on cubic polynomials, has capabilities which in the other systems mentioned would require the use of either a composite surface or of a single patch based on polynomials of degree higher than cubic. This advantage is reinforced by the local modification property and the automatic second-order continuity of the B-spline surface which were mentioned above.

7.3 LOFTED SURFACES

Before the days of computers, large objects such as ship hulls or aircraft fuselages were usually designed in terms of parallel plane cross-sections at a number of longitudinal stations. Once these were specified, longitudinal curves were constructed to blend them together into a single three-dimensional shape. Traditionally, these curves were drawn full size, with the assistance of mechanical splines. The only part of a shipyard where space was available for this to be done was the loft, and hence this kind of blending procedure became known as **lofting.** It was perhaps more of an art than a science; there was not (and still is not) any quantitative measure of the 'fairness' of a lofted curve. The acceptance or rejection of such curves depended largely on a loftsman's aesthetic judgement, and two different loftsmen would almost certainly draw significantly different 'fairest' curves to meet the same set of constraints.

A development of the manual lofting process came with the piecewise use of plane analytic curves (conics in particular). This technique was used by much of the British aircraft industry during the fifties and early sixties. The necessary calculations were performed on desk calculators before computers became widely used.

With the advent of powerful computers it became possible to automate the lofting process. One system which resulted was Numerical Master Geometry (Section 7.2.4), but this uses spline curves to define both cross-sectional shapes and longitudinal curves and has hence been dealt with earlier as a tensor-product method. Another system is CONSURF, developed by Ball (1974, 1975, 1977) at the former British Aircraft Corporation (Warton), which operates more in the spirit of the traditional lofting procedure. CONSURF is based upon two specialised forms of the rational cubic curve, the linear parameter segment and the generalised conic segment described earlier in Section 5.2.2. In the simplest case all cross-sections are plane and parallel, and these will be described in terms of generalised conic segments. In the longitudinal direction linear parameter segments are usually used, though the generalised conic segment may also be used in this direction in highly curved regions such as fuselage nose-cones. With the curve network established, an appropriate modification of (7.38) is used to define each surface patch.

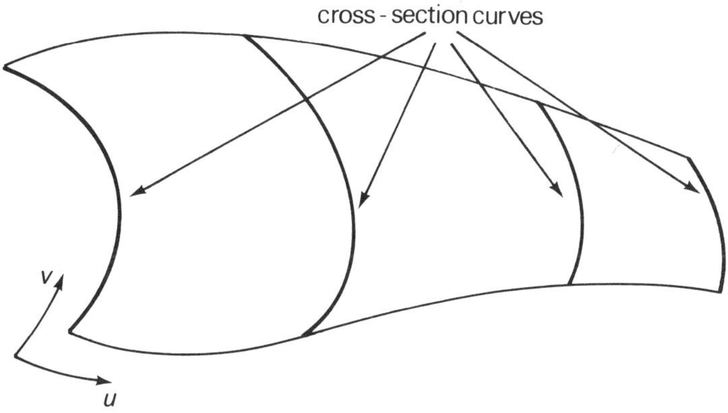

Figure 7.11 – A Strip of CONSURF Patches

Let us briefly consider the construction of a longitudinal strip of patches as shown in Figure 7.11. We will assume that all the cross-sectional curves lie in parallel planes (though Ball, 1975, describes how this condition may be dispensed with). We recall from Section 5.2.2 that a generalised conic segment may be defined in terms of its end-points, its end tangents and its p-ratio. CONSURF enables the surface strip to be defined by stipulating that its cross-section at any longitudinal station will be a generalised conic segment. We therefore need four vector-valued functions and one scalar-valued function of u, as follows:

(i) $\mathbf{r}_0(u)$ and $\mathbf{r}_3(u)$ are the equations of the lower and upper boundaries in the diagram; for any value of u they give the two end-points required. These curves will be composite rational cubics, composed of linear parameter segments with continuity of tangent direction at the joins between them.

(ii) $\mathbf{q}_1(u)$ and $\mathbf{q}_2(u)$ must also be specified. These are composite *slope-defining curves*, again made up of linear parameter segments. Together with $\mathbf{r}_0(u)$ and $\mathbf{r}_3(u)$ they are used to define the end tangents of cross-sectional curves for any value of u by

$$\mathbf{r}_{v,0}(u) = \mathbf{q}_1(u) - \mathbf{r}_0(u) \ , \quad \mathbf{r}_{v,3}(u) = \mathbf{r}_3(u) - \mathbf{q}_2(u)$$

along the lower and upper boundaries respectively.

(iii) The p-ratio function $p(u)$ is constructed from the known values of p at the transverse patch boundaries using an interpolatory technique (Akima, 1970).

Once these five functions have been specified, $\mathbf{r}_0$, $\mathbf{r}_3$, $\mathbf{r}_{v,0}$, $\mathbf{r}_{v,3}$ and p can be evaluated for any value of u. As shown in Section 5.2.2, we can then find scalars λ and μ such that the points

$$\mathbf{r}_1 = \mathbf{r}_0 + \lambda\mathbf{r}_{v,0} \ , \quad \mathbf{r}_2 = \mathbf{r}_3 - \mu\mathbf{r}_{v,3} \ ,$$

together with $\mathbf{r}_0$ and $\mathbf{r}_3$, define the characteristic polygon of a generalised conic segment whose p-ratio is p. This segment is the cross-section curve for the specified value of u.

This is a very simplified account of CONSURF. For further details the reader should consult Ball's papers, bearing in mind that we have modified his formulation slightly to preserve uniformity of notation in this book. Ball states that the mathematical complexities of the system are concealed within the program, so that the user can work entirely in terms of simple geometrical concepts. The cross-section curves, longitudinal curve segments and slope-defining segments which he has to specify are constructed solely in terms of end-points, end tangents and the curve fullness determining parameter p. Once a surface strip has been defined, the system prints out the values of λ and μ at various stations along the strip. These provide a check for the user; any irregular variation of them in the longitudinal direction indicates that the computed surface is unsatisfactory, and probably contains unwanted wrinkles.

Another notable lofting system is BSURF, which was developed by the Boeing aircraft company for numerical control applications. This uses the APT numerical control language to define a number of plane sections of a body. These can either be analytic curves or point-defined curves generated by the interpolatory routine TABCYL, which was mentioned in Section 6.4. The sections are then faired in using closely-spaced curves in the longitudinal direction. Appropriate cutter offsets are also calculated for machining in this direction.

A third lofting procedure, developed at Lockheed-Georgia for the preliminary design of aircraft fuselages, is known as **surface moulding** (Flanagan and Hefner, 1967). It has the interesting feature that all cross-sections are composed of arcs of plane **super-ellipses** defined by

$$\frac{x^n}{a^n} + \frac{y^n}{b^n} = 1 \ .$$

For $n = 2$ this equation is that of an ordinary ellipse with semiaxes a and b, but for $n > 2$ the fullness of the curve is greater, and as $n \to \infty$ the curve approaches a rectangular configuration in the quadrant $x > 0, y > 0$, as shown in Figure 7.12.

As we have seen, lofting is basically a problem of interpolating a surface between two fixed curves. Traditionally, it has been a lengthy and meticulous process, because curves lofted through two sets of mutually perpendicular cross-sections were generally found not to intersect, that is not to lie in the same surface. The resolution of such incompatibilities often involved a great deal of trial and error. The use of computers and of modern surface-defining systems has removed much of this tedium. Further, it is now possible to define the lofted surface as a whole, rather than merely to construct lofted curves which lie on the required surface.

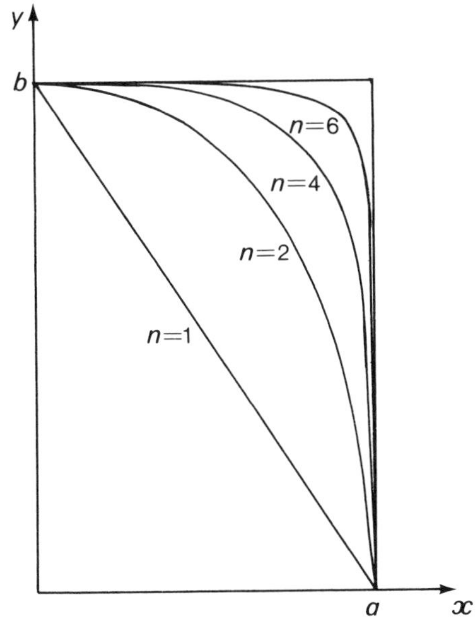

Figure 7.12 – Super-elliptic Arc used to define Cross-Sections in Surface Moulding.

Analytically, the lofted surface is simpler than the Coons surface, because it uses blending in only one direction rather than two. Referring back to Section 7.1, we note that (7.1) and (7.2) define simple lofted surfaces which were used as components in the construction of a Coons patch. Similarly, the first terms on the right of (7.4) and (7.12) give lofted surfaces defined in terms of more general blending functions. In the second case the slopes of the surface across the two fixed curves are specified. Lofting methods are discussed further in Chapter 8, in the context of cross-sectional design.

7.4 NON-PARAMETRIC SURFACES

The disadvantages of the Cartesian *vis-a-vis* the parametric approach to curve and surface design have been discussed in Section 4.1.5 and elsewhere. However, non-parametric methods may be used effectively in the definition of surfaces which are single-valued and have no large gradients with respect to some flat plane of reference. Where they are suitable, they have the advantage of requiring less computation than parametric methods (see, for instance, Dimsdale and Burkley, 1976).

The simplest Cartesian surface-fitting algorithms fit scalar-valued polynomials in x and y of the type

$$z(x,y) = \sum_{i=0}^{p} \sum_{j=0}^{q} a_{ij} x^i y^j$$

to values of z which are specified at the intersection points of a fixed rectangular mesh in the (x,y)-plane, as in Figure 7.13. The mesh spacing need not be uniform.

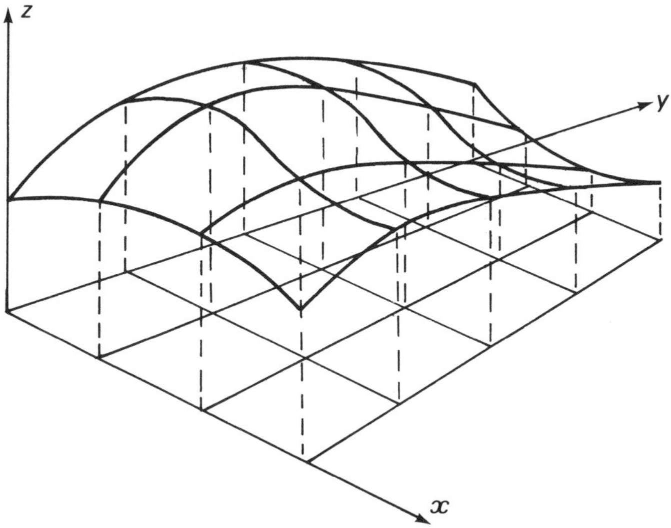

Figure 7.13 – A Surface defined on a Fixed Rectangular Mesh.

We can use an extension of Lagrange's method to find a single polynomial which will interpolate all the data points if we are prepared to take p and q sufficiently large. But the surface so generated is likely to exhibit unwanted oscillations, which often result when the Lagrangean method is used to fit a large number of points (see Section 6.2.1). For this reason, most practical systems fit a single polynomial of low degree in x and y to each separate mesh cell and apply gradient continuity conditions across patch boundaries to generate a smooth composite surface.

Two systems which have been developed along these lines are FMESH (described in Bézier, 1972) and GEMESH (Moore, 1959, Dollries, 1963). The first was developed in Japan by Inaba for the Fujitsu company, and the second by the General Electric Company in the United States. Both use a bicubic patch function

$$z(x,y) = \sum_{i=0}^{3} \sum_{j=0}^{3} a_{ij} x^i y^j \,. \tag{7.44}$$

As in the case of the parametric bicubic patches discussed earlier, the coefficients a_{ij} may be evaluated in terms of positional and derivative information at the patch corners. In FMESH values of z, z_x and z_y are used, and four further items of data are required to specify each patch unambiguously; these may be, for example, four specified points in the interior of the patch. GEMESH requires corner values of z, z_x, z_y and z_{xy} for the determination of each patch. Gradient continuity between patches results, as with parametric bicubic patches, when these four quantities are matched at contiguous patch corners. In FMESH the twist or cross-derivatives z_{xy} are not required explicitly, but their values can be inferred from the alternative information supplied. The main problem with GEMESH was that values of z_{xy} were usually approximated in terms of the z-values at neighbouring patch corners, and were therefore inaccurate. This led to undesirable surface wrinkles in the directions of patch diagonals (Shu, Hori *et al.*, 1970).

As in the parametric case, we can largely avoid the problem of the corner twist derivatives by using bicubic splines. More computation is needed, but the resulting surface has the added advantage of continuity of second derivatives everywhere. Suppose, then, that we have a rectangular mesh of $M \times N$ cells, with function values $z_{mn} = z(x_m, y_n)$ specified at all mesh intersections, where $m = 0, 1, 2, \ldots, M$ and $n = 0, 1, 2, \ldots, N$. We wish to construct a bicubic surface patch, which can conveniently be written as

$$z(x,y) = \sum_{i=0}^{3} \sum_{j=0}^{3} a_{ij} \left(\frac{x - x_m}{h_m}\right)^i \left(\frac{y - y_n}{k_n}\right)^j \,, \tag{7.45}$$

over the mesh cell $x_m \leqslant x \leqslant x_{m+1}$, $y_n \leqslant y \leqslant y_{n+1}$. In this equation $h_m = x_{m+1} - x_m$ and $k_n = y_{n+1} - y_n$. It may be shown that the coefficients a_{ij} in (7.45) are related to the patch corner points and derivatives by

$$\mathbf{A} = \mathbf{C}\,\mathbf{H}_m\,\mathbf{Q}\,\mathbf{K}_n\,\mathbf{C}^T \,, \tag{7.46}$$

in which $\mathbf{A} = [a_{ij}]$, $\mathbf{C}$ is the constant matrix defined in (7.17), $\mathbf{H}_m$, $\mathbf{K}_n$ are given by (7.41), while

$$\mathbf{Q} = \begin{bmatrix} z_{mn} & z_{m,n+1} & z_{y,mn} & z_{y,m,n+1} \\ z_{m+1,n} & z_{m+1,n+1} & z_{y,m+1,n} & z_{y,m+1,n+1} \\ z_{x,mn} & z_{x,m,n+1} & z_{xy,mn} & z_{xy,m,n+1} \\ z_{x,m+1,n} & z_{x,m+1,n+1} & z_{xy,m+1,n} & z_{xy,m+1,n+1} \end{bmatrix} .$$

Equation (7.46) is the scalar analogue of (7.42). The derivative elements in **Q** are calculated by constructing spline curves through the data points, as for the parametric case, though here they will be plane splines. These curves are computed along all the mesh lines in both the x and y directions, and from them the gradients z_x and z_y at the data points are obtained. We can then spline the gradients z_y along the x-boundaries of the surface to obtain z_{xy} at the data points on these boundaries. We use the resulting values to spline z_x along all the mesh lines in the y-direction, and so finally obtain z_{xy} at all the interior mesh points. We then know **Q** for each mesh cell and can use (7.46) and (7.45) to define each surface patch. It is not difficult to show that since all the composite patch boundary lines have continuity of second derivatives (being spline curves) then z_{xx} and z_{yy} are continuous over the surface as a whole. We observe that, as in Sections 7.2.1 and 7.2.4, the entire surface can be computed from the data points, cross-boundary derivatives around its boundary (needed to establish the splines fitting the data) and twist derivatives at its corners (needed to establish the splines fitting the first-derivative values). The arbitrary assumption of, say, $z_{xy} = 0$ at these four points is likely to have only a local effect of any significance, while the cross-boundary gradients can be specified without much difficulty in most applications.

7.5 SPLINE-BLENDED SURFACES

We now return to parametric coordinates and examine a generalisation of the Coons patch technique. As we have seen, the Coons approach enables us to construct a surface patch to fit each mesh cell of a specified topologically rectangular curve network. By choosing the right kind of patch, we can ensure that the resulting composite surface has any desired order of derivative continuity, in addition to simple positional continuity.

However, it is also possible to construct a single non-composite surface which interpolates the entire curve network, and which can fulfil continuity conditions of any order required. To illustrate this, we consider once again (7.5), which is the equation of a patch fitting four specified boundary curves. We may write this equation in the form

$$\mathbf{r}(u,v) = \sum_{i=0}^{1} \alpha_i(u)\mathbf{r}(i,v) + \sum_{j=0}^{1} \alpha_j(v)\mathbf{r}(u,j) - \sum_{i=0}^{1}\sum_{j=0}^{1} \alpha_i(u)\alpha_j(v)\mathbf{r}(i,j) \; , \tag{7.47}$$

where $0 \leqslant u, v \leqslant 1$. Now suppose that, instead of the four boundary curves $\mathbf{r}(u,j)$, $j = 0, 1$ and $\mathbf{r}(i,v)$, $i = 0, 1$ we have a complete curve network $\mathbf{r}(u,j)$, $j = 0, 1, \ldots, N$ and $\mathbf{r}(i,v)$, $i = 0, 1, \ldots, M$. Here $0 \leqslant u \leqslant M$ and $0 \leqslant v \leqslant N$, and the curve network has its intersections at the $(M + 1) \times (N + 1)$ array of points $\mathbf{r}(i,j)$. By analogy with (7.47), we might hope to fit the entire system of curves by setting

$$\mathbf{r}(u,v) = \sum_{i=0}^{M} \alpha_i(u)\mathbf{r}(i,v) + \sum_{j=0}^{N} \alpha_j(v)\mathbf{r}(u,j) - \sum_{i=0}^{M}\sum_{j=0}^{N} \alpha_i(u)\alpha_j(v)\mathbf{r}(i,j)\ . \tag{7.48}$$

Reasoning by analogy with the argument given in Section 7.1, it is not difficult to see that the appropriate conditions on the blending functions α_i are

$$\alpha_i(t) = \begin{cases} 0\ , & t \neq i \\ 1\ , & t = i \end{cases}\ , \qquad i = 0, 1, \ldots, \max\,(M,N)\,.$$

We can easily find a set of functions satisfying these conditions; classical interpolation theory will suffice, for instance. A convenient choice, which avoids the need to work with high-degree polynomials, is to choose the α_i as cubic splines based on the integer knot set 0, 1, 2, . . . , max (M,N). The resulting surface interpolates the curve network as a whole, and provided all the original curves are continuous up to second derivatives, the entire surface will have this same order of continuity because of the properties of our spline blending functions. If a higher order of continuity is required, a higher degree of spline may be used for the α_i.

Surfaces of this kind differ from the composite surfaces studied earlier in two respects. Firstly, the surface is now not composite, except in the sense that its blending functions (being splines) are composite. Secondly the attainment of the desired order of derivative continuity now results not from the form of the patch equation as with (7.12), for instance, but from the choice of blending functions.

These spline-blended surfaces were first described by Gordon (1969). It should be noted that the original network may consist of curves of any kind, composite or otherwise, provided that they have second-order continuity. This interpolation method is therefore more flexible than the tensor-product techniques discussed earlier.

7.6 DEGENERATE PATCHES

The surface patches considered in this chapter have all been topologically rectangular, that is defined in terms of four boundary curves. We sometimes need to use a topologically triangular patch, however, as shown in Figure 7.14. In this case, in the context of a system which basically employs four-sided patches, we may obtain a three-sided patch by allowing the length of one boundary curve to tend to zero. Such a patch is said to be **degenerate**.

Consider the heavily outlined three-sided patch in Figure 7.14. Here the boundary $\mathbf{r}(0,v)$ has zero length. Thus $\mathbf{r}(0,v)$ is constant for $0 \leqslant v \leqslant 1$, and $\mathbf{r}_v(0,v) = \mathbf{r}_{vv}(0,v) = \ldots = \mathbf{0}$. For this patch to be well-defined, the direction of its surface normal at the point marked A must be uniquely determinable. Usually,

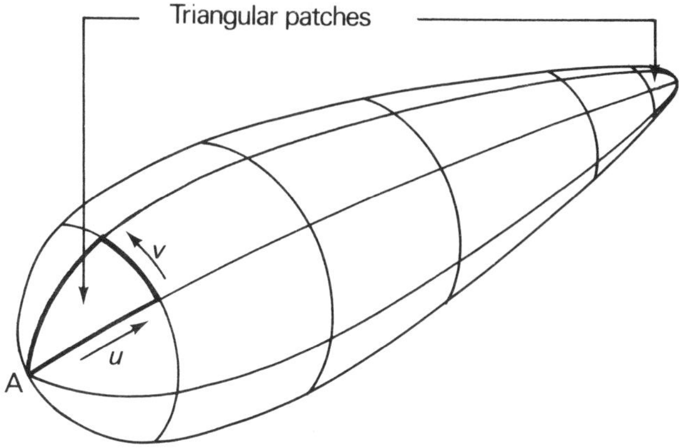

Figure 7.14 – Triangular patches in the Context of a Topologically Rectangular Patch System.

we would calculate the direction of the surface normal on the curve $\mathbf{r}(0,v)$ by forming $\mathbf{r}_u(0,v) \times \mathbf{r}_v(0,v)$, but since in our case $\mathbf{r}_v(0,v)$ is zero, the vector product has zero magnitude and we are unable to establish its direction. This being so, we must use a limiting process based on Taylor series, as follows. When u is small, we can write

$$\mathbf{r}_u(u,v) = \mathbf{r}_u(0,v) + O(u)$$

and

$$\mathbf{r}_v(u,v) = \mathbf{r}_v(0,v) + u\mathbf{r}_{uv}(0,v) + O(u^2)$$

$$= u\mathbf{r}_{uv}(0,v) + O(u^2) \, .$$

It follows that

$$\mathbf{r}_u(u,v) \times \mathbf{r}_v(u,v) = u\mathbf{r}_u(0,v) \times \mathbf{r}_{uv}(0,v) + O(u^2) \, .$$

This gives a vector normal to the surface at the point (u,v). The unit normal at this point is therefore

$$\mathbf{n}(u,v) = \frac{u\mathbf{r}_u(0,v) \times \mathbf{r}_{uv}(0,v) + O(u^2)}{|u\mathbf{r}_u(0,v) \times \mathbf{r}_{uv}(0,v) + O(u^2)|} \, .$$

Then as $u \to 0$ we obtain

$$\lim_{u \to 0} \mathbf{n}(u,v) = \frac{\mathbf{r}_u(0,v) \times \mathbf{r}_{uv}(0,v)}{|\mathbf{r}_u(0,v) \times \mathbf{r}_{uv}(0,v)|} \, .$$

Provided the limit exists, this gives the surface normal direction at the point A, which is uniquely defined only if the right-hand side is independent of v. This will be so provided that $\mathbf{r}_u(0,v)$ and $\mathbf{r}_{uv}(0,v)$ lie in a common plane, which is tangent to the patch at the point A, for all v in the interval $0 \leqslant v \leqslant 1$. This condition is necessary and sufficient for the degenerate patch to be properly defined.

We illustrate the application of the foregoing condition in a practical system by considering a degenerate cubic Bézier patch. The patch equation is (7.23),

$$\mathbf{r}(u,v) = \mathbf{UMBM}^T\mathbf{V} \ .$$

On differentiating and setting $u = 0$ we find

$$\mathbf{r}_u(0,v) = (0 \quad 1 \quad 0 \quad 0)\mathbf{MBM}^T(1 \quad v \quad v^2 \quad v^3)^T$$

and

$$\mathbf{r}_{uv}(0,v) = (0 \quad 1 \quad 0 \quad 0)\mathbf{MBM}^T(0 \quad 1 \quad 2v \quad 3v^2)^T \ .$$

But $$(0 \quad 1 \quad 0 \quad 0)\mathbf{MB} = 3(\mathbf{r}_{10} - \mathbf{r}_{00}, \mathbf{r}_{11} - \mathbf{r}_{01}, \mathbf{r}_{12} - \mathbf{r}_{02}, \mathbf{r}_{13} - \mathbf{r}_{03}), \quad (7.49)$$

and therefore both $\mathbf{r}_u(0,v)$ and $\mathbf{r}_{uv}(0,v)$ are linear combinations of the four vectors in the right-hand side of (7.49), with coefficients which are functions of v. Then if $\mathbf{r}_u(0,v)$ and $\mathbf{r}_{uv}(0,v)$ are to lie in the same plane for all v, the four vectors on which they depend must also all lie in that plane. Since in the degenerate case under discussion $\mathbf{r}_{00} = \mathbf{r}_{01} = \mathbf{r}_{02} = \mathbf{r}_{03}$, we conclude that the five polyhedron vertices $\mathbf{r}_{00}$, $\mathbf{r}_{10}$, $\mathbf{r}_{11}$, $\mathbf{r}_{12}$ and $\mathbf{r}_{13}$ must be coplanar, as shown in Figure 7.15.

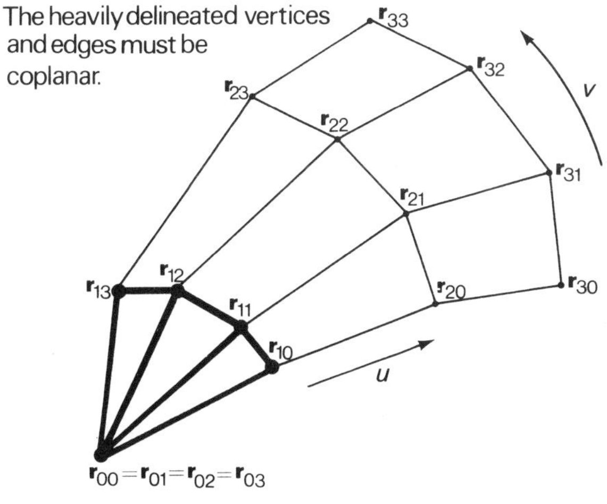

Figure 7.15 – Characteristic Polyhedron of a Degenerate Bézier Patch.
Coplanar vertices and vectors are heavily delineated,

It is perhaps worth mentioning that composite surfaces based entirely on triangular elements are in common use for the numerical solution of certain variational problems by the finite element method. Details may be found in Chapter 4 of Mitchell and Wait (1977). At the time of writing the authors know of no practical system for the design of engineering surfaces which is based on three-sided patches though some of the necessary groundwork is covered in Barnhill, Birkhoff and Gordon (1973). See also Barnhill (1974) and Gregory (1974).

7.7 CURVES ON PARAMETRIC SURFACES; PATCH DISSECTION

Suppose that y is defined as a function of x by $\phi(x,y) = 0$, and that the graph of y against x in two-dimensional Cartesian coordinates contains an arc which lies inside the unit square $0 \leqslant x,\ y \leqslant 1$. In this case, $\phi(u,v) = 0$ represents a curve lying on the parametric surface patch $\mathbf{r} = \mathbf{r}(u,v)$, $0 \leqslant u,\ v \leqslant 1$. If the relation between u and v can be expressed parametrically by $u = u(t)$, $v = v(t)$, then the equation of the curve can easily be found in terms of the new parameter t. We illustrate by considering the curve defined by

$$u = \sum_{k=0}^{3} \alpha_k t^k, \quad v = \sum_{k=0}^{3} \beta_k t^k$$

(where the α_k and β_k are scalars) on the Ferguson patch

$$\mathbf{r}(u,v) = \sum_{i=0}^{3}\sum_{j=0}^{3} \mathbf{a}_{ij} u^i v^j \ .$$

The equation of the curve results when the relations for u and v are substituted into the patch equation. We obtain

$$\begin{aligned}\mathbf{r}(t) &= \sum_{i=0}^{3}\sum_{j=0}^{3} \mathbf{a}_{ij}\left(\sum_{k=0}^{3} \alpha_k t^k\right)^i \left(\sum_{k=0}^{3} \beta_k t^k\right)^j \\ &= \sum_{p=0}^{18} \mathbf{b}_p t^p, \end{aligned} \tag{7.50}$$

where the vectors $\mathbf{b}_p$ are linear combinations of the $\mathbf{a}_{ij}$. Note that even in this comparatively simple case the degree of the curve is a rather alarming eighteen!

In principle, we can use curves of this kind to subdivide patches into smaller units. For example, if ABCD in Figure 7.16 is a patch whose equation is known, and EF is a curve on the patch surface defined as above, then we can obtain equations for the two sub-patches ABFE and EFCD. It must be arranged for the parameter t in (7.50) to run from 0 to 1 across the patch; if t is replaced by u we then have the equation of EF regarded as a patch boundary. The v-boundaries AE, ED, BF and FC may be obtained by curve splitting, as described in Section 6.5. Then ABFE and EFCD can be represented as generalised Ferguson patches of degree eighteen in the u-direction and three in the v-direction.

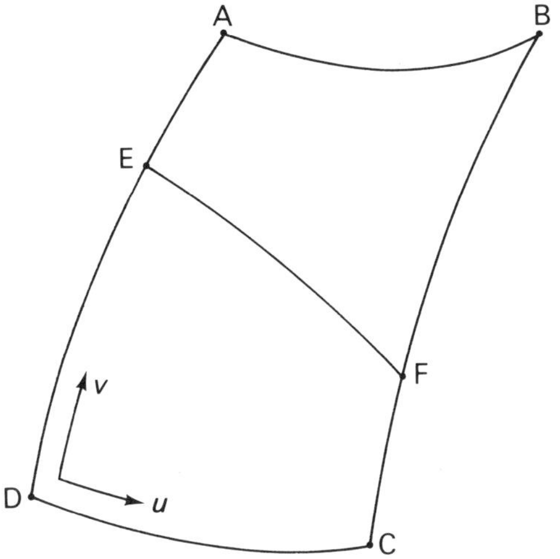

Figure 7.16 – Subdivision of a Surface Patch

The algebraic details of such a patch dissection will be rather complicated, in general. However, they are straightforward in the simple case when the curve on the surface is defined by $u = t, v = v_0 =$ constant. The line EF is then expressed by a parametric cubic in t, and it divides the original patch into two standard Ferguson patches.

The FMILL/APTLFT numerically controlled machining program makes use of curves of this latter type. The desired surface is first defined as outlined in Section 7.2.1 in terms of a number of patches. Then dissecting lines are generated running the length of the entire surface, closely spaced at equal parametric intervals in the transverse direction. Next closely spaced points are specified on each of these lines, at equal parametric intervals in the longitudinal direction. At each point in the resulting array $\mathbf{r}_u \times \mathbf{r}_v$ is calculated to find the direction of the surface normal. Lastly, these normals are used to calculate the corresponding cutter tip positions as shown later in Section 9.3. The cutter is programmed to move so that its tip follows a straight line between each pair of successive computed positions. On reaching either end of the surface the cutter moves transversely onto the next line and returns in the opposite direction. In this way the entire surface is machined. Further details of FMILL/APTLFT may be found in Gould (1972).

An interactive surface design system based on patch dissection has been described by Ghezzi and Tisato (1973). The general shape of an object is first roughly defined in terms of a small number of large patches. Then in regions where refinement of detail is needed the patches are subdivided, possibly several times. Any requisite modifications are made to the smaller patches which result; these modifications have only a local effect and do not alter the overall shape initially defined.

Chapter 8
Cross-sectional designs

8.1 LINEAR AXIS DESIGNS USING BÉZIER PATCHES

There are many products which possess a 'natural' axis, about which they generally have some degree of symmetry. These products are usually designed by defining a number of cross-sections normal to this axis. The shape of the surface between the defined cross-sections must then be determined by using some interpolation scheme, usually guided by the provision of one or more longitudinal cross-sections or **profiles** (see Figure 8.1), as described in Faux (1978).

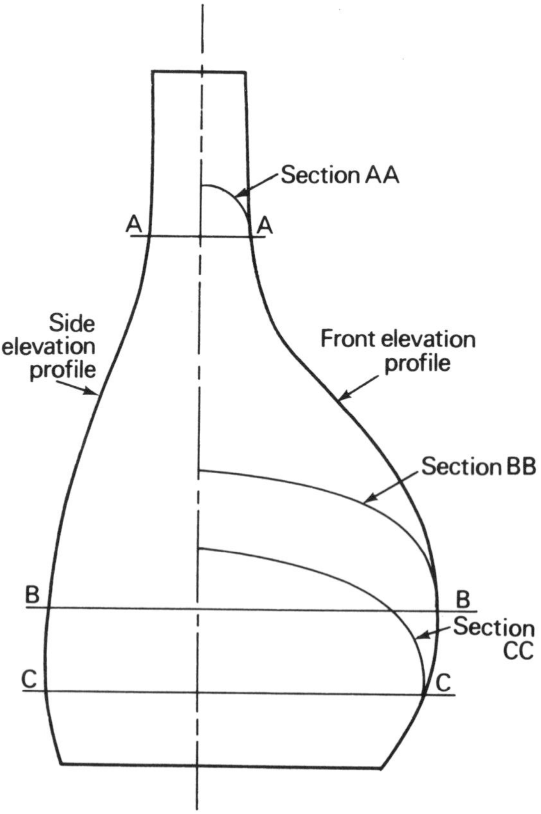

Figure 8.1

The individual cross-sections may be defined by explicit, implicit or parametric equations, and the interpolation between them may also take any of these three forms. For the reasons given in Section 4.1.5, we will use the parametric method in this section, but it should *not* be inferred that this choice is the best in all circumstances.

When the axis is rectilinear, we shall choose the z-axis as the axis of the design, so that the individual cross-sections will be members of the family of curves $x = x(u,v)$, $y = y(u,v)$. Here the parameter v distinguishes the individual cross-sections, and will be called the **profile parameter**. The **cross-section parameter** u describes the individual points on a given cross-section curve.

The variation of the cross-section along the z-axis may then be described by defining z as a function of the parameter v, and the surface is then described by the equation

$$\mathbf{r} = x(u,v)\mathbf{i} + y(u,v)\mathbf{j} + z(v)\mathbf{k} . \qquad (8.1)$$

If we wish to refer to the cross-sectional component of any position vector $\mathbf{r}$, we will denote this by $\mathbf{s} = x\mathbf{i} + y\mathbf{j}$. Then, for example, the symbols $\mathbf{s}_1$ and $\mathbf{s}(u)$ denote the corresponding components of the vectors $\mathbf{r}_1$ and $\mathbf{r}(u)$. Then equation (8.1) may be abbreviated to

$$\mathbf{r} = \mathbf{s}(u,v) + z(v)\mathbf{k} . \qquad (8.2)$$

Cross-sectional designs may, of course, be carried out using the bicubic or birational patches discussed in Sections 5.1 and 5.2. Since the parametric curves of constant v are the cross-section curves, it follows that the characteristic polygons of these curves are plane polygons in the cross-section planes.

Taking the Bézier bicubic patch method as an example, the cross-section curve for a given value of v is composed of segments having equations of the form

$$\mathbf{s}(u,v) = (1 - u)^3\mathbf{s}_0(v) + 3u(1 - u)^2\mathbf{s}_1(v) + 3u^2(1 - u)\mathbf{s}_2(v) + u^3\mathbf{s}_3(v)$$

and $$z = z(v) , \qquad (8.3)$$

where $\mathbf{s}_0(v)$, $\mathbf{s}_1(v)$, $\mathbf{s}_2(v)$ and $\mathbf{s}_3(v)$ indicate the positions of the vertices of the characteristic polygon of the cross-section at $z = z(v)$.

If the cross-section curves possess some symmetry, such as a rotational symmetry about the axis or a reflection symmetry about a plane containing the axis, then it is only necessary to design part of the curve, the remainder of the curve being generated using the co-ordinate transformations described in Chapter 3. Thus, for example, the Bézier curve in Figure 8.2 has a 4-fold rotational symmetry about the axis, and we need only define the segment in the first quadrant by writing $\mathbf{s}_0 = a\mathbf{i}$, $\mathbf{s}_1 = a\mathbf{i} + b\mathbf{j}$, $\mathbf{s}_2 = b\mathbf{i} + a\mathbf{j}$, $\mathbf{s}_3 = a\mathbf{j}$. The remaining quadrants are obtained by reflections in the Ox and Oy axes.

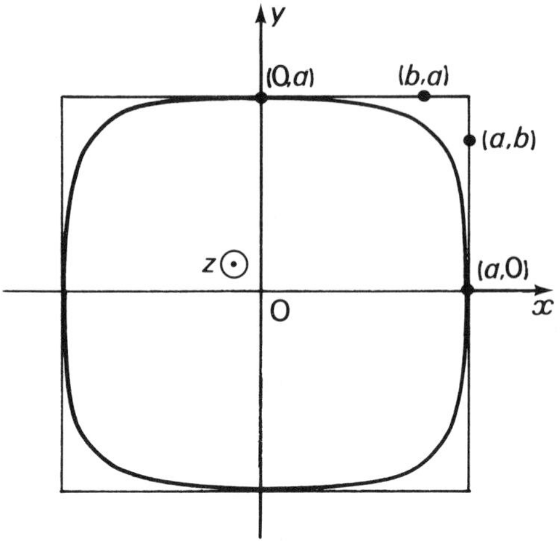

Figure 8.2

The longitudinal variation of the cross-sections in equation (8.3) is controlled by the characteristic polygons of the curves $\mathbf{r}_0(v) = \mathbf{s}_0(v) + z(v)\mathbf{k}$ and so on.

8.1.1 Design with proportional cross-sections

If the cross-sections have constant shape but varying size, then all the cross-sections are proportional to one basic cross-section, which we will choose to be the section with parameter $v = 0$. Thus

$$\mathbf{s}(u,v) = \alpha(v)\mathbf{s}(u,0) , \tag{8.4}$$

where $\alpha(v)$ and $z(v)$ together describe the longitudinal profile. It follows that the vertices of the characteristic polygons of the cross-sections are scaled in the same manner, so that $\mathbf{s}_0(v) = \alpha(v)\mathbf{s}_0(0)$ and so on (see Figure 8.3). We denote the vertices of the basic polygon by $\mathbf{s}_{00}$, $\mathbf{s}_{10}$, $\mathbf{s}_{20}$ and $\mathbf{s}_{30}$.

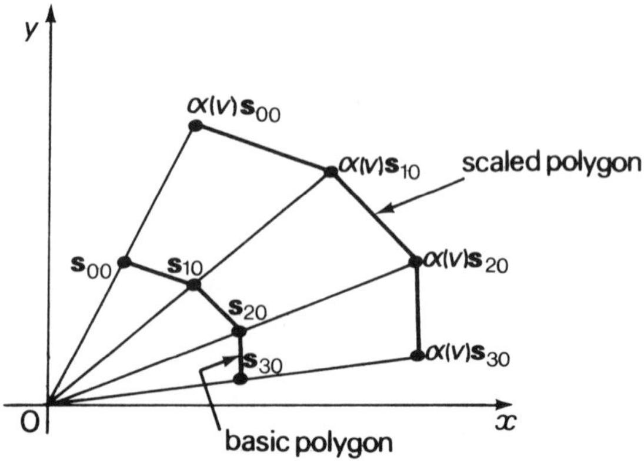

Figure 8.3

The scaling curve $\alpha = \alpha(v)$, $z = z(v)$ may now be described by a two-dimensional Bézier curve in the α-z plane, with its own characteristic polygon $P_0P_1P_2P_3$, as shown in Figure 8.4. Thus

$$\alpha(v) = (1 - v)^3\alpha_0 + 3v(1 - v)^2\alpha_1 + 3v^2(1 - v)\alpha_2 + v^3\alpha_3$$

and $$z(v) = (1 - v)^3 z_0 + 3v(1 - v)^2 z_1 + 3v^2(1 - v)z_2 + v^3 z_3 \ . \quad (8.5)$$

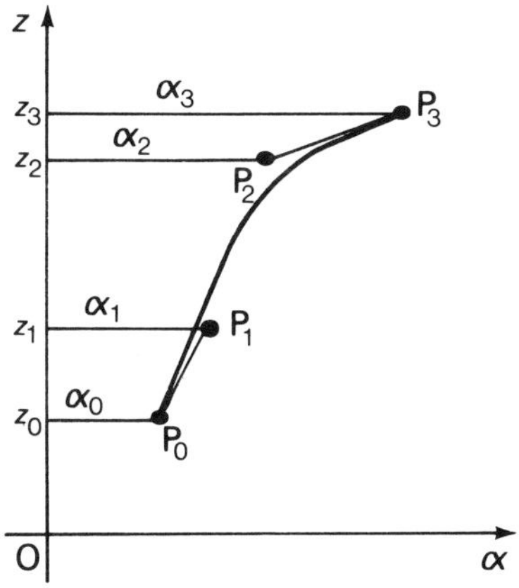

Figure 8.4

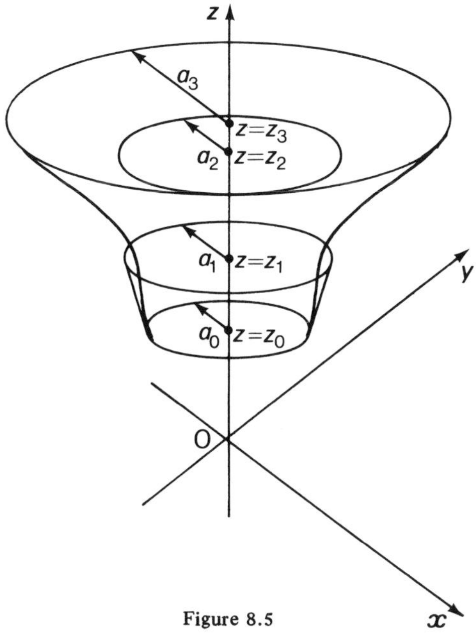

Figure 8.5

For example, if the basic cross-section is an approximate unit circle obtained by combining a number of cubic segments, the surface corresponding to Figure 8.4 is the bell-shape shown in Figure 8.5.

Then radii R of the cross-sections at $z = z_0$ and $z = z_3$ are then α_0 and α_3 respectively. At $z = z_0$ the surface is tangential to the cone defined by the circles $R = \alpha_0, z = z_0$ and $R = \alpha_1, z = z_1$. A similar tangent cone at $z = z_3$ is defined by the circles $R = \alpha_2, z = z_2$ and $R = \alpha_3, z = z_3$, as shown in Figure 8.5.

The 16 vertices of the characteristic polyhedron of any given patch are then given by

$$\mathbf{B} = \begin{bmatrix} \mathbf{r}_{00} & \mathbf{r}_{01} & \mathbf{r}_{02} & \mathbf{r}_{03} \\ \mathbf{r}_{10} & \mathbf{r}_{11} & \mathbf{r}_{12} & \mathbf{r}_{13} \\ \mathbf{r}_{20} & \mathbf{r}_{21} & \mathbf{r}_{22} & \mathbf{r}_{23} \\ \mathbf{r}_{30} & \mathbf{r}_{31} & \mathbf{r}_{32} & \mathbf{r}_{33} \end{bmatrix} = \begin{bmatrix} \mathbf{s}_{00} \\ \mathbf{s}_{10} \\ \mathbf{s}_{20} \\ \mathbf{s}_{30} \end{bmatrix} [\alpha_0 \quad \alpha_1 \quad \alpha_2 \quad \alpha_3] + \begin{bmatrix} \mathbf{k} \\ \mathbf{k} \\ \mathbf{k} \\ \mathbf{k} \end{bmatrix} [z_0 \quad z_1 \quad z_2 \quad z_3] \,. \tag{8.6}$$

8.1.2 Blend between two cross-sections, using one profile curve only

The sections at $v = 0$ and $v = 1$ may, however, be different in shape as well as size. A simple family of intermediate shapes is given by the equation

$$\mathbf{s}(u,v) = (1 - \alpha(v))\mathbf{s}(u,0) + \alpha(v)\mathbf{s}(u,1) \,, \tag{8.7}$$

where $\alpha(0) = 0$ and $\alpha(1) = 1$. Since only one blending function $\alpha(v)$ is used, the intermediate surface is determined when one profile is described.

For any given parameter $v = v_0$, the section is the locus of points which divide the lines joining corresponding points on the curves $\mathbf{s}(u,0)$ and $\mathbf{s}(u,1)$ in the constant ratio $1 - \alpha(v_0):\alpha(v_0)$ as shown in Figure 8.6. The characteristic polygons are interpolated in the same way. Let the polygon for the cross-section $\mathbf{s}(u,0)$ have the vertices $\mathbf{s}_{00}$, $\mathbf{s}_{10}$, $\mathbf{s}_{20}$ and $\mathbf{s}_{30}$ and that for $\mathbf{s}(u,1)$ the vertices $\mathbf{s}_{03}$, $\mathbf{s}_{13}$, $\mathbf{s}_{23}$ and $\mathbf{s}_{33}$.

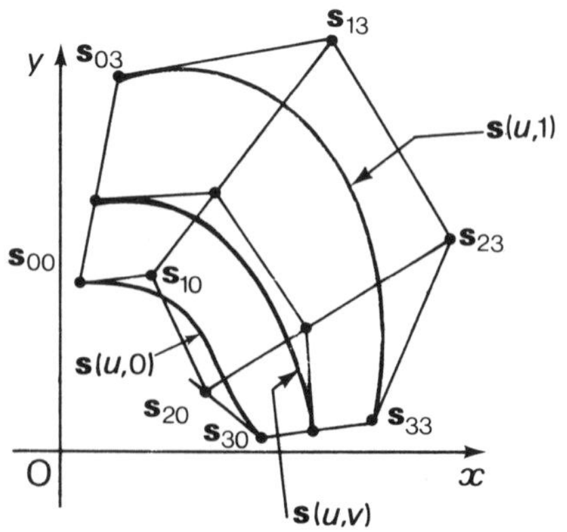

Figure 8.6

If the functions $\alpha(v)$ and $z(v)$ are given by the equations (8.5), with $\alpha_0 = 0$ and $\alpha_3 = 1$, then the vertices of the characteristic polyhedron have the form

$$\mathbf{B} = \begin{bmatrix} \mathbf{r}_{00} & \mathbf{r}_{01} & \mathbf{r}_{02} & \mathbf{r}_{03} \\ \mathbf{r}_{10} & \mathbf{r}_{11} & \mathbf{r}_{12} & \mathbf{r}_{13} \\ \mathbf{r}_{20} & \mathbf{r}_{21} & \mathbf{r}_{22} & \mathbf{r}_{23} \\ \mathbf{r}_{30} & \mathbf{r}_{31} & \mathbf{r}_{32} & \mathbf{r}_{33} \end{bmatrix} = \begin{bmatrix} \mathbf{s}_{00} & \mathbf{s}_{03} \\ \mathbf{s}_{10} & \mathbf{s}_{13} \\ \mathbf{s}_{20} & \mathbf{s}_{23} \\ \mathbf{s}_{30} & \mathbf{s}_{33} \end{bmatrix} \begin{bmatrix} 1 & 1-\alpha_1 & 1-\alpha_2 & 0 \\ 0 & \alpha_1 & \alpha_2 & 1 \end{bmatrix} + \begin{bmatrix} \mathbf{k} \\ \mathbf{k} \\ \mathbf{k} \\ \mathbf{k} \end{bmatrix} [z_0 \ z_1 \ z_2 \ z_3] \ . \tag{8.8}$$

This simple form requires the user to supply the plane characteristic polygons for the cross-sections at each end of the patch, together with the polygon for the α-z variation of the chosen profile (for example, Figure 8.4). There are just 22 pieces of data per patch. When the cross-sections are composed of several segments, the corresponding patches must all use the same values of α_1, α_2, z_0, z_1, z_2 and z_3, so that only one profile may be predefined using this simple form.

Collins and Gould (1974) have used a similar method to blend cross-sections defined in terms of the polar co-ordinates defined in Section 1.2.1. The two cross-sections are then described by the functions $r = r_0(\theta)$ and $r = r_1(\theta)$ in terms of the same parameter θ. Intermediate cross-sections have the polar equation

$$r(\theta,v) = [1 - \alpha(v)]r_0(\theta) + \alpha(v)r_1(\theta) \ . \tag{8.9}$$

The blending function can then be determined if the longitudinal cross-section by any plane $\theta = \theta_0$ is known. Thus if the section by the plane $\theta = 0$ is given by $r = R(v)$, $z = z(v)$, then the blending function is

$$\alpha(v) = \frac{R(v) - r_0(0)}{r_1(0) - r_0(0)} \ . \tag{8.10}$$

The advantage of this method is that it is straightforward and easily visualised. Note that we must choose the cross-sections and profile curve such that $r_1(0) \neq r_0(0)$.

8.1.3 Blend between two cross-sections, using two profile curves

A more general blend of the two given cross-section curves, in which profiles are provided by both sides of the patch, is given by

$$\mathbf{B} = \left[\begin{array}{cc|cc} \mathbf{s}_{00} & \mathbf{s}_{03} & \multicolumn{2}{c}{0} \\ \mathbf{s}_{10} & \mathbf{s}_{13} & & \\ \hline \multicolumn{2}{c|}{0} & \mathbf{s}_{20} & \mathbf{s}_{23} \\ & & \mathbf{s}_{30} & \mathbf{s}_{33} \end{array}\right] \left[\begin{array}{cc|cc} 1 & 1-\alpha_{01} & 1-\alpha_{02} & 0 \\ 0 & \alpha_{01} & \alpha_{02} & 1 \\ \hline 1 & 1-\alpha_{31} & 1-\alpha_{32} & 0 \\ 0 & \alpha_{31} & \alpha_{32} & 1 \end{array}\right] + \begin{bmatrix} \mathbf{k} \\ \mathbf{k} \\ \mathbf{k} \\ \mathbf{k} \end{bmatrix} [z_0 \ z_1 \ z_2 \ z_3] \ . \tag{8.11}$$

The interpolation for the first two points of the cross-section polygons is determined by α_{01}, α_{02}, z_0, z_1, z_2 and z_3, whereas that for the last two points is governed by $\alpha_{31}, \alpha_{32}, z_0, z_1, z_2$ and z_3. An example is shown in Figures 8.7, 8.8. It can be seen that only one profile can be chosen freely (the **principal profile**), and whilst the other **auxiliary profile** can have given tangents at each end, its fullness is fixed because the same z values have been used for both profiles. This is a consequence of the choice of the equation (8.1), in which z is a function of v only.

Although more general surfaces may be obtained by relaxing this constraint, we have not done so here because we have been concerned to illustrate *simple* methods of definition in terms of a small number of plane curves.

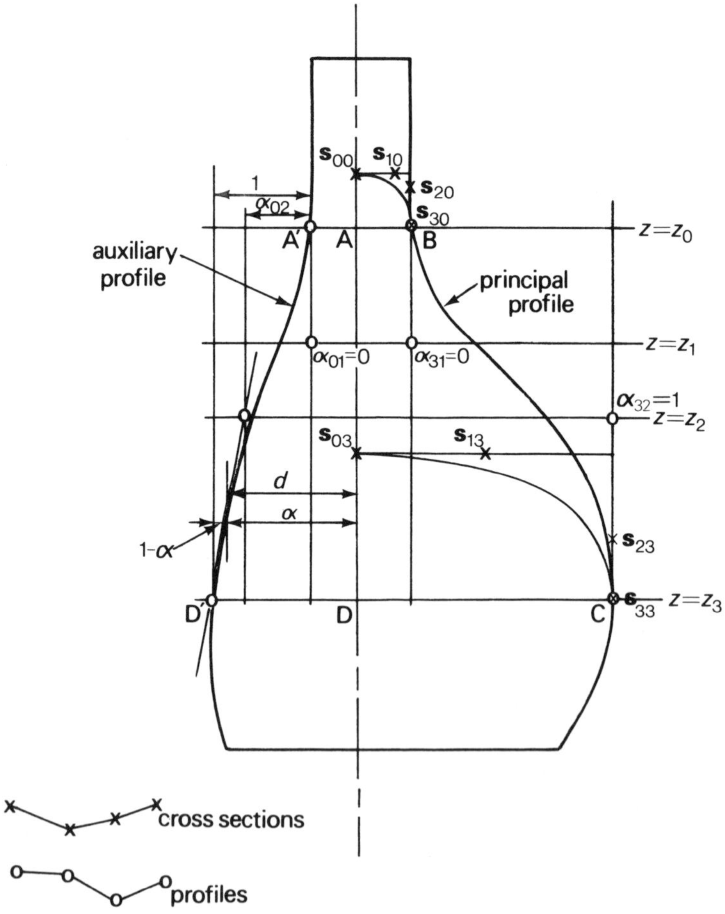

Figure 8.7 – Characteristic Polygonals of one Patch (schematic only).

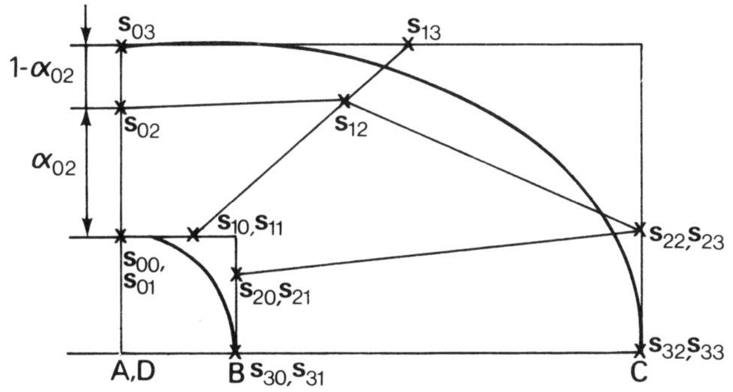

Figure 8.8

A more general surface based on two sections and two profile curves takes the form

$$\mathbf{r}(u,v) = \alpha_1(v)\mathbf{s}_1(u) + \alpha_2(v)\mathbf{s}_2(u) + z(v)\mathbf{k} \,, \tag{8.12}$$

which is a natural extension of the proportional cross-section design of Section 8.1.1.

All the cross-sections belong to the family of shapes based on the two basic section curves $\mathbf{s} = \mathbf{s}_1(u)$ and $\mathbf{s} = \mathbf{s}_2(u)$. This equation lends itself more readily to the design of multipatch surfaces than does (8.11). Provided that the section curves are smooth, as well as the blending curves defined by $\alpha_1(v), z(v)$ and $\alpha_2(v), z(v)$, the resulting surface will also be smooth.

The functions $\alpha_1(v)$, $\alpha_2(v)$ and $z(v)$ will most conveniently be calculated from two longitudinal profiles specified in co-ordinate terms.

Extension to a family of cross-sections based on three or more basic sections is possible. However, the oscillation problem familiar in the interpolation of point data will then become serious. If more complex cross-section families are required, the spline-blended surfaces of Section 7.5 are recommended.

8.2 LINEAR AXIS DESIGNS USING GENERALISED BÉZIER PATCHES

Although Bézier polynomial equations are normally used to describe the Cartesian coordinates of a curve, they may be used to describe the axial variation of any property of the cross-sectional curves.

For example, the family of cross-sections

$$\mathbf{s} = a \cos u \, \mathbf{i} + a \sin u \, \mathbf{j} \tag{8.13}$$

represents a family of circles of variable radius a.

If the radius varies linearly with z, the surface generated is a cone. This relationship may be described parametrically by the equations

$$a = a_0(1 - v) + a_1 v$$

$$z = z_0(1 - v) + z_1 v$$

The surface of revolution shown in Figure 8.5 may be described by equation (8.12), together with

$$a = (1 - v)^3\alpha_0 + 3v(1 - v)^2\alpha_1 + 3v^2(1 - v)\alpha_2 + v^3\alpha_3 \tag{8.14}$$

and

$$z = (1 - v)^3 z_0 + 3v(1 - v)^2 z_1 + 3v^2(1 - v)z_2 + v^3 z_3 \ .$$

The Bézier polynomials may be used to describe the variation of more than one parameter. Thus, for example, the cross-section curves

$$\mathbf{s} = (a - b\cos 4u)\cos u\,\mathbf{i} + (a - b\cos 4u)\sin u\,\mathbf{j} \tag{8.15}$$

describe, for small values of b/a, circles to which some squareness has been added. For larger values of b/a, clover-leaf patterns are obtained. Hence, by varying a and b along the z-axis, a smooth surface such as that shown in Figure 8.9 may be obtained.

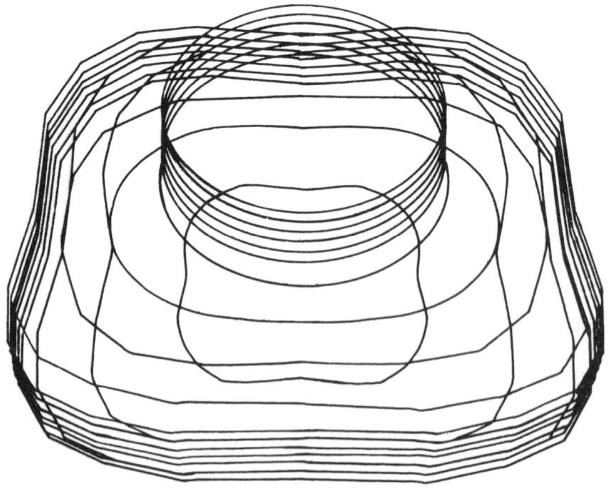

Figure 8.9

The surface in Figure 8.9 was produced by using Bézier cubics to define the axial variation of a and b as follows:

$$a = (1-v)^3 a_0 + 3v(1-v)^2 a_1 + 3v^2(1-v)a_2 + v^3 a_3,$$

$$b = (1-v)^3 b_0 + 3v(1-v)^2 b_1 + 3v^2(1-v)b_2 + v^3 b_3, \quad (8.16)$$

and $$z = (1-v)^3 z_0 + 3v(1-v)^2 z_1 + 3v^2(1-v)z_2 + v^3 z_3 .$$

However it should be noted that, just as only one profile could be freely designed using the methods of the previous section, the variation of a and b cannot be determined independently. One parameter must be chosen as the **principal parameter**, since the other must use the same values of z_0, z_1, z_2 and z_3, and will be called the **auxiliary parameter**.

The family of cross-sections need not be described parametrically. The family of conics

$$(1-\lambda)(x^2 + y^2 - 1) + \lambda(x-1)(y-1) = 0 \quad (8.17)$$

is a one-parameter family of curves from which a surface may be generated by determining the variation of λ with z.

Thus we may, for example, write

$$\lambda = (1-v)^3\lambda_0 + 3v(1-v)^2\lambda_1 + 3v^2(1-v)\lambda_2 + v^3\lambda_3$$

and $$z = (1-v)^3 z_0 + 3v(1-v)^2 z_1 + 3v^2(1-v)z_2 + v^3 z_3, \quad (8.18)$$

which gives a surface of the kind shown in Figure 8.10.

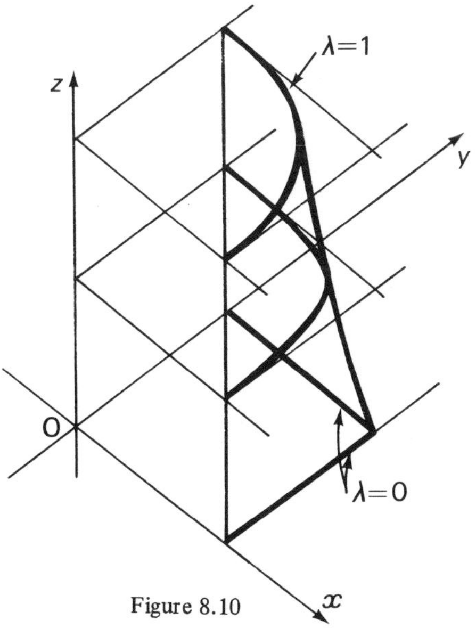

Figure 8.10

These generalisations of the Bézier curves are examples of the 'indirect design' described by Wielinga (1974).

8.3 CROSS-SECTIONAL DESIGNS BASED ON PROPORTIONAL DEVELOPMENT

For many years, a graphical technique known as proportional development has been used in the shipbuilding, aircraft and automobile industries. In this method, four boundary curves are developed into a surface by using linear proportions of the boundary curves as seen in the direction of the design axis.

In Figure 8.11, the curves DHC and AGB are two cross-sections in planes normal to the z-axis. We wish to generate an intermediate cross-section EJF.

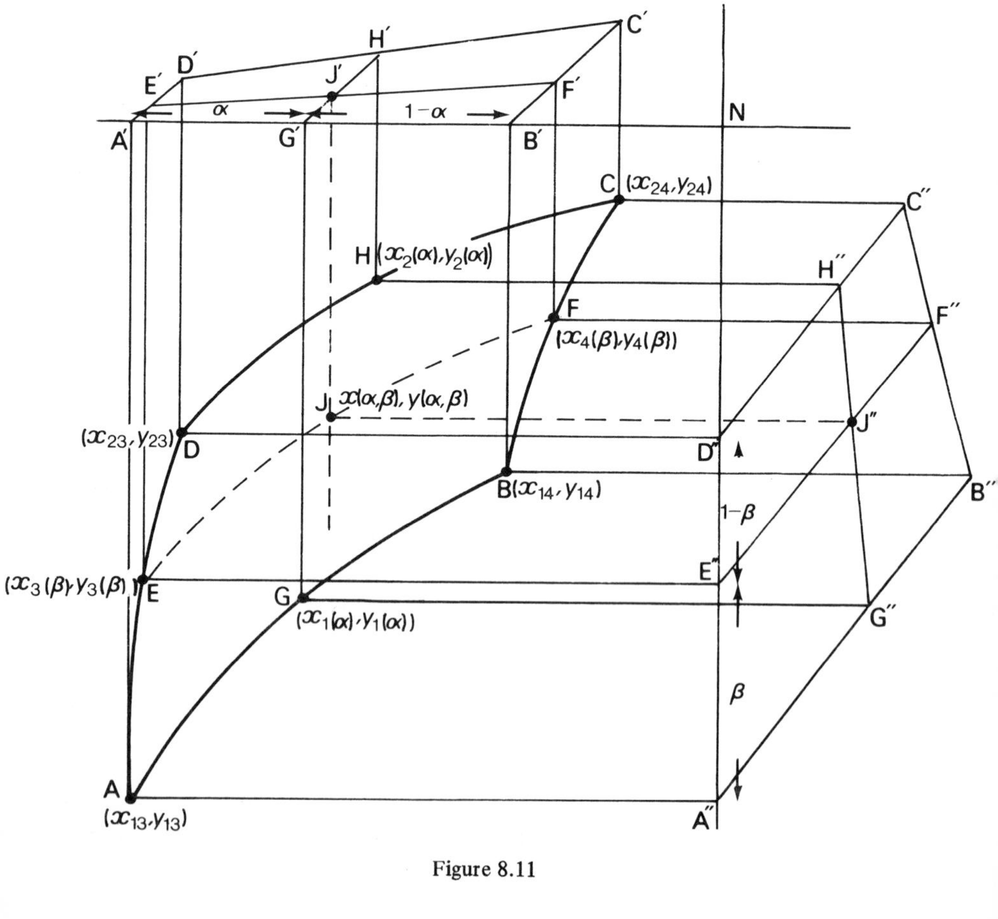

Figure 8.11

We first determine corresponding points G and H on AB and DC as follows. Draw the horizontal line A′B′ with A′ and B′ vertically above A and B. Draw A′D′ and B′C′ as parallel lines meeting the verticals DD′ and CC′ at D′ and C′ respectively (any convenient angle may be chosen for D′A′B′ and C′B′N). The point corresponding to G is obtained by drawing the vertical line GG′, then the line G′H′ parallel to A′D′, and finally the vertical H′H to meet DC at H. We label the points G and H with the parameter α, where A′G′:G′B′ = α:1−α. (Note that E′J′:J′F′ = D′H′:H′C′ = α:1−α also, since the lines A′D′, G′H′ and B′C′ are all parallel.)

Then the curves AB and DC have been linearly parametrised in x, since G has coordinates

$$x_G = x_1(\alpha) = (1 - \alpha)x_{13} + \alpha x_{14} ,$$

and $$y_G = y_1(\alpha) ,$$

where x_{13} and x_{14} are the x-coordinates of A and B, and $y_1(\alpha)$ is a function determined by the shape of the curve AGB. A similar linear parametrisation applies to the curve DC.

By similar means, we parametrise AED and BFC linearly in y, using β as the parameter. We obtain the following equations for the boundary curves in terms of the coordinates marked in Figure 8.11:

$$\begin{aligned} &\text{AGB:} \quad x = x_1(\alpha) = (1 - \alpha)x_{13} + \alpha x_{14}, \quad y = y_1(\alpha) ; \\ &\text{DHC:} \quad x = x_2(\alpha) = (1 - \alpha)x_{23} + \alpha x_{24}, \quad y = y_2(\alpha) \\ &\text{and} \\ &\text{AED:} \quad x = x_3(\beta), \quad y = y_3(\beta) = (1 - \beta)y_{13} + \beta y_{23} ; \\ &\text{BFC:} \quad x = x_4(\beta), \quad y = y_4(\beta) = (1 - \beta)y_{14} + \beta y_{24} . \end{aligned} \tag{8.19}$$

The intermediate points such as J on the new cross-section EJF are constructed as follows. The line H′G′ is the line representing points of parameter α. The particular point representing J is obtained by projecting E and F vertically to E′ and F′. The point J′ is then the intersection between E′F′ and G′H′. The x-coordinate of J is thus given by $x_J = (1 - \alpha)x_E + \alpha x_F$, and using (8.19) to describe E and F, we obtain

$$\begin{aligned} x_J &= x(\alpha,\beta) = (1 - \alpha)x_3(\beta) + \alpha x_4(\beta) , \\ \text{and similarly} \quad y_J &= y(\alpha,\beta) = (1 - \beta)y_1(\alpha) + \beta y_2(\alpha) . \end{aligned} \tag{8.20}$$

Since the curves AGB, EJF and DHC are cross-sections by planes normal to the z-axis, and are identified by the parameter β, the shape of the surface is determined by the relationship between z and β.

Thus, finally, the surface has the equation

$$\mathbf{r} = \mathbf{r}(\alpha,\beta) = x(\alpha,\beta)\mathbf{i} + y(\alpha,\beta)\mathbf{j} + z(\beta)\mathbf{k} \qquad (8.21)$$

where $x(\alpha,\beta)$ and $y(\alpha,\beta)$ take the form of (8.20). Comparison with (8.1) shows that this is a special case of the designs considered in Section 8.1.

If, for example, the functions $x_3(\beta)$, $x_4(\beta)$, $y_1(\alpha)$, $y_2(\alpha)$ and $z(\beta)$ are cubic polynomials, then we obtain a special case of the Bézier bicubic surface in which x is linear in α and y linear in β. This means that the four characteristic polygons across the patch have equal x-increments between points, as shown for polygon ABCD in Figure 8.12. Similarly the four longitudinal polygons have equal y increments between the points, as with polygon AEFG in the figure. These constraints limit the scope for providing continuity between patches, because the tangent vector lengths are predetermined by the boundary curves.

Although not strictly the subject of this chapter, it is worth remarking here that proportional development is possible even when the curves DHC and AGB are not cross-section curves in parallel planes, but general space curves. This more general case is described by Duncan and Mair (1974). The proportions in z are then obtained from a second view. Duncan and Mair have described a computer program which implements this more general development in connection with their Polyhedral Machining method.

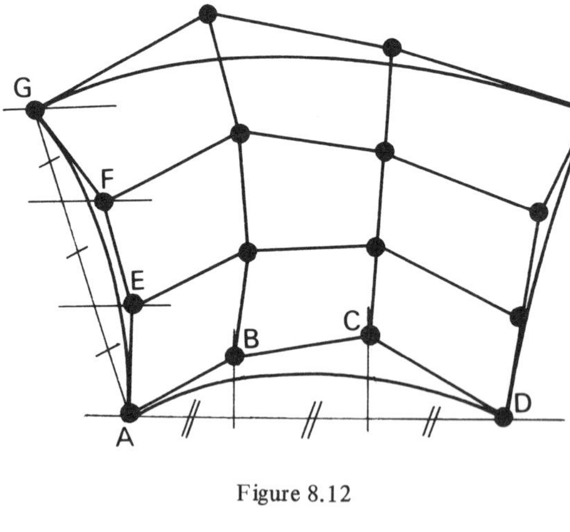

Figure 8.12

8.4 CROSS-SECTIONAL DESIGN BASED ON A CURVED SPINE

In the design of pipework and ducting, interest often centres on the cross-sections normal to the direction of flow. If the pipes or ducts are curved, it is usual to design their surfaces in terms of a number of cross-sections normal to some 'mean flow line'.

If this curve (often known as the spine) has the parametric form $\mathbf{r} = \mathbf{r}_s(v)$, then we may use the normal vectors $\mathbf{N}(v)$ and $\mathbf{B}(v)$ defined in Section 4.2.4 as unit vectors in a local coordinate system in the normal plane at each point of the spine.

If the cross-section in this normal plane is described parametrically by $\mathbf{s} = f(u,v)\mathbf{N}(v) + g(u,v)\mathbf{B}(v)$, we may write the equation of the general point on the surface in the form

$$\mathbf{r} = \mathbf{r}(u,v) = \mathbf{r}_s(v) + f(u,v)\mathbf{N}(v) + g(u,v)\mathbf{B}(v) \;, \qquad (8.22)$$

as shown in Figure 8.13. This is a straightforward generalisation of equation (8.1), in which the spine was a straight line $\mathbf{r}_s(v) = z(v)\mathbf{k}$.

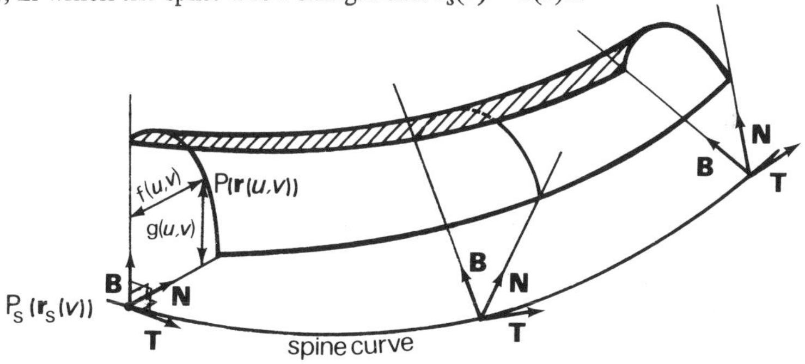

Figure 8.13

However, this 'natural' axis system has two deficiencies: firstly, the axis system rotates about the spine as it proceeds, making it difficult to visualise the meaning of the components $f(u,v)$ and $g(u,v)$, and secondly, the axis system is undefined if the spine has any straight segments.

The 'Duct' system described by Gossling (1976) is based on local axes in the normal plane which are chosen to ensure that one axis is always in the horizontal plane, thus largely overcoming both objections to (8.22). The surface equation then takes the form

$$\mathbf{r} = \mathbf{r}(u,v) = \mathbf{r}_s(v) + f(u,v)\mathbf{n}_1(v) + g(u,v)\mathbf{n}_2(v) \qquad (8.23)$$

where $\quad \mathbf{n}_1 = (\dot{\mathbf{r}}_s \times \mathbf{k})/|\dot{\mathbf{r}}_s \times \mathbf{k}|, \qquad \mathbf{n}_2 = (\mathbf{n}_1 \times \dot{\mathbf{r}}_s)/|\dot{\mathbf{r}}_s|$.

Then the unit vector $\mathbf{n}_1$ is parallel to the Oxy plane, and lies in the normal plane. The unit vector $\mathbf{n}_2$ lies in the normal plane and is perpendicular to $\mathbf{n}_1$.

One feature of both (8.22) and (8.23) is that the surface normal vector has components dependent on the second derivative of the spine curve equation. Thus continuity of surface normal depends on curvature continuity of the spine curve unless $f(u,v)$ and $g(u,v)$ are suitably chosen.

The 'Duct' system uses a Bézier curve as the spine, so that curvature continuity cannot be guaranteed, as discussed in Section 6.3.3, and the program is arranged to minimise the resulting discontinuities of surface normal direction.

The functions $f(u,v)$ and $g(u,v)$ are effectively defined by surface patches of Bézier bicubic type, as shown in Figure 7.4. Because the curves v = constant must be plane curves, it follows that the polygon points $(\mathbf{r}_{00},\mathbf{r}_{10},\mathbf{r}_{20},\mathbf{r}_{30})$, $(\mathbf{r}_{01},\mathbf{r}_{11},\mathbf{r}_{21},\mathbf{r}_{31})$, $(\mathbf{r}_{02},\mathbf{r}_{12},\mathbf{r}_{22},\mathbf{r}_{32})$ and $(\mathbf{r}_{03},\mathbf{r}_{13},\mathbf{r}_{23},\mathbf{r}_{33})$ must be 4 sets of coplanar points.

The first and last sets are the characteristic polygons of actual cross-section curves, and will be chosen to fit the data provided by the designer on these two cross-sections. They will automatically be plane curves. Unless the designer chooses otherwise, 'Duct' will fit a series of segments of approximately elliptic shape, maintaining slope continuity across the joins.

The points $\mathbf{r}_{01}$, $\mathbf{r}_{02}$ and $\mathbf{r}_{31}$, $\mathbf{r}_{32}$ complete the characteristic polygons of the longitudinal edges of the patch. Their positions on the planes of the points $(\mathbf{r}_{01},\mathbf{r}_{11},\mathbf{r}_{21},\mathbf{r}_{31})$ and $(\mathbf{r}_{02},\mathbf{r}_{12},\mathbf{r}_{22},\mathbf{r}_{32})$ are calculated by the program to minimise the discontinuity of the surface normal.

The interior points of the polyhedron are arranged such that $\mathbf{r}_{uv} = \mathbf{0}$ at the patch corners, the large number of patches ensuring that the resulting surface is satisfactory.

The reader is referred to Gossling (1976) for further details of the method, including the handling of degenerate patches.

8.5 AREAS AND VOLUMES OF LINEAR AXIS DESIGNS

The volume of a cross-sectional design of the form given in equation (8.2) may be computed by integrating the area $A(v)$ of the cross-sections with respect to z.

Thus for a given Bézier patch of the type discussed in Section 8.1, the volume subtended at the axis is given by

$$V = \int_{z_0}^{z_3} A(v)\,\mathrm{d}z = \int_0^1 A(v)\,\dot{z}(v)\,\mathrm{d}v\,. \tag{8.24}$$

The area $A(v)$ may be calculated analytically as follows. The vector area $\mathbf{A}(v)$ is taken from equation (4.46). Provided that the parameter u increases in a right-handed screw relationship to the positive z-axis, the *magnitude* of the area $\mathbf{A}(v)$ is then given by

$$A(v) = \tfrac{1}{2}\int_0^1 \mathbf{k}\,.\,\mathbf{s}\times\mathbf{s}_u\,\mathrm{d}u\,. \tag{8.25}$$

The vector $\mathbf{s}(u,v)$ may be expressed in terms of the polyhedron vertex vectors $\mathbf{s}_{ij}$ of Section 8.1 by

$$\mathbf{s}(u,v) = \sum_i \sum_j \sum_k \sum_\ell u^i m_{ij} \mathbf{s}_{jk} m_{\ell k} v^\ell , \tag{8.26}$$

in which m_{ij} is the ij element of the Bézier coefficient matrix $\mathbf{M}$ given in equation (5.6), and the summations are each over the range 0 to 3.

Similarly,
$$\mathbf{s}_u(u,v) = \sum_p{}' \sum_q \sum_r \sum_s p\, u^{p-1} m_{pq} \mathbf{s}_{qr} m_{sr} v^s , \tag{8.27}$$

in which the symbol Σ' indicates summation from 1 to 3 only.

These expressions may be substituted into (8.25) to obtain the equations

$$A(v) = \sum_k \sum_\ell \sum_r \sum_s m_{\ell k} m_{sr} v^{\ell+s} d_{kr} , \tag{8.28}$$

where
$$d_{kr} = \tfrac{1}{2} \sum_i \sum_j \sum_p{}' \sum_q \int_0^1 \mathbf{k} \cdot m_{ij} \mathbf{s}_{jk} \times m_{pq} \mathbf{s}_{qr}\, p\, u^{i+p-1}\, du . \tag{8.29}$$

Equation (8.29) may be simplified to give

$$d_{kr} = \tfrac{1}{2} \sum_i \sum_p{}' \mathbf{k} \cdot \frac{(\mathbf{a}_{ik} \times \mathbf{a}_{pr})\, p}{i + p} , \tag{8.30}$$

where
$$\mathbf{a}_{ik} = \sum_j m_{ij} \mathbf{s}_{jk} . \tag{8.31}$$

Equations (8.28), (8.30) and (8.31) may be used to obtain the area of any cross-section rapidly and accurately.

For the purposes of the volume calculation, we simplify (8.28) by substituting $q_{\ell s} = \sum_k \sum_r m_{\ell k} d_{kr} m_{sr}$ and obtain

$$A(v) = \sum_\ell \sum_s q_{\ell s} v^{\ell+s} . \tag{8.32}$$

The expression for $z(v)$ is of the Bézier form $z(v) = \sum_i \sum_j v^i m_{ij} z_j$, so that the derivative $\dot{z}(v)$ is given by

$$\dot{z}(v) = \sum_i{}' \sum_j i\, v^{i-1} m_{ij} z_j = \sum_i{}' i\, v^{i-1} c_i , \tag{8.33}$$

where

$$c_i = \sum_j m_{ij} z_j \; . \tag{8.34}$$

It remains to substitute (8.32) and (8.33) into the expression (8.24) to obtain the volume V.

After integration, we obtain the equation

$$V = \sum_i{}' \sum_\ell \sum_s \frac{i\, c_i\, q_{\ell s}}{i + \ell + s} \; . \tag{8.35}$$

Similar analysis is possible for higher order polynomial patches, but the calculations for rational polynomial patches usually require numerical integrations, and direct numerical calculation of the volumes is probably just as fast.

Chapter 9

Computing methods for surface design and manufacture

9.1 INTERSECTIONS OF CURVES AND SURFACES

9.1.1 Introductory remarks

The intersection curve between two surfaces is determined by the solution of non-linear equations, unless the two surfaces are planes. Except for some simpler surfaces such as quadrics, for which analytical methods may be used to simplify the numerical work (Levin, 1976), it is usually necessary to attack the problem directly by numerical methods.

The calculation of the intersection curve C of two surfaces S_1 and S_2 may be regarded either as a problem involving the solution of simultaneous (usually non-linear) equations, or as a minimisation problem in which the squared distance $|\mathbf{r}_1 - \mathbf{r}_2|^2$ between variable points $P_1(\mathbf{r}_1)$ on S_1 and $P_2(\mathbf{r}_2)$ on S_2 is minimised by adjusting the points $\mathbf{r}_1$ and $\mathbf{r}_2$.

Although the simultaneous equation approach is not fool-proof without proper safeguards, we shall describe it in some detail because it clarifies the difficulties which may arise in solving intersection problems by any method. Moreover, there is much in common between this approach and the iterative method used by the APT III ARELEM program for the calculation of tool paths for numerically controlled (n.c) machines.

As has been remarked in Section 1.1.8, in connection with the intersections of plane curves, the problem is somewhat simplified when one surface is defined implicitly and one parametrically. If the implicit surface has the equation $f(\mathbf{r}) = 0$, and the parametric surface is defined by $\mathbf{r} = \mathbf{r}_1(u,v)$, we can see that the intersection curve takes the form $F(u,v) = 0$.

On the other hand, the intersection of two implicit surfaces $f_1(\mathbf{r}) = 0$ and $f_2(\mathbf{r}) = 0$ involves the simultaneous solution of two equations in three variables.

Finally, if the two surfaces are defined parametrically by $\mathbf{r} = \mathbf{r}_1(u_1,v_1)$ and $\mathbf{r} = \mathbf{r}_2(u_2,v_2)$, their intersection is defined by $\mathbf{r}_1(u_1,v_1) - \mathbf{r}_2(u_2,v_2) = \mathbf{0}$. This vector equation corresponds to three scalar equations in four variables.

In each case, there is one more variable than there are equations, since the points of a curve have, in general, one degree of freedom. In solving these equations, we produce a sequence of points along the curve by imposing a further constraint $g(\mathbf{r}) = 0$ at each step. We therefore solve a sequence of three-surface intersection problems.

If we consider the hybrid problem $f(\mathbf{r}) = 0$ and $\mathbf{r} = \mathbf{r}_1(u,v)$ once more, we see that the step constraint $g(\mathbf{r}) = 0$ is equivalent to an additional equation $G(u,v) = 0$, so that we must solve

$$F(u,v) = 0$$

and

$$G(u,v) = 0 \tag{9.1}$$

simultaneously at each step.

For two implicit surfaces $f_1(\mathbf{r}) = 0$ and $f_2(\mathbf{r}) = 0$, the additional equation $g(\mathbf{r}) = 0$ provides us with three equations in three unknowns:

$$f_1(\mathbf{r}) = 0 \ ,$$

$$f_2(\mathbf{r}) = 0 \ ,$$

and

$$g(\mathbf{r}) = 0 \ . \tag{9.2}$$

We should obviously look to see whether any of the surfaces can be translated into parametric form, because this would enable us to reduce the problem to the form (9.1). For example, planes may be parametrised as described in Section 4.1.4, and quadrics may be represented as rational quadratic functions by stereographic projection onto a plane, as described by Sommerville (1934). However, some arbitrary decisions have to be made in setting up the parametric forms because these are not unique.

In the case of two parametrically defined surfaces $\mathbf{r} = \mathbf{r}_1(u_1,v_1)$ and $\mathbf{r}_2 = \mathbf{r}_2(u_2,v_2)$, the constraint $g(\mathbf{r}) = 0$ may be translated into a constraint $G_1(u_1,v_1) = 0$, or alternatively $G_2(u_2,v_2) = 0$. These equivalent constraints may be combined and solved with the surface equations. Thus

$$\mathbf{r}_1(u_1,v_1) - \mathbf{r}_2(u_2,v_2) = \mathbf{0}$$

and

$$\lambda G_1(u_1,v_1) + \mu G_2(u_2,v_2) = 0 \ , \tag{9.3}$$

where λ and μ may be chosen to simplify the constraint equation. Great advantage would be had if one surface could be translated into implicit form. Unfortunately, the most commonly used parametric surfaces, the bicubic patches, cannot be expressed in this way.

Before considering the step constraint and solution methods in more detail, we note that these intersection curves have been considered without reference to their application. When we wish to calculate the projections of these curves, or compute offset curves for numerical control purposes, it may often be advantageous to combine the two processes. Thus, for example, Sabin (1975) has shown that the central projection of the intersection curve between two quadrics onto a given picture plane is most economically performed directly in picture co-ordinates, provided that a variety of views is not required and care is taken to avoid incorrect plotting near double points in the projected curve. Moreover, the choice of step constraint is properly made with reference to picture co-ordinates in this case.

9.1.2 Solution of the equations

Although the hybrid problem (9.1) apparently produces the simplest equations, there are often difficulties which are best illustrated by considering equations (9.2), where the solution is carried out in the space variables.

In Appendix A4 we show that a system of equations such as (9.2) can be solved iteratively by the Newton-Raphson method (which we will henceforth refer to as **Newton's method** for brevity) provided that a good initial approximation is known and that the Jacobian matrix **J** is non-singular at each approximation point.

In the terms of equations (9.2) a single step of Newton's method may be expressed as $\mathbf{r}_{i+1} = \mathbf{r}_i + \delta\mathbf{r}_i$. Here $\mathbf{r}_i$ is the ith iterated solution, and the correction term $\delta\mathbf{r}_i$ is given by

$$\mathbf{J}(\mathbf{r}_i)\delta\mathbf{r}_i = -\mathbf{F}(\mathbf{r}_i) , \tag{9.4}$$

in which the Jacobian matrix **J** and the vector **F** are given respectively by

$$\mathbf{J}(\mathbf{r}) = \begin{bmatrix} \nabla f_1{}^T \\ \hline \nabla f_2{}^T \\ \hline \nabla g^T \end{bmatrix}, \qquad \mathbf{F}(\mathbf{r}) = \begin{bmatrix} f_1 \\ f_2 \\ g \end{bmatrix} .$$

The geometrical interpretation of equation (9.4) is that $\mathbf{r}_{i+1}$ lies at the intersection of three offset planes which are parallel to the tangent planes to the surfaces $f_1(\mathbf{r}) = f_1(\mathbf{r}_i)$, $f_2(\mathbf{r}) = f_2(\mathbf{r}_i)$ and $g(\mathbf{r}) = g(\mathbf{r}_i)$ at the point $\mathbf{r} = \mathbf{r}_i$.

In Figure 9.1, we illustrate the situation by a two-dimensional view of the surface families $f_1(\mathbf{r})$ = constant and $f_2(\mathbf{r})$ = constant. The offset planes OP_1 and OP_2 pass close to the intersections M and N of the normals from P_i to the corresponding surfaces $f_1 = 0$ and $f_2 = 0$ if $f_1(\mathbf{r}_i)$ and $f_2(\mathbf{r}_i)$ are sufficiently small. We assume that g is chosen such that the intersection of its offset plane with the common line of OP_1 and OP_2 is well defined.

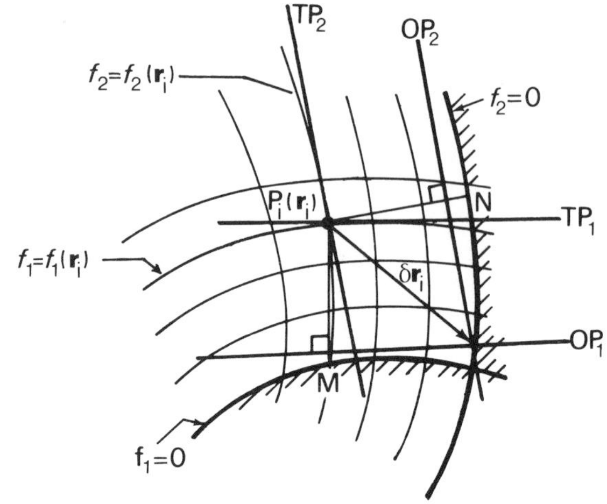

Figure 9.1

However, if the two offset planes OP_1 and OP_2 are parallel at $\mathbf{r} = \mathbf{r}_i$, they either meet at infinity or coincide. In either case, the simple Newton procedure fails. The usual reason is that the initial approximation was too far from the correct solution, and this situation can often be avoided by using information from previous steps along the contour.

A more serious problem occurs when the surfaces S_1 and S_2 have the same normal at one or more points along their line of intersection. However close the initial approximation, the offset planes are then coincident or almost coincident and their intersection is undefined or ill-defined. The Jacobean matrix $\mathbf{J}$ is then singular or 'ill-conditioned'. This is not surprising, because the intersection curve of the surfaces themselves is ill-conditioned in the sense that a small distortion of the surfaces makes a large change in the intersection curve.

If we want our procedures to be robust enough to cope with these problems, we must resort to more sophisticated methods such as the modified Newton method described by Deuflhard (1974).

Writing $\mathbf{J}_i$ for $\mathbf{J}(\mathbf{r}_i)$ and $\mathbf{F}_i$ for $\mathbf{F}(\mathbf{r}_i)$, we may denote the Newton iteration step as $\delta\mathbf{r}_i = -\mathbf{J}_i^{-1}\mathbf{F}_i$. Deuflhard's modification consists of taking a suitable step in the steepest descent direction $-\nabla T$ of the **level function** $T = \frac{1}{2}(\mathbf{J}_i^{-1}\mathbf{F}_i)^T \mathbf{J}_i^{-1}\mathbf{F}_i$. For this particular level function, $-\nabla T = -\mathbf{J}_i^{-1}\mathbf{F}_i$, so that the steepest descent direction is the same as that of the Newton step. However, the length of the step taken is now determined by a linear search along this direction to locate the point at which T is minimum.

When the Jacobian is ill-conditioned or singular, Deuflhard uses the pseudo-inverse $\mathbf{J}_i^{+}$ (see Penrose, 1955) in place of the inverse $\mathbf{J}_i^{-1}$ in the level function T. Ben-Israel (1966) has shown that the resulting method converges locally to a point $\mathbf{r}$ satisfying the equation

$$\mathbf{J}^T(\mathbf{r})\,\mathbf{F}(\mathbf{r}) = \mathbf{0} \ .$$

Thus the point $\mathbf{r}$ is either a solution point of $\mathbf{F}(\mathbf{r}) = \mathbf{0}$, *or* a point where $\mathbf{J}(\mathbf{r})$ is singular, when the three surface normals are coplanar, as we have seen. To resolve this difficulty, we need to look at the second derivatives. We will discuss the use of second derivatives in Section 9.1.4, in connection with minimisation algorithms.

The reader is referred to Deuflhard (1974) for further details of this method, and for general discussion of systems of non-linear equations to Rabinowitz (1970) and Dennis (1976).

The same problem occurs in a more concealed fashion with equations (9.1). If the surfaces $f(\mathbf{r}) = 0$ and $\mathbf{r} = \mathbf{r}_1(u,v)$ have the same normal at some point, then ∇f is parallel to $\mathbf{r}_u \times \mathbf{r}_v$ there. It follows that $\dfrac{\partial F}{\partial u} = \nabla F.\mathbf{r}_u = 0$ and $\dfrac{\partial F}{\partial v} = 0$ in the same way. Thus the Jacobian of equations (9.1) is singular whatever the form of $G(u,v)$.

In this section, we have assumed that a sequence of constraints $g(\mathbf{r}) = 0$ can be found which ensure convergence of the Newton iteration, and provide a subdivision of the intersection curve which is suited to our requirements. The next section considers the stepping procedures in more detail.

9.1.3 Step length determination

The choice of step length naturally depends on the application and on the interpolation scheme to be employed subsequently. We will assume linear interpolation here, as used in vector plotting devices and many numerically controlled machines. The intersection curve is then approximated by a sequence of linear segments, as shown in Figure 9.2. The optimum choice for parabolic and circular arc interpolation will not be discussed, although the following notes will be of some use in dealing with these also.

When linear interpolation is used, it is usual to set limits on the maximum deviation δ between the true intersection curve and the chord between successive points, measured normal to the chord as shown in Figure 9.2.

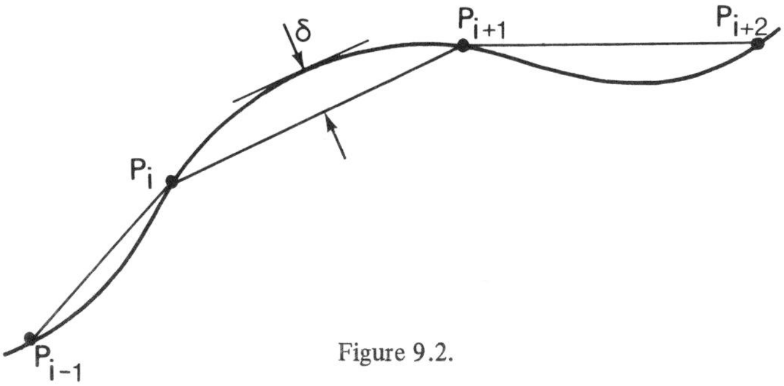

Figure 9.2.

For small steps, it is reasonable to approximate the curve by its osculating circle, so that the local curvature of the intersection curve may be used to determine the step length. By Pythagoras' theorem, we may show that

$$L^2 = 4\delta(2\rho - \delta) \tag{9.5}$$

where L is the step length, and ρ is the radius of curvature (see Figure 9.3).

The curvature of the intersection of the two surfaces S_1 and S_2 may be expressed in terms of the normal curvatures κ_{n1} and κ_{n2} of the curve on S_1 and S_2 respectively, together with the unit normals $\mathbf{n}_1$ and $\mathbf{n}_2$ to the two surfaces (see Section 4.2.9).

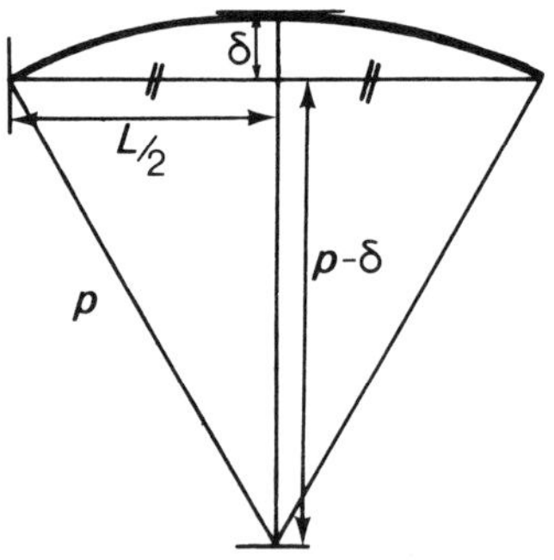

Figure 9.3

Let the curve have the equation $\mathbf{r} = \mathbf{r}(s)$, and the corresponding tangent, normal and binormal vectors be **T**, **N** and **B**.

The tangent vector is perpendicular to both surface normals, so that

$$\mathbf{T} = \pm \frac{\mathbf{n}_1 \times \mathbf{n}_2}{|\mathbf{n}_1 \times \mathbf{n}_2|}\,. \tag{9.6}$$

Thus $\kappa\mathbf{B} = -\kappa\mathbf{N} \times \mathbf{T} = \mp\kappa\mathbf{N} \times (\mathbf{n}_1 \times \mathbf{n}_2)/|\mathbf{n}_1 \times \mathbf{n}_2|$, and we may expand the triple vector product to obtain

$$\kappa\mathbf{B} = \frac{\pm(\kappa_{n1}\mathbf{n}_2 - \kappa_{n2}\mathbf{n}_1)}{|\mathbf{n}_1 \times \mathbf{n}_2|}\,. \tag{9.7}$$

If the angle between the surface normals is θ, the curvature is given by

$$\kappa^2 = \frac{\kappa_{n1}{}^2 - 2\kappa_{n1}\kappa_{n2}\cos\theta + \kappa_{n2}{}^2}{\sin^2\theta}\,. \tag{9.8}$$

Note that the curvature tends to infinity when the normals become parallel, unless $\kappa_{n1} = \kappa_{n2}$, in which case $\kappa = \kappa_{n1} = \kappa_{n2}$.

A first approximation to the normal curvature of a curve C on a surface S may be obtained from the position of two adjacent points of C and the surface normal of S at one of them. Consider two adjacent points P and Q on the curve $\mathbf{r} = \mathbf{r}(s)$, having arc length parameters s and $s + \delta s$ (see Figure 9.4).

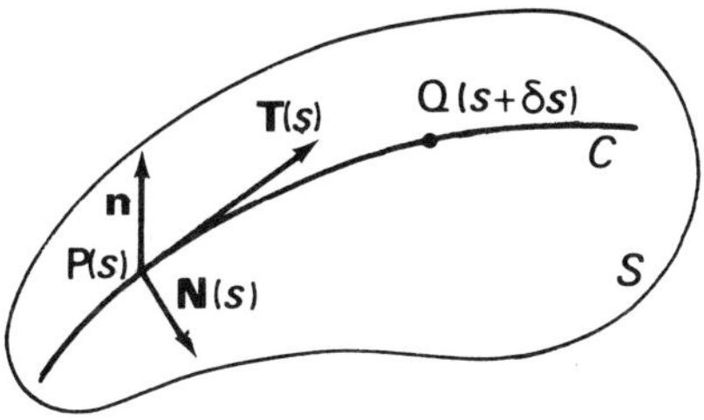

Figure 9.4

By a Taylor expansion of $\mathbf{r}(s)$, we have

$$\mathbf{r}(s + \delta s) = \mathbf{r}(s) + \delta s\, \dot{\mathbf{r}}(s) + \frac{\delta s^2}{2}\ddot{\mathbf{r}}(s) + O(\delta s^3) ,$$

in which the dots denote differentiation with respect to s.

Thus, by the Frenet-Serret formulae,

$$\mathbf{r}(s + \delta s) - \mathbf{r}(s) = \mathbf{T}\delta s + \kappa\mathbf{N}\frac{\delta s^2}{2} + O(\delta s^3) ,$$

where the vectors $\mathbf{T}$ and $\kappa\mathbf{N}$ are evaluated at s.

Since the curve lies on the surface S, the tangent vector $\mathbf{T}$ is perpendicular to the surface normal $\mathbf{n}$, and we may thus obtain

$$\mathbf{n} . (\mathbf{r}(s + \delta s) - \mathbf{r}(s)) = \kappa\mathbf{n}.\mathbf{N}\frac{\delta s^2}{2} + O(\delta s^3) = \kappa_n\frac{\delta s^2}{2} + O(\delta s^3).$$

By approximating the arc length δs by the chord length and neglecting the remainder, we obtain the equation

$$\kappa_n \simeq \frac{\mathbf{n}.[\mathbf{r}(s + \delta s) - \mathbf{r}(s)]}{|\mathbf{r}(s + \delta s) - \mathbf{r}(s)|^2} . \tag{9.9}$$

Alternatively, the curvature κ can be computed directly in terms of $\mathbf{r}(s)$, $\mathbf{r}(s + \delta s)$ and $\mathbf{T}(s)$. Since $\mathbf{T}$ must be obtained from (9.6), the actual coding is no more efficient than careful implementation of (9.8) and (9.9), although the expression for κ is formally much simpler:

$$\kappa = 2|(\mathbf{r}(s + \delta s) - \mathbf{r}(s)) \times \mathbf{T}|/|\mathbf{r}(s + \delta s) - \mathbf{r}(s)|^2 \ .$$

We have chosen to develop the result in terms of normal curvatures in order to show the sensitivity of the curvature κ to the angle between the two surface normals. As equation (9.8) shows, the curvature may become very large when the angle is small, so that the step length must be very small. Thus we have two complications arising when the normals are parallel or nearly parallel: difficulty in applying Newton's method because the Jacobian is ill-conditioned, and difficulty in defining the step length.

The specification of the surface $g(\mathbf{r}) = 0$ which determines the step length may be made in a number of ways. Although a sphere of radius L would appear to be the obvious choice, the equation of a plane is linear, and the resulting computational simplicity has led to its use in numerical control programs such as APT III (IITRI, 1967) and HAPT-3D (Hyodo, 1973).

The plane may be chosen to be normal to the current tangent to the intersection curve. On the basis of the osculating circle approximation, the distance d from the plane to the current point is taken as $d = L(1 - \delta/\rho)$, where L is taken from equation (9.5). If the current tangent vector is $\mathbf{T}_0$, and the current position vector $\mathbf{r}_0$, then

$$g(\mathbf{r}) = (\mathbf{r} - \mathbf{r}_0).\mathbf{T}_0 - d \qquad (9.10)$$

as shown in Figure 9.5.

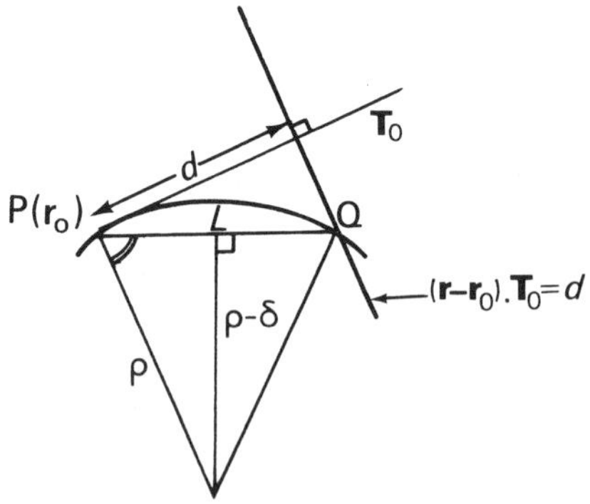

Figure 9.5

In numerical control programs, where cutter offset calculations are required, the process is usually iterative, and the plane $g(\mathbf{r}) = 0$ is arranged to be normal to the tangent at the final point Q rather than at P. More will be said about these programs in Section 9.3.

We will close this section by considering the intersection between a given surface and a given plane. This important problem is encountered when surface cross-sections are required for surface checking at the design, manufacture and inspection stages, and for design of ribs supporting sheet metal surfaces.

We will consider the case of a bicubic patched surface, although similar remarks will apply to other parametric surfaces. Since the function $\mathbf{r}_1(u,v)$ for the given surface is now bicubic in u and v, and the equation of the section plane $f(\mathbf{r}) = \mathbf{n}_2 . \mathbf{r} - p$ is linear in $\mathbf{r}$, the intersection curve is determined by a cubic equation in u and v. If the steps are determined by a planar constraint of the form (9.10), we obtain two simultaneous bicubic equations

$$F(u,v) \equiv \mathbf{n}_2 . \mathbf{r}_1(u,v) - p = 0 \ ,$$

and

$$G(u,v) \equiv (\mathbf{r}_1(u,v) - \mathbf{r}_0) . \mathbf{T}_0 - d = 0 \ ,$$

for each step along the intersection curve.

9.1.4 Solution of surface intersection problems using least-squares minimisation algorithms

The surface intersection equations (9.2) can alternatively be expressed as a minimisation problem. We now search for a point $P^*(\mathbf{r}^*)$ which minimises the function $\phi(\mathbf{r}) = \frac{1}{2}(f_1{}^2(\mathbf{r}) + f_2{}^2(\mathbf{r}) + g^2(\mathbf{r}))$. It is clear that the global minimum of this function occurs when $f_1(\mathbf{r}) = f_2(\mathbf{r}) = g(\mathbf{r}) = 0$. The theory of such **least-squares minimisation** has been described by Powell (1972), and the reader is referred to that work for further details of the theory outlined below.

At a **local minimum point** P^* of $\phi(\mathbf{r})$, it must be true that the function is a minimum along any curve $\mathbf{r} = \mathbf{r}(u)$ which passes through P^*. It follows that $\dfrac{d\phi}{du} = 0$ and $\dfrac{d^2\phi}{du^2} > 0$ for *every* curve $\mathbf{r} = \mathbf{r}(u)$ passing through P^*.

By the chain rule, we may express these conditions in terms of the gradient and second derivative of ϕ with respect to x, y and z.

Thus

$$\frac{d\phi}{du} = \frac{\partial\phi}{\partial x}\frac{dx}{du} + \frac{\partial\phi}{\partial y}\frac{dy}{du} + \frac{\partial\phi}{\partial z}\frac{dz}{du} = \dot{\mathbf{r}}^T\,\nabla\phi \tag{9.11}$$

and

$$\frac{d^2\phi}{du^2} = \frac{\partial^2\phi}{\partial x^2}\left(\frac{dx}{du}\right)^2 + \frac{\partial^2\phi}{\partial y^2}\left(\frac{dy}{du}\right)^2 + \frac{\partial^2\phi}{\partial z^2}\left(\frac{dz}{du}\right)^2 + 2\frac{\partial^2\phi}{\partial x\partial y}\frac{dx}{du}\frac{dy}{du}$$

$$+ 2\frac{\partial^2\phi}{\partial y\partial z}\frac{dy}{du}\frac{dz}{du} + 2\frac{\partial^2\phi}{\partial z\partial x}\frac{dz}{du}\frac{dx}{du} + \frac{\partial\phi}{\partial x}\frac{d^2x}{du^2} + \frac{\partial\phi}{\partial y}\frac{d^2y}{du^2} + \frac{\partial\phi}{\partial z}\frac{d^2z}{du^2}$$

or
$$\frac{d^2\phi}{du^2} = \dot{\mathbf{r}}^T\mathbf{H}\dot{\mathbf{r}} + \ddot{\mathbf{r}}^T\nabla\phi \ , \tag{9.12}$$

where $\nabla\phi$ is the **gradient** of ϕ (see Section 4.3.2), and **H** is the **Hessian** matrix of second derivatives of ϕ.

At a **stationary point** of ϕ, $\frac{d\phi}{du} = 0$ for any tangent direction $\dot{\mathbf{r}}$. Thus $\dot{\mathbf{r}}^T\nabla\phi = 0$ for any $\dot{\mathbf{r}}$, so that the condition for a stationary point is that $\nabla\phi = \mathbf{0}$.

At a **local minimum point**, we must also have $\frac{d^2\phi}{du^2} > 0$ for any non-zero tangent vector, so that $\dot{\mathbf{r}}^T\mathbf{H}\dot{\mathbf{r}} > 0$ for any $\dot{\mathbf{r}} \neq \mathbf{0}$. (The minimum point is a stationary point, so that the term $\ddot{\mathbf{r}}^T\nabla\phi$ is zero.) Any matrix **A** such that $\mathbf{v}^T\mathbf{A}\mathbf{v} > 0$ for all vectors $\mathbf{v} \neq \mathbf{0}$ is called a **positive definite matrix**. It follows that the Hessian must be positive definite at a minimum point of ϕ. If $\dot{\mathbf{r}}^T\mathbf{H}\dot{\mathbf{r}} < 0$ for all $\dot{\mathbf{r}} \neq \mathbf{0}$, the Hessian is **negative definite**, $\frac{d^2\phi}{du^2} < 0$, and the point P concerned is a **local maximum point.** If $\dot{\mathbf{r}}^T\mathbf{H}\dot{\mathbf{r}}$ is sometimes positive and sometimes negative, the Hessian is called **indefinite** and the point P is a **saddle point** at which ϕ has a maximum on some curves and a minimum on others. An algorithm which does not distinguish between the different kinds of stationary point may well terminate on a saddle point.

Using the notation introduced in equation (9.4), we may express the gradient of ϕ as

$$\nabla\phi = \mathbf{J}^T\mathbf{F} \ . \tag{9.13}$$

Then the stationary points of ϕ occur at the solution of $\mathbf{J}^T\mathbf{F} = \mathbf{0}$. (Thus Deuflhard's modified Newton method described in Section 9.1.2 terminates at a stationary point of ϕ, which may not be a solution of the original equations.)

The Hessian matrix of ϕ can be expressed as

$$\mathbf{H} = \mathbf{J}\mathbf{J}^T + \mathbf{A} \ ,$$

where **A** has the elements

$$a_{ij} = f_1 \frac{\partial^2 f_1}{\partial x_i \partial x_j} + f_2 \frac{\partial^2 f_2}{\partial x_i \partial x_j} + g \frac{\partial^2 g}{\partial x_i \partial x_j} \ . \tag{9.14}$$

The algorithms used for the minimisation of ϕ are effectively Newton methods for the equations $\nabla\phi = 0$, with suitable modifications. Since the Jacobian matrix of these equations is given by the Hessian **H**, the basic iteration of these methods is based on $\mathbf{H}\delta\mathbf{r} = -\nabla\phi$, or

$$(\mathbf{JJ}^T + \mathbf{A})\delta\mathbf{r} = -\mathbf{J}^T\mathbf{F} \ . \tag{9.15}$$

In the vicinity of a solution point $\mathbf{r}^*$, the values of f_1, f_2 and g should be small, so that $\mathbf{A}$ may usually be neglected in comparison with $\mathbf{JJ}^T$. The exceptions occur when the surfaces do not have an exact intersection point, or when the surface normals are nearly coplanar, so that $\mathbf{J}$ is ill-conditioned. If these difficulties can be avoided, neglecting $\mathbf{A}$ in (8.15) results in the equations $\mathbf{J}\delta\mathbf{r} = -\mathbf{F}$, and we obtain the ordinary Newton equations (9.4) for the intersection problem.

For general surface intersections, the neglect of $\mathbf{A}$ is not justified, and the full Hessian $\mathbf{JJ}^T + \mathbf{A}$ should be used.

Although the increment $\delta\mathbf{r}$ resulting from equation (9.15) may be used directly to define the next approximation to $\mathbf{r}^*$, it is more usual to use the *direction* of $\delta\mathbf{r}$ as a search direction along which a linear minimisation of ϕ is performed. Provided that the Hessian is positive definite, this direction will always result in a reduction in ϕ. However, if the Hessian is not positive definite, the direction may not be a descent direction. Marquardt (1963) and Gill and Murray (1974, 1976) have described two methods which ensure that a descent direction is always chosen by adjusting the direction prescribed by (9.15) towards the steepest descent direction $-\nabla\phi$.

Gill and Murray (1976) also describe a technique for dealing with saddle points. At a saddle point, equation (9.15) has the solution $\delta\mathbf{r} = \mathbf{0}$, so that the iteration will normally terminate. Gill and Murray give a method of detecting the indefinite nature of the Hessian and computing a descent direction to enable the calculation to proceed.

Butterfield (1978) has described a minimisation procedure for the intersection of two *parametric* surfaces $\mathbf{r} = \mathbf{r}_1(u_1,v_1)$ and $\mathbf{r} = \mathbf{r}_2(u_2,v_2)$. The user chooses one variable to determine a particular point on the intersection curve, and the algorithm minimises the function $|\mathbf{r}_1(u_1,v_1) - \mathbf{r}_2(u_2,v_2)|^2$ with respect to the remaining three variables. The algorithm uses the full Hessian matrix, and can deal with saddle points.

9.2 OFFSET SURFACES

When a surface is machined using a ball-ended cutter, the tool centre moves on another parallel surface, offset from the original by an amount equal to the cutter radius R.

For example, in the FMILL–APTLFT system referred to in Section 7.2.1, FMILL produces a sequence of points and normals for the Ferguson surface, and APTLFT adds the offset R along the surface normal at each each point, together with a displacement R along the tool axis to determine the position of the cutter tip (for tool setting purposes it is convenient to the use the cutter tip rather than the tool centre as the reference point). The new points determine the linear segments of the cutter path, so that the machining accuracy is a function of the cutter radius and the spacing of the curves and stringers specified by the user in the data for the FMILL program.

Offset surfaces are also required, for example, to make allowance for the skin thickness in sheet metal surfaces. We will therefore mention some general points about parallel surfaces before considering the numerical control programs in more detail in Section 9.3.

We will consider the parametric surface $\mathbf{r} = \mathbf{r}_1(u,v)$ and the corresponding offset surface at distance d along the normal $\mathbf{n}_1(u,v)$ given by

$$\mathbf{r}' = \mathbf{r}_1(u,v) + d\mathbf{n}_1(u,v). \tag{9.16}$$

Willmore (1959) shows that the normals of corresponding points on the surfaces $\mathbf{r}$ and $\mathbf{r}'$ have the same direction, but have the opposite sense if d exceeds *one* of the principal radii of curvature. If $\mathbf{G}$ and $\mathbf{G}'$ are the first fundamental matrices of the two surfaces, then

$$|\mathbf{G}'|\mathbf{n}' = |\mathbf{G}|\,(1 - d\kappa_a)\,(1 - d\kappa_b)\mathbf{n}\ , \tag{9.17}$$

where κ_a and κ_b are the principal curvatures of the original surface.

The principal curvatures of the offset surface are κ_a' and κ_b' where

$$\kappa_a' = \frac{\kappa_a}{1 - d\kappa_a}\ , \qquad \kappa_b' = \frac{\kappa_b}{1 - d\kappa_b} \tag{9.18}$$

Thus, when $d = \kappa_a^{-1}$ or $d = \kappa_b^{-1}$, the curvature of the parallel surface is infinite, and the surface has a ridge at such points. This occurs when the offset is made on the concave side of the surface. In machining a surface, the cutter radius must obviously not exceed the minimum radius of curvature in the surface, otherwise gouging of the surface will occur. If sections of the offset surface are plotted, the danger points will appear as cusps or even loops, as described by Flutter (1973).

9.3 CUTTER PATHS FOR NUMERICAL CONTROL

The comparatively simple cutter path calculations described in Section 9.2 are inadequate for general n.c. programming for three reasons. Firstly, the user has to choose suitably spaced curves on the offset surface to ensure that the machined surface meets the specified tolerances. Secondly, the use of more complex cutter shapes for practical reasons makes analytical determination of the offsets impracticable. Most seriously, however, the surfaces are not unbounded, a common problem being to machine a surface which is bounded on one or more sides by surfaces which must also be accurately machined.

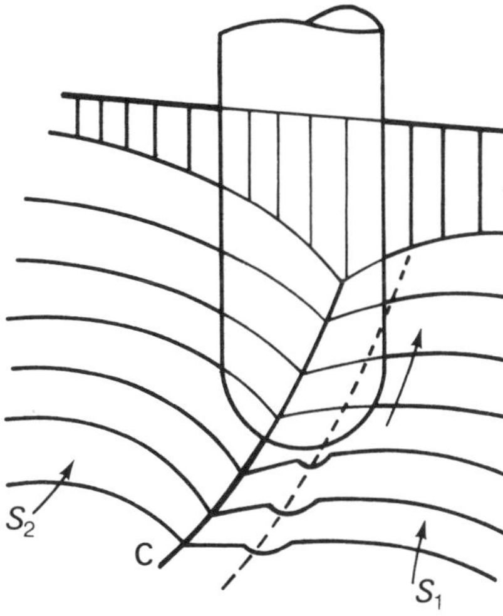

Figure 9.6

Referring to Figure 9.6, we have surface S_1 which is being currently machined, and a surface S_2 which is to be machined later. To cut along the boundary curve C, we must move the cutter in such a way as to maintain the correct machining tolerances with respect to S_1 (known as the **part surface**) and S_2 (the **drive surface**). These tolerances define, in effect, a pair of offset surfaces for each of S_1 and S_2, so that the accuracy of the boundary curve is specified only indirectly (see Figure 9.7).

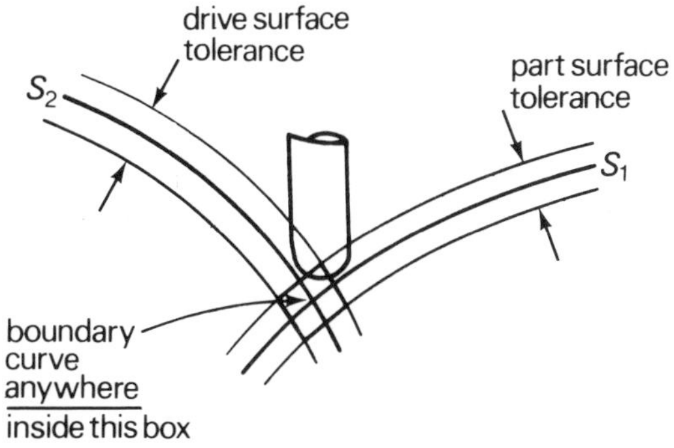

Figure 9.7

Note that the terms part and drive surface are reversed when we machine S_2 with S_1 as the boundary.

In order to machine S_1 completely, a number of cutter paths must be defined whose spacing represents a compromise between excessive cusp height and long machining times.

In APT III (IITRI, 1967), the cutter passes are defined by means of a series of intermediate drive surfaces, terminated by a pass along the boundary. If the surface S_1 is also bounded by a third surface S_3, each cutter pass must terminate with the cutter also within tolerance of surface S_3. This surface is referred to as a **check surface**. The cutter path for each pass is approximated by a series of linear segments. The end points of these segments are each defined by the position of the cutter tip when the tool is within tolerance of the part surface, the current drive surface and one of a sequence of **pseudo check surfaces** whose spacing determines the lengths of the steps taken along the path. The rôle of these pseudo check surfaces is similar to that of the constraint surfaces $g(\mathbf{r}) = 0$ in the intersection problem discussed in Sections 9.1.1 to 9.1.3.

Whereas the choice of the constraint surface there is determined by the accuracy required of the intersection curve, our main concern here is to maintain the required tolerances relative to the part, drive and, possibly, check surfaces. If the accuracy of the intersection curve is important, the tolerances may have to be carefully chosen when the angle between the surface normals is small.

The procedure in the APT III ARELEM routines follows a similar technique to that described earlier for intersection curve generation.

In order to define the cutter path tangent by equation (9.6), we need to define unit normals $\mathbf{n}_1$ and $\mathbf{n}_2$ on the surfaces S_1 and S_2. However, we are not now following the exact intersection curve, and $\mathbf{n}_1$ and $\mathbf{n}_2$ are taken in the direction of the **common normals** P_1N_1 and P_2N_2 shown in Figure 9.8. Thus, for example, P_1N_1 is normal to the part surface S_1 and the tool surface, and P_1N_1 represents the shortest distance between the tool and the part surface for any given position $T(\mathbf{r}_t)$ of the cutter tip.

The points P_1, N_1, P_2 and N_2 are calculated iteratively. Starting at N, where the tool normal has direction $\mathbf{n}$, the intersection P of the tool normal with the surface S_1 is found, and the angle between the tool normal PN and the surface normal QP is calculated. If the angle is less than about 0.01 radian, the points P and N are accepted. If not, N is modified using estimates of the curvature of the surface at P.

For any given surface, APT III requires a subroutine DDST to locate the intersection P for a given point N and direction $\mathbf{n}$, and a routine UNRMAL to calculate the surface normal at P. The routine DDST is discussed further in Section 9.4.

The APT III procedure uses the common normals $\mathbf{n}_1$ and $\mathbf{n}_2$ at each stage to define the 'tangent direction' of the cutter path for any given tool position T.

If T is not within tolerance, this is only an approximation to the cutter path tangent, which is itself ill-defined due to the tolerances allowed for S_1 and S_2.

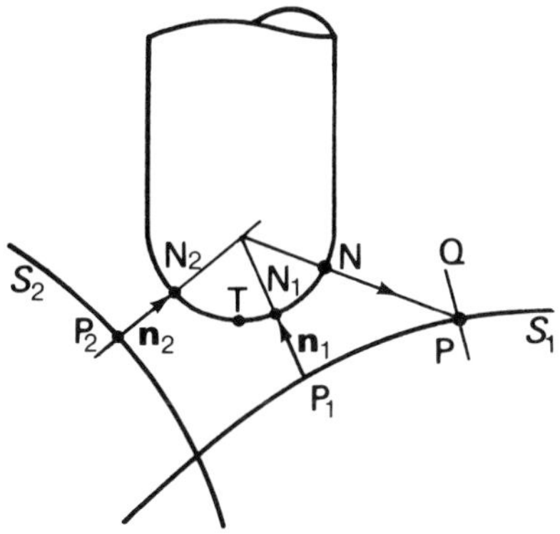

Figure 9.8

Starting with the tool within tolerance of the part and drive surfaces, the following imaginary tool movements are calculated by the computer in order to obtain the next tool point. A step is first taken along the current tangent direction $\mathbf{n}_1 \times \mathbf{n}_2$, the step length being related to the current estimate of curvature by equation (9.5). Since the new tool position $T'(\mathbf{r}_t')$ is probably out of tolerance, an iterative procedure is used to bring it back onto the correct cutter path. At each iteration, the increment $\delta\mathbf{r}_t$ in the cutter position is obtained from the equations.

$$\mathbf{n}_1'.\delta\mathbf{r}_t = s_1 \quad ,$$

$$\mathbf{n}_2'.\delta\mathbf{r}_t = s_2 \quad ,$$

and

$$(\mathbf{n}_1' \times \mathbf{n}_2').\delta\mathbf{r}_t = 0 \quad , \tag{9.19}$$

where $\mathbf{n}_1'$ and $\mathbf{n}_2'$ are common normals at T′, and s_1 and s_2 are the normal distances from the tool to the surfaces S_1 and S_2. The first two equations represent the tangent planes at P_1' and P_2' in Figure 9.9, and the last equation is a plane through T′ whose normal is the 'tangent direction' $\mathbf{n}_1' \times \mathbf{n}_2'$. This last plane is a pseudo check surface whose function was described earlier. The tool is moved into contact with the tangent planes as shown in Figure 9.9, and the process is repeated until the normal distances s_1 and s_2 are within the tolerances specified by the user.

To take maximum advantage of the tolerance allowed on S_1 and S_2 to reduce the number of cutter points, the tool is actually moved into contact with parallel surfaces offset towards the convex sides of the surfaces S_1 and S_2, and the distances s_1 and s_2 are adjusted to bring the tool close to the edge of the tolerance band on this side in each case.

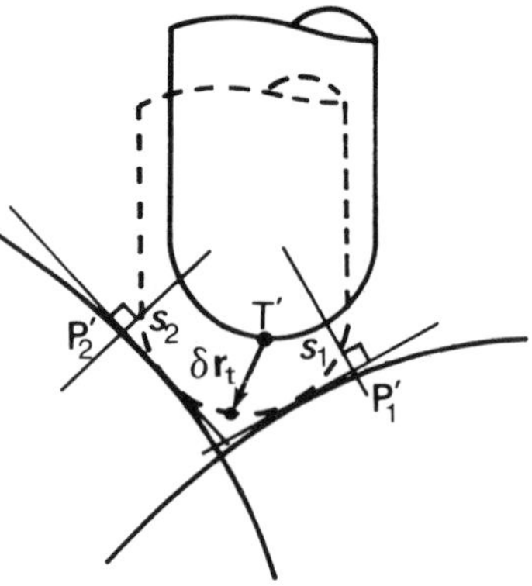

Figure 9.9

The normals $\mathbf{n}_1'$ and $\mathbf{n}_2'$ are updated at each iteration, so that the pseudo check surface is eventually normal to the cutter path at the *end* of the step.

Having found a new point on the cutter path, the *actual* step length is calculated and compared with the tolerance requirements for the part and drive surfaces using equations (9.5) and (9.9). If the step is too long, it is reduced and the process is repeated. Otherwise, the new tool point T* is accepted, and is output to the control tape as the tool point following the point T from which we started.

When the tool is in the vicinity of the boundary check surface the third equation in (9.13) is replaced by

$$\mathbf{n}_3'.\delta\mathbf{r}_t' = s_3 \tag{9.14}$$

where $\mathbf{n}_3'$ and s_3 are defined in the same way as $\mathbf{n}_1'$, $\mathbf{n}_2'$, s_1 and s_2.

Although the APT ARELEM appears clumsy, it must be remembered that the program is designed to deal with a wide variety of cutter shapes and surface definitions.

9.4 INTERSECTION BETWEEN A LINE AND A SURFACE

In the APT ARELEM program, it is necessary to provide a routine for each surface type which locates the nearest intersection between a given line $\mathbf{r} = \mathbf{r}_0 + \lambda\mathbf{n}$ and the surface.

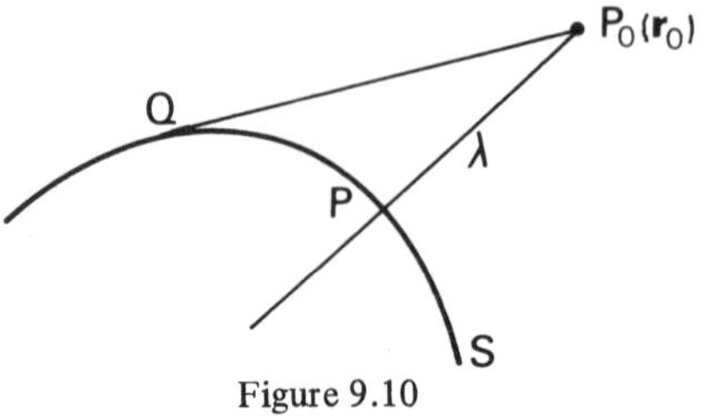

Figure 9.10

If the surface is defined implicitly by the equation $f(\mathbf{r}) = 0$, we may simply substitute for $\mathbf{r}$ and obtain an equation $f(\mathbf{r}_0 + \lambda\mathbf{n}) = F(\lambda) = 0$. Unless the surface is a plane, this is a non-linear equation for λ, and we may use the Newton method to compute the smallest positive root we require for our purpose. The iteration will fail if $\frac{dF}{d\lambda}$ is too small at any step. Since $\frac{dF}{d\lambda} = \frac{d\mathbf{r}}{d\lambda}.\nabla f = \mathbf{n}.\nabla f$, we see that failure occurs when $\mathbf{n}$ is nearly perpendicular to the surface normal ∇f, as shown at point Q in Figure 9.10. The correction $\delta\lambda$ then becomes very large.

In recent versions of APT, parametric surfaces have been added, so that a corresponding method is needed for these. If the surface S has the equation $\mathbf{r} = \mathbf{r}_1(u,v)$, we must solve the three equations

$$F(\lambda,u,v) \equiv \mathbf{r}_0 + \lambda\mathbf{n} - \mathbf{r}_1(u,v) = \mathbf{0} \tag{9.21}$$

for λ, u and v.

The Jacobian of these equations is given by

$$\mathbf{J}^T = \left[\frac{\partial F}{\partial \lambda} \left| \frac{\partial F}{\partial u} \right| \frac{\partial F}{\partial v}\right] = \left[\mathbf{n}|\mathbf{r}_{1u}|\mathbf{r}_{1v}\right] ,$$

so that the Newton method may fail when $|\mathbf{J}| = \mathbf{n}.(\mathbf{r}_{1u} \times \mathbf{r}_{1v})$ is small, again because $\mathbf{n}$ is in fact parallel to S at the intersection.

The equation (9.21) can alternatively be solved by minimising the function $|\mathbf{r}_0 + \lambda\mathbf{n} - \mathbf{r}_1(u,v)|^2$ with respect to λ, u and v in much the same way as described in 9.1.4 for the intersection problem. The minimisation method has the advantage of overcoming the problem which occurs when the line $\mathbf{r} = \mathbf{r}_0 + \lambda\mathbf{n}$ either just touches the surface or has no intersection with it, provided that the full Hessian matrix is used as discussed by Butterfield (1978).

If the parametric surface consists of several patches, the iteration may extend over more than one patch, and it is useful to limit the search by calculating polyhedral bounds on the patches so that those patches whose bounds are not intersected by the line may be rejected without further calculations. Such methods have been discussed by Dimsdale (1977).

Another useful application of the algorithm described in this section is in the projection of a curve onto a surface. For example, in the design of motor cars it is often desired to project feature lines which have been designed as plane curves in side elevation onto the curved surface of the vehicle.

In such cases, a number of points on the plane curve are projected onto the surface using one of the techniques described above. A suitable spline technique may then be used to provide a smooth curve through these points.

9.5 THE DEVELOPMENT OF DEVELOPABLE SURFACES

In the manufacture of developable surfaces such as those used for aircraft wings, it is necessary to calculate the plane developments of these surfaces and to locate fixing points on the plane sheet corresponding to given points on the curved surface.

We will first consider the development of the tangent plane generated surfaces described in Section 4.2.13. The development is based on the fact that the curvature of a curve on the developed surface is equal to the curvature of the projection of the original curve onto the tangent plane, known as the **geodesic curvature** κ_g (see Figure 9.11).

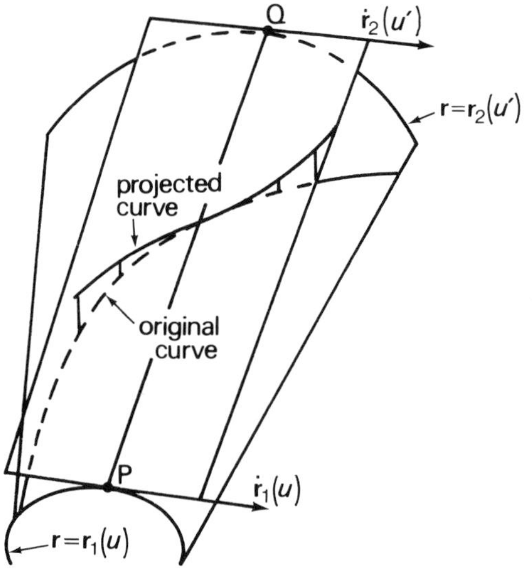

Figure 9.11

The geodesic curvature is shown by Willmore (1959) to be given by

$$\kappa_g = \frac{\mathbf{n}.(\dot{\mathbf{r}} \times \ddot{\mathbf{r}})}{\dot{s}^3} \tag{9.21}$$

where **n** is the unit normal to the surface, and the curve has the equation $\mathbf{r} = \mathbf{r}(u)$.

Let the surface be generated by the tangent planes of the two curves $\mathbf{r} = \mathbf{r}_1(u)$, and $\mathbf{r} = \mathbf{r}_2(u)$. Then for any point P with parameter u on the primary curve $\mathbf{r} = \mathbf{r}_1(u)$ the corresponding generator meets the secondary curve at a point Q with parameter u' where $[\mathbf{r}_1(u) - \mathbf{r}_2(u')] \,.\, \dot{\mathbf{r}}_1(u) \times \dot{\mathbf{r}}_2(u') = 0$, as given in equation (4.44) in Section 4.2.13. The equation of the surface is then given by

$$\mathbf{r} = (1 - v)\mathbf{r}_1(u) + v\mathbf{r}_2(u').$$

If, for example, the curves are parametric cubics, we must solve a quintic equation for u' for any given u. We first solve this equation for u', and compute the direction and length of the generator. We are then able to calculate the surface normal **n**, which enables us to obtain κ_g from equation (9.21). We now know the curvature of the developed curve, and may obtain the curve itself by integrating equations (1.38), rewritten in terms of parameter u.

Since we also know the length and direction of the generator at each point P, we may obtain the development of point Q by noting that the angle between the primary curve tangent and the generator is unchanged during the development.

In a similar manner, we may locate intermediate points along the generators, so that fixtures defined by u and v on the curved surface may be located on the development.

For the convolute $\mathbf{r} = \mathbf{r}_0(u) + v\mathbf{T}_0(u)$, it is easy to show that $\kappa_g = \kappa_0$, the curvature of the original curve. The development of these surfaces is therefore a special case of the procedure described above.

9.6 PIECEWISE LINEAR APPROXIMATION OF A PARAMETRIC CURVE

In Section 9.1.3, we discussed the approximation of a curve $f_1(\mathbf{r}) = f_2(\mathbf{r}) = 0$ by a sequence of linear segments suitable for draughting machines or other machines using linear interpolation between data points. Because the intersection curve could only be evaluated point by point by solving simultaneous equations, we used approximations to the curvature to define a step length L which ensured that the maximum normal distance δ between the curve and the chord did not exceed the specified tolerance (see Figure 9.2).

If, on the other hand, we wish to plot a curve whose parametric equation $\mathbf{r} = \mathbf{r}(u)$ is known, we can use more reliable and straightforward algorithms.

We will calculate here the maximum normal distance δ for the chord joining the points $u = 0$ and $u = 1$. For a polynomial or rational polynomial curve, the results of Section 5.3 can be used to transform any segment $u_0 \leqslant u \leqslant u_1$ into $0 \leqslant u' \leqslant 1$.

Let the chord AB joining the points A ($u = 0$) and B ($u = 1$) be denoted by the vector **c**, as shown in Figure 9.12. Let $\vec{NP} = \mathbf{p}(u)$ represent the perpendicular from the chord AB to the general point P($\mathbf{r}(u)$) of the curve.

Then
$$\mathbf{p} = \mathbf{r}(u) - \mathbf{r}(0) - \lambda\mathbf{c} \ , \tag{9.22}$$

where
$$\mathbf{c}.\mathbf{p} = 0 \ .$$

It follows that $\mathbf{c}.[\mathbf{r}(u) - \mathbf{r}(0)] = \lambda|\mathbf{c}|^2$, so that

$$\mathbf{p} = \mathbf{r}(u) - \mathbf{r}(0) - \frac{[\mathbf{r}(u) - \mathbf{r}(0)].\mathbf{c}}{|\mathbf{c}|^2}\mathbf{c} \ .$$

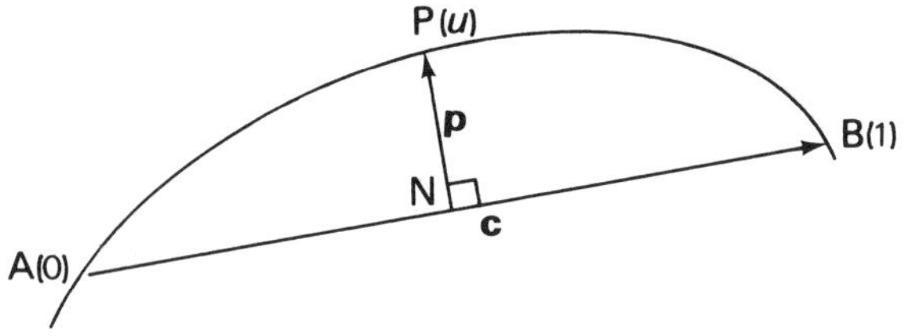

Figure 9.12

In matrix notation,

$$\mathbf{p} = \mathbf{P}[\mathbf{r}(u) - \mathbf{r}(0)] \ , \tag{9.23}$$

where $\mathbf{P}$ is the **projection matrix** $\mathbf{I} - \dfrac{\mathbf{c}\mathbf{c}^T}{\mathbf{c}^T\mathbf{c}}$.

The vector $\mathbf{p}$ is zero at $u = 0$ and $u = 1$. By the remainder theorem, it follows that $\mathbf{p}$ has the factor $u(u - 1)$ for polynomial curves.

For the quadratic Bézier curve,

$$\mathbf{p} = 2u(1 - u)\mathbf{P}(\mathbf{r}_1 - \mathbf{r}_0) \ , \tag{9.24}$$

and for the cubic Bézier curve,

$$\mathbf{p} = 3u(1 - u)\mathbf{P}[(1 - u)(\mathbf{r}_1 - \mathbf{r}_0) + u(\mathbf{r}_2 - \mathbf{r}_0)] \ . \tag{9.25}$$

At the maximum perpendicular, the curve tangent is perpendicular to $\mathbf{p}$, so that in general

$$\dot{\mathbf{r}}^T\mathbf{p} = \dot{\mathbf{r}}^T\mathbf{P}[\mathbf{r}(u) - \mathbf{r}(0)] = 0 \ . \tag{9.26}$$

Although this is a polynomial equation of degree $2n - 1$ for a curve of degree n, the factors $u(u - 1)$ occur as in (9.24) and (9.25), so that we have to solve an equation of degree $2n - 3$ only.

For the quadratic Bézier curve (a parabola), there is just one root, and this always has the value $u = ½$. Since there is only one root, it must correspond to a maximum.

For the cubic Bézier curve, there are three stationary points at the roots of the equation

$$[(1 - u)^2(\mathbf{r}_1 - \mathbf{r}_0) + 2u(1 - u)(\mathbf{r}_2 - \mathbf{r}_1) + u^2(\mathbf{r}_3 - \mathbf{r}_2)]^T\mathbf{P} \times$$
$$\times \ [(1 - u) \times (\mathbf{r}_1 - \mathbf{r}_0) + u(\mathbf{r}_2 - \mathbf{r}_0)] = 0 \ . \tag{9.27}$$

At a true maximum point, we must have $\ddot{\mathbf{r}}^T \mathbf{p} < 0$, so that the condition to be applied at each root of (9.27) is

$$[(1 - u)(\mathbf{r}_2 - 2\mathbf{r}_1 + \mathbf{r}_0) + u(\mathbf{r}_3 - 2\mathbf{r}_2 + \mathbf{r}_1)]^T\mathbf{P} \times$$
$$\times [(1 - u)(\mathbf{r}_1 - \mathbf{r}_0) + u(\mathbf{r}_2 - \mathbf{r}_0)] < 0 \; . \quad (9.28)$$

These results, together with the transformations given in Section 5.3, enable us to calculate the maximum perpendicular distance for any segment $u_0 \leqslant u \leqslant u_1$ of the curve. We may use these in an iterative procedure to compute a sequence of segments of maximum length which are within the tolerance specified. The time required to evaluate the maximum perpendicular distance is a limitation of the method.

Butterfield (1978) has suggested fitting a quadratic spline to a number of points on the plane projections of a curve. As we have seen, a parabolic segment has its maximum perpendicular distance at $u = \frac{1}{2}$ for the segment $0 \leqslant u \leqslant 1$. In general, the maximum occurs at $u = \frac{1}{2}(u_0 + u_1)$ for a subsegment $u_0 \leqslant u \leqslant u_1$. Butterfield generalises this result to deal with the piecewise quadratic spline. This leads to a rapid evaluation of the maximum for general parametric curves.

A quite different and more general approach is that adopted by Cox (1971) in dealing with convex functions of the form $y = f(x)$. Cox uses least-squares minimisation to obtain the best-fitting polygonal arc whose vertices are *not* constrained to lie on the curve. However, a straightforward extension to parametric curve approximation is not possible because there is no direct correspondence between points on the polygon and points on the curve.

Appendix 1

Elementary matrix algebra

A1.1 PRELIMINARY DEFINITIONS

A **matrix** is a rectangular array of elements which obeys the laws of matrix algebra given below. The elements are usually numbers.

A matrix with m rows and n columns is described as an **m×n matrix**. A matrix with one row is called a **row vector** and a matrix with one column a **column vector**.

A matrix is usually denoted by a bold capital letter such as **A**. The element in row r and column s of matrix **A** is denoted by a_{rs}, the elements usually being named by the small letter corresponding to the matrix name. If the matrix is a row or column vector, a single suffix is used to describe its elements.

Thus the matrix $\mathbf{B} = \begin{bmatrix} 3 & -1 & 4 \\ 2 & 1 & 5 \end{bmatrix}$ is a 2×3 matrix with, for example, $b_{13} = 4$. The matrix $\mathbf{A} = [2 \quad 1 \quad -1]$ is a 1×3 matrix (or row vector) with $a_1 = 2$, $a_2 = 1, a_3 = -1$.

A1.2 THE LAWS OF MATRIX ALGEBRA

Two $m\times n$ matrices **A** and **B** are **equal** if all corresponding elements are equal. Thus **A** = **B** if $a_{rs} = b_{rs}$ for any r and s in the range $1 \leqslant r \leqslant m$, $1 \leqslant s \leqslant n$. There can be no equality between matrices of different sizes.

The **sum** of two $m\times n$ matrices **A** and **B** is the $m\times n$ matrix formed by the sums of corresponding elements of **A** and **B**. Thus **C** = **A** + **B** if $c_{rs} = a_{rs} + b_{rs}$ for any r and s in the range $1 \leqslant r \leqslant m$, $1 \leqslant s \leqslant n$. For example, if

$$\mathbf{A} = \begin{bmatrix} 2 & 1 & -1 \\ 1 & 3 & 2 \end{bmatrix} \text{ and } \mathbf{B} = \begin{bmatrix} 3 & -1 & 4 \\ -2 & 1 & -5 \end{bmatrix}, \text{ then } \mathbf{A} + \mathbf{B} = \begin{bmatrix} 5 & 0 & 3 \\ -1 & 4 & 3 \end{bmatrix}.$$

Matrices of different sizes may *not* be added.

For any $m\times n$ matrix **A**, we may define a corresponding matrix '–**A**' whose typical element is $-a_{rs}$. Then **A** + (–**A**) = **0**, where **0** is an $m\times n$ **null matrix** whose elements are all zero.

The **difference** of two $m\times n$ matrices **A** and **B** is defined by the equation

$\mathbf{A}-\mathbf{B}=\mathbf{A}+(-\mathbf{B})$. For example, if $\mathbf{A}=\begin{bmatrix}2 & 1 & 3\\ -1 & 4 & 5\end{bmatrix}$ and $\mathbf{B}=\begin{bmatrix}-1 & -3 & 4\\ 4 & 2 & 3\end{bmatrix}$,
then $\mathbf{A}-\mathbf{B}=\begin{bmatrix}3 & 4 & -1\\ -5 & 2 & 2\end{bmatrix}$.

The matrix obtained by multiplying every element of the $m \times n$ matrix $\mathbf{A}$ by the same scalar (number) λ is denoted by $\lambda\mathbf{A}$. Thus if $\mathbf{B}=\lambda\mathbf{A}$, then $b_{rs}=\lambda a_{rs}$ for any r and s in the range $1 \leqslant r \leqslant m$ and $1 \leqslant s \leqslant n$. This operation is known as **scalar multiplication.**

From these laws, the following properties of matrix addition and scalar multiplication are easily proved:

i) $\mathbf{A}+\mathbf{B}=\mathbf{B}+\mathbf{A}$,

ii) $\mathbf{A}+(\mathbf{B}+\mathbf{C})=(\mathbf{A}+\mathbf{B})+\mathbf{C}$,

iii) $\lambda(\mathbf{A}+\mathbf{B})=\lambda\mathbf{A}+\lambda\mathbf{B}$,

iv) $(\lambda+\mu)\mathbf{A}=\lambda\mathbf{A}+\mu\mathbf{A}$,

v) $\lambda(\mu\mathbf{A})=(\lambda\mu)\mathbf{A}=\mu(\lambda\mathbf{A})$,

provided that $\mathbf{A}$, $\mathbf{B}$ and $\mathbf{C}$ are matrices of the same size, and λ and μ are scalars.

A1.3 THE PRODUCT OF TWO MATRICES

In defining the product of two matrices, we bear in mind the applications for which matrix algebra is intended and in particular the study of simultaneous linear equations and linear transformations.

If x_1,x_2 are related to y_1,y_2,y_3, which in turn are related to z_1,z_2,z_3,z_4 by the equations

$$\begin{aligned} x_1 &= a_{11}y_1 + a_{12}y_2 + a_{13}y_3 \ , \\ x_2 &= a_{21}y_1 + a_{22}y_2 + a_{23}y_3 \ , \end{aligned} \tag{A1.1}$$

and

$$\begin{aligned} y_1 &= b_{11}z_1 + b_{12}z_2 + b_{13}z_3 + b_{14}z_4 \ , \\ y_2 &= b_{21}z_1 + b_{22}z_2 + b_{23}z_3 + b_{24}z_4 \ , \\ y_3 &= b_{31}z_1 + b_{32}z_2 + b_{33}z_3 + b_{34}z_4 \ , \end{aligned} \tag{A1.2}$$

then we may substitute (A1.2) into (A1.1) to obtain the direct relationships between x_1, x_2 and z_1, z_2, z_3, z_4.

Thus $x_1 = (a_{11}b_{11} + a_{12}b_{21} + a_{13}b_{31})z_1 + (a_{11}b_{12} + a_{12}b_{22} + a_{13}b_{32})z_2$

$+ (a_{11}b_{13} + a_{12}b_{23} + a_{13}b_{33})z_3 + (a_{11}b_{14} + a_{12}b_{24} + a_{13}b_{34})z_4,$

and $x_2 = (a_{21}b_{11} + a_{22}b_{21} + a_{23}b_{31})z_1 + (a_{21}b_{12} + a_{22}b_{22} + a_{23}b_{32})z_2$

$+ (a_{21}b_{13} + a_{22}b_{23} + a_{23}b_{33})z_3 + (a_{21}b_{14} + a_{22}b_{24} + a_{23}b_{34})z_4.$

(A1.3)

If we express this transformation in the same form as (A1.1) and (A1.2) by writing

$$x_1 = c_{11}z_1 + c_{12}z_2 + c_{13}z_3 + c_{14}z_4$$

(A1.4)

and

$$x_2 = c_{21}z_1 + c_{22}z_2 + c_{23}z_3 + c_{24}z_4,$$

then the coefficients c_{rs} can be obtained directly from (A1.3).

Thus, for example, $c_{13} = a_{11}b_{13} + a_{12}b_{23} + a_{13}b_{33}$.

If we consider the matrices $\mathbf{A} = \begin{bmatrix} a_{11} & a_{12} & a_{13} \\ a_{21} & a_{22} & a_{23} \end{bmatrix}$, $\mathbf{B} = \begin{bmatrix} b_{11} & b_{12} & b_{13} & b_{14} \\ b_{21} & b_{22} & b_{23} & b_{24} \\ b_{31} & b_{32} & b_{33} & b_{34} \end{bmatrix}$ and $\mathbf{C} = \begin{bmatrix} c_{11} & c_{12} & c_{13} & c_{14} \\ c_{21} & c_{22} & c_{23} & c_{24} \end{bmatrix}$ of the coefficients describing these transformations, we see that c_{13} is obtained by multiplying corresponding elements of row 1 of **A** and column 3 of **B** and summing their products. In general, element $c_{rs} = a_{r1}b_{1s} + a_{r2}b_{2s} + a_{r3}b_{3s} = \sum_{t=1}^{3} a_{rt}b_{ts}$. The matrix **C** is said to be the **matrix product** of **A** and **B**, denoted by **AB**. Thus the result of two successive transformations of the form (A1.1) and (A1.2) is obtained by multiplying the corresponding matrices **A** and **B**. Note that we took *rows* of **A** and *columns* of **B** in forming the product **AB**. The *order* of the matrices is important, as we shall see in Section A1.4.

In general, the product **AB** of an $m \times n$ matrix **A** and an $n \times p$ matrix **B** is an $m \times p$ matrix **C** whose elements are given by

$$c_{rs} = \sum_{t=1}^{n} a_{rt}b_{ts}. \tag{A1.5}$$

For example, if $\mathbf{A} = \begin{bmatrix} 1 & -1 \\ -3 & 2 \end{bmatrix}$ and $\mathbf{B} = \begin{bmatrix} 2 & 1 & 7 \\ 3 & 2 & 4 \end{bmatrix}$, then

$$\mathbf{C} = \begin{bmatrix} 1\times2 + (-1)\times3 & 1\times1 + (-1)\times2 & 1\times7 + (-1)\times4 \\ (-3)\times2 + 2\times3 & (-3)\times1 + 2\times2 & (-3)\times7 + 2\times4 \end{bmatrix}$$

$$= \begin{bmatrix} -1 & -1 & 3 \\ 0 & 1 & -13 \end{bmatrix}.$$

Note that the number of columns of **A** must equal the number of rows of **B**, otherwise the product is not defined.

If we denote $\begin{bmatrix} x_1 \\ x_2 \end{bmatrix}$, $\begin{bmatrix} y_1 \\ y_2 \\ y_3 \end{bmatrix}$ and $\begin{bmatrix} z_1 \\ z_2 \\ z_3 \\ z_4 \end{bmatrix}$ by **X**, **Y** and **Z**, then the transformations (A1.1) and (A1.2) may be written more concisely as $\mathbf{X} = \mathbf{AY}$ and $\mathbf{Y} = \mathbf{BZ}$ using the definition (A1.5) for the products. Remember that, for example, x_1 is a short-hand for the element x_{11} of **X**, since **X** has only one column.

Then the combination of the two transformations is given by

$$\mathbf{X} = \mathbf{AY} = \mathbf{A}(\mathbf{BZ}) = (\mathbf{AB})\mathbf{Z} = \mathbf{CZ}.$$

A1.4 PROPERTIES OF MATRIX PRODUCTS

Most properties of matrix products are similar to those of ordinary numbers. Thus, provided that the sizes of the matrices **A**, **B** and **C** are suitable for multiplication, we have the results

i) $(\mathbf{AB})\mathbf{C} = \mathbf{A}(\mathbf{BC})$,

ii) $\mathbf{A}(\mathbf{B} + \mathbf{C}) = \mathbf{AB} + \mathbf{AC}$,

iii) $(\mathbf{A} + \mathbf{B})\mathbf{C} = \mathbf{AC} + \mathbf{BC}$,

iv) $\mathbf{A}(\lambda\mathbf{B}) = \lambda(\mathbf{AB}) = (\lambda\mathbf{A})\mathbf{B}$.

Further important properties can be derived by defining the **unit** or **identity** matrix $\mathbf{I}_m$. This is an $m \times m$ (square) matrix whose elements i_{rs} satisfy the equations

$$i_{rs} = \begin{cases} 1 & \text{when } r = s \\ 0 & \text{when } r \neq s\,. \end{cases}$$

Thus

$$\mathbf{I}_3 = \begin{bmatrix} 1 & 0 & 0 \\ 0 & 1 & 0 \\ 0 & 0 & 1 \end{bmatrix},$$

and we see that the elements of the **principal diagonal** (top left to bottom right) are unity, and all other elements are zero.

Now if **A** is an $m \times n$ matrix, we have the further properties

v) $\mathbf{I}_m \mathbf{A} = \mathbf{A}$,

vi) $\mathbf{A}\mathbf{I}_n = \mathbf{A}$.

The sizes of unit and null matrices are often apparent from their context and they are then written simply as **I** and **0** respectively. Finally, if **A** is an $m \times n$ matrix and **0** an $n \times p$ null matrix, then

vii) $\mathbf{A0} = \mathbf{0}$.

All these properties agree with those of ordinary numbers if we regard **I** as the analogue of unity and **0** as the analogue of zero. However, it is important to bear in mind the following properties which do not show the same agreement.

First, the products **AB** and **BA** may not be equal. For example:

i) $$\begin{bmatrix} 1 & 1 \\ 3 & 2 \end{bmatrix} \begin{bmatrix} 3 & 1 & 2 \\ 4 & 1 & 5 \end{bmatrix} = \begin{bmatrix} 7 & 2 & 7 \\ 17 & 5 & 16 \end{bmatrix},$$

but $\begin{bmatrix} 3 & 1 \\ 4 & 1 \end{bmatrix} \begin{bmatrix} 2 & 1 & 1 \\ 5 & 3 & 2 \end{bmatrix}$ is not defined.

ii) $$\begin{bmatrix} 2 & 1 & 0 \\ 0 & 1 & 2 \end{bmatrix} \begin{bmatrix} 4 & 1 \\ -1 & 2 \\ 0 & -3 \end{bmatrix} = \begin{bmatrix} 7 & 4 \\ -1 & -4 \end{bmatrix},$$

whereas $$\begin{bmatrix} 4 & 1 \\ -1 & 2 \\ 0 & -3 \end{bmatrix} \begin{bmatrix} 2 & 1 & 0 \\ 0 & 1 & 2 \end{bmatrix} = \begin{bmatrix} 8 & 5 & 2 \\ -2 & 1 & 4 \\ 0 & -3 & -6 \end{bmatrix}.$$

iii) $$\begin{bmatrix} 1 & 1 \\ 2 & 1 \end{bmatrix} \begin{bmatrix} 1 & 2 \\ 3 & 1 \end{bmatrix} = \begin{bmatrix} 4 & 3 \\ 5 & 5 \end{bmatrix},$$

whilst $$\begin{bmatrix} 1 & 2 \\ 3 & 1 \end{bmatrix} \begin{bmatrix} 1 & 1 \\ 2 & 1 \end{bmatrix} = \begin{bmatrix} 5 & 3 \\ 5 & 4 \end{bmatrix}.$$

If $\mathbf{AB} = \mathbf{BA}$ for two particular $n \times n$ matrices $\mathbf{A}$ and $\mathbf{B}$, then $\mathbf{A}$ and $\mathbf{B}$ are said to **commute.**

It can be seen that care must be taken with the order of matrix products. In particular, when forming a compound transformation of variables of the kind described in Section A1.3, it is important to multiply the matrices in the correct order.

The other unusual property of matrix multiplication is that we may have $\mathbf{AB} = \mathbf{0}$ (the null matrix) even though both $\mathbf{A}$ and $\mathbf{B}$ are non-null matrices. Thus, for example,

$$\begin{bmatrix} 1 & 2 \\ 2 & 4 \end{bmatrix} \begin{bmatrix} 2 & -2 \\ -1 & 1 \end{bmatrix} = \begin{bmatrix} 0 & 0 \\ 0 & 0 \end{bmatrix}.$$

The consequence is that if $\mathbf{AB} = \mathbf{AC}$ and $\mathbf{A} \neq \mathbf{0}$, we cannot conclude that $\mathbf{B} = \mathbf{C}$. For although $\mathbf{A}(\mathbf{B} - \mathbf{C}) = \mathbf{0}$ and $\mathbf{A} \neq \mathbf{0}$, we can still have $\mathbf{B} - \mathbf{C} \neq \mathbf{0}$.

A1.5 NON-SINGULAR MATRICES

For a given $n \times n$ matrix $\mathbf{A}$, it may be possible to find an **inverse matrix** $\mathbf{A}^{-1}$ for which $\mathbf{AA}^{-1} = \mathbf{I}_n$ and $\mathbf{A}^{-1}\mathbf{A} = \mathbf{I}_n$. The matrix $\mathbf{A}$ is then said to be **non-singular**.

If $\mathbf{A}$ is non-singular, then the equation $\mathbf{AB} = \mathbf{AC}$ may be multiplied at the front (pre-multiplied) by $\mathbf{A}^{-1}$ to give $\mathbf{A}^{-1}\mathbf{AB} = \mathbf{A}^{-1}\mathbf{AC}$, whence $\mathbf{IB} = \mathbf{IC}$, and finally $\mathbf{B} = \mathbf{C}$. Thus we may effectively cancel $\mathbf{A}$ in this case because it is non-singular.

However, we have seen that it is possible to find non-zero matrices $\mathbf{A}$ and $\mathbf{B}$ such that $\mathbf{AB} = \mathbf{0}$. But $\mathbf{A0} = \mathbf{0}$, so that $\mathbf{AB} = \mathbf{A0}$. If $\mathbf{A}$ were non-singular, this would imply, by cancellation, that $\mathbf{B} = \mathbf{0}$. The example in Section A1.6 shows that this is not necessarily true. Thus the matrix $\mathbf{A} = \begin{bmatrix} 1 & 2 \\ 2 & 4 \end{bmatrix}$ in the example is a **singular** matrix.

When the variables $\mathbf{X} = \begin{bmatrix} x_1 \\ x_2 \\ x_3 \end{bmatrix}$ and $\mathbf{Y} = \begin{bmatrix} y_1 \\ y_2 \\ y_3 \end{bmatrix}$ are related by the transformation $\mathbf{AX} = \mathbf{Y}$, and $\mathbf{A}$ is a non-singular matrix, then $\mathbf{A}^{-1}\mathbf{AX} = \mathbf{A}^{-1}\mathbf{Y}$ and hence $\mathbf{X} = \mathbf{A}^{-1}\mathbf{Y}$. Thus the transformation can be reversed by use of the inverse matrix $\mathbf{A}^{-1}$.

The methods of evaluating inverse matrices can be found in most books on Linear Algebra, and also in modern general mathematics texts such as Goult *et al.* (1973).

A1.6 THE TRANSPOSE OF A MATRIX

For any $m \times n$ matrix **A**, the **transpose** of **A**, denoted by $\mathbf{A}^T$, is an $n \times m$ matrix obtained from **A** by interchanging the rows and columns. More precisely, if α_{rs} is the element in row r and column s of $\mathbf{A}^T$ then

$$\alpha_{rs} = a_{sr} \text{ for } 1 \leqslant r \leqslant n, \quad 1 \leqslant s \leqslant m.$$

A square matrix **A** such that $\mathbf{A}^T = \mathbf{A}$ is said to be **symmetric**.

A square matrix **B** such that $\mathbf{B}^T = -\mathbf{B}$ is said to be **skew-symmetric**.

Examples

$$\text{Let } \mathbf{A} = \begin{bmatrix} 5 & 1 & 2 \\ 2 & 3 & 4 \end{bmatrix}, \text{ then } \mathbf{A}^T = \begin{bmatrix} 5 & 2 \\ 1 & 3 \\ 2 & 4 \end{bmatrix}.$$

$$\text{The matrix } \mathbf{B} = \begin{bmatrix} 1 & 2 & -1 \\ 2 & 2 & 5 \\ -1 & 5 & 4 \end{bmatrix} \text{ is symmetric,}$$

$$\text{whereas} \quad \mathbf{C} = \begin{bmatrix} 0 & 1 & -2 \\ -1 & 0 & 3 \\ 2 & -3 & 0 \end{bmatrix} \text{ is skew-symmetric.}$$

The transpose matrices of sums and products obey the following rules:

(i) $$(\mathbf{A} + \mathbf{B})^T = \mathbf{A}^T + \mathbf{B}^T$$

(ii) $$(\mathbf{AB})^T = \mathbf{B}^T\mathbf{A}^T .$$

A1.7 ORTHOGONAL MATRICES

A matrix **P** is said to be orthogonal if $\mathbf{PP}^T = \mathbf{I}$, so that its inverse is identical with its transpose. For example, it is easily verified that the matrix $\begin{bmatrix} \cos\theta & \sin\theta & 0 \\ -\sin\theta & \cos\theta & 0 \\ 0 & 0 & 1 \end{bmatrix}$ is orthogonal.

A1.8 MATRIX REPRESENTATION OF THE SCALAR AND VECTOR PRODUCTS

In Section 2.4 we have defined the scalar and vector products which are useful in solving geometrical problems. In applying these to the derivation of co-ordinate transformations we need to relate the dot and cross product notation to the matrix representation used in Chapter 3.

If we denote all our vectors by 3X1 matrices so that $\mathbf{a} = [a_1\ a_2\ a_3]^T$, then we see by our definition of matrix products that the scalar product $\mathbf{a}.\mathbf{b} = a_1b_1 + a_2b_2 + a_3b_3$ is simply

$$[a_1 \quad a_2 \quad a_3] \begin{bmatrix} b_1 \\ b_2 \\ b_3 \end{bmatrix}.$$

Then $\mathbf{a}^T\mathbf{b}$ is equivalent to the scalar product $\mathbf{a}.\mathbf{b}$.

Note that we are not using capitals even though we are treating the vectors as matrices. Thus we are using the vector notation $\mathbf{a}$ to denote a 3X1 matrix or column 3-vector and the notation $\mathbf{a}^T$ to denote a row 3-vector.

The expressions $(\mathbf{a}.\mathbf{b})\mathbf{c}$ which arise in triple vector products can be written as $(\mathbf{a}^T\mathbf{b})\mathbf{c}$, and since $\mathbf{a}^T\mathbf{b}$ is a scalar, this can be written as $\mathbf{c}(\mathbf{a}^T\mathbf{b})$ or $(\mathbf{c}\mathbf{a}^T)\mathbf{b}$. Here $\mathbf{c}\mathbf{a}^T$ is a 3X3 matrix. In detail, we have

$$(\mathbf{a}^T\mathbf{b})\mathbf{c} = (a_1b_1 + a_2b_2 + a_3b_3) \begin{bmatrix} c_1 \\ c_2 \\ c_3 \end{bmatrix} = \begin{bmatrix} (a_1b_1 + a_2b_2 + a_3b_3)c_1 \\ (a_1b_1 + a_2b_2 + a_3b_3)c_2 \\ (a_1b_1 + a_2b_2 + a_3b_3)c_3 \end{bmatrix},$$

whereas

$$(\mathbf{c}\mathbf{a}^T)\mathbf{b} = \begin{bmatrix} c_1 \\ c_2 \\ c_3 \end{bmatrix} [a_1 \quad a_2 \quad a_3] \begin{bmatrix} b_1 \\ b_2 \\ b_3 \end{bmatrix} = \begin{bmatrix} c_1a_1 & c_1a_2 & c_1a_3 \\ c_2a_1 & c_2a_2 & c_2a_3 \\ c_3a_1 & c_3a_2 & c_3a_3 \end{bmatrix} \begin{bmatrix} b_1 \\ b_2 \\ b_3 \end{bmatrix},$$

$$= \begin{bmatrix} c_1a_1b_1 + c_1a_2b_2 + c_1a_3b_3 \\ c_2a_1b_1 + c_2a_2b_2 + c_2a_3b_3 \\ c_3a_1b_1 + c_3a_2b_2 + c_3a_3b_3 \end{bmatrix} = (\mathbf{a}^T\mathbf{b})\mathbf{c}\ .$$

The vector product $\mathbf{a} \times \mathbf{b}$ is a vector with components

$$\begin{bmatrix} a_2b_3 - a_3b_2 \\ a_3b_1 - a_1b_3 \\ a_1b_2 - a_2b_1 \end{bmatrix};$$

its equivalent in matrix terms is

$$\begin{bmatrix} 0 & -a_3 & a_2 \\ a_3 & 0 & -a_1 \\ -a_2 & a_1 & 0 \end{bmatrix} \begin{bmatrix} b_1 \\ b_2 \\ b_3 \end{bmatrix} = \mathbf{A}\,\mathbf{b}.$$

Although this converts **a** into the 3X3 matrix **A** and **b** into a vector, we shall see that this distinction between the treatment of **a** and **b** is justified by the applications.

A1.9 PARTITIONED MATRICES

It is often convenient to subdivide matrices into smaller blocks which are given their own names. For example, the matrix $\mathbf{A} = \begin{bmatrix} 2 & 1 & 4 \\ -2 & 5 & 7 \end{bmatrix}$ may be divided into $\mathbf{B} = \begin{bmatrix} 2 & 1 \\ -2 & 5 \end{bmatrix}$ and $\mathbf{C} = \begin{bmatrix} 4 \\ 7 \end{bmatrix}$, and the subdivision expressed by the notation $\mathbf{A} = [\mathbf{B} \mid \mathbf{C}]$.

In a similar way, $\mathbf{A} = \left[\begin{array}{c|c} \mathbf{A}_1 & \mathbf{A}_2 \\ \hline \mathbf{A}_3 & \mathbf{A}_4 \end{array}\right]$ is the matrix formed by stacking the matrices $\mathbf{A}_1$, $\mathbf{A}_2$, $\mathbf{A}_3$ and $\mathbf{A}_4$ together in the manner shown. The sizes of the matrices must be matched, so that, for example, the number of rows in $\mathbf{A}_1$ and $\mathbf{A}_2$ must be the same.

Examples

(i) The matrix **M** formed by stacking the column vectors **a**, **b** and **c** side by side is written as

$$\mathbf{M} = [\mathbf{a} \mid \mathbf{b} \mid \mathbf{c}] \;.$$

The matrix $\mathbf{M}^T$ may then be written as $\mathbf{M}^T = \left[\begin{array}{c} \mathbf{a}^T \\ \hline \mathbf{b}^T \\ \hline \mathbf{c}^T \end{array}\right]$.

(ii) If $\mathbf{A} = \begin{bmatrix} 1 & 2 \\ 3 & 7 \end{bmatrix}$ and $\mathbf{B} = \begin{bmatrix} -3 & 4 \\ 1 & -2 \end{bmatrix}$, then the matrix $\mathbf{M} = \left[\begin{array}{c|c} \mathbf{A} & \mathbf{I} \\ \hline \mathbf{O} & \mathbf{B} \end{array}\right]$ denotes

$$\left[\begin{array}{cc|cc} 1 & 2 & 1 & 0 \\ 3 & 7 & 0 & 1 \\ \hline 0 & 0 & -3 & 4 \\ 0 & 0 & 1 & -2 \end{array}\right],$$ since **I** is the identity matrix, and **0** a null matrix.

Appendix 2
Determinants

DEFINITIONS

The 2X2 **determinant** Δ of the square matrix $\begin{bmatrix} a_{11} & a_{12} \\ a_{21} & a_{22} \end{bmatrix}$ is written $\begin{vmatrix} a_{11} & a_{12} \\ a_{21} & a_{22} \end{vmatrix}$ and is defined by the equation

$$\Delta = \begin{vmatrix} a_{11} & a_{12} \\ a_{21} & a_{22} \end{vmatrix} = a_{11}a_{22} - a_{12}a_{21} \ . \qquad \text{(A2.1)}$$

The 3X3 **determinant** Δ of the square matrix $\begin{bmatrix} a_{11} & a_{12} & a_{13} \\ a_{21} & a_{22} & a_{23} \\ a_{31} & a_{32} & a_{33} \end{bmatrix}$ is written $\begin{vmatrix} a_{11} & a_{12} & a_{13} \\ a_{21} & a_{22} & a_{23} \\ a_{31} & a_{32} & a_{33} \end{vmatrix}$ and is defined by

$$\Delta = \begin{vmatrix} a_{11} & a_{12} & a_{13} \\ a_{21} & a_{22} & a_{23} \\ a_{31} & a_{32} & a_{33} \end{vmatrix} = a_{11} \begin{vmatrix} a_{22} & a_{23} \\ a_{32} & a_{33} \end{vmatrix} - a_{12} \begin{vmatrix} a_{21} & a_{23} \\ a_{31} & a_{33} \end{vmatrix} + a_{13} \begin{vmatrix} a_{21} & a_{22} \\ a_{31} & a_{32} \end{vmatrix} . \qquad \text{(A2.2)}$$

The coefficients of a_{11}, a_{12} and a_{13} in (A2.2) are obtained as follows. For each element of the determinant we may define a **minor** which is obtained by striking out the row and column in which the element is found. Thus, for example, the minor of a_{21} is $\begin{vmatrix} a_{12} & a_{13} \\ a_{32} & a_{33} \end{vmatrix}$. There is also a sign associated with each element. For element a_{ij}, the sign is $(-1)^{i+j}$, so that the signs appear as an alternating patchwork $\begin{vmatrix} + & - & + \\ - & + & - \\ + & - & + \end{vmatrix}$. The product of the minor and the sign gives the **co-factor** of the element a_{ij}, denoted by A_{ij}. Then equation (A2.2) may be written briefly as

$$\Delta = a_{11}A_{11} + a_{12}A_{12} + a_{13}A_{13} \ . \qquad \text{(A2.3)}$$

The determinant may, in fact, be expressed in terms of any row and column in the same way. Thus, for example, $\Delta = a_{12}A_{12} + a_{22}A_{22} + a_{32}A_{32}$.

The following properties of these determinants may be simply verified.

(i) The determinant of a matrix is equal to the determinant of its transpose. Denoting the determinant of a matrix **A** by $|\mathbf{A}|$, we have $|\mathbf{A}| = |\mathbf{A}^T|$.

(ii) If two rows or columns of **A** are interchanged, the sign of the determinant is changed. Thus if the modified matrix is called **B**, we have $|\mathbf{B}| = -|\mathbf{A}|$.

(iii) If matrix **B** is obtained by multiplying one row or column of **A** by a constant k, then $|\mathbf{B}| = k|\mathbf{A}|$.

(iv) If two rows or columns of **A** are identical, then $|\mathbf{A}| = 0$.

(v) If **B** is obtained from **A** by adding a multiple of one row or column of **A** to another, then $|\mathbf{B}| = |\mathbf{A}|$.

(vi) If **A** and **B** are both $n \times n$ matrices, then the determinant of the product **AB** is given by $|\mathbf{AB}| = |\mathbf{A}|\,|\mathbf{B}|$.

(vii) The elements of the inverse **C** of a matrix **A** can be shown to be given by $\mathbf{c}_{rs} = \mathbf{A}_{sr}/|\mathbf{A}|$ (see Goult *et al*, 1973, p.321). Then the inverse does not exist if $|\mathbf{A}| = 0$, in which case the matrix **A** is singular.

Determinants of higher orders can be defined in a similar way, but those just defined will be sufficient for our purpose.

It is also possible to permit one row or column of a determinant to consist of vectors, it being understood that all multiplications in the evaluation are scalar multiplications when vectors are involved.

Example

$$\begin{vmatrix} \mathbf{i} & \mathbf{j} & \mathbf{k} \\ a_x & a_y & a_z \\ b_x & b_y & b_z \end{vmatrix} = (a_y b_z - a_z b_y)\mathbf{i} - (a_x b_z - a_z b_x)\mathbf{j} + (a_x b_y - a_y b_x)\mathbf{k} \ .$$

Appendix 3

Important properties of polynomials

We here review some of the basic features of polynomials, which are the functions underlying many of the methods described earlier.

The following result is fundamental:

THEOREM: A polynomial of degree n,

$$p(x) = a_n x^n + a_{n-1}x^{n-1} + \ldots . + a_1 x + a_0 \ ,$$

can be factorised uniquely in the form

$$p(x) = a_n(x - \alpha_1)(x - \alpha_2) \ldots . (x - \alpha_n) \ ,$$

as the product of precisely n linear factors.

The proof of this theorem depends upon what is known as the Fundamental Theorem of Algebra, which has no known purely algebraic proof. Details will be found in any textbook on functions of a complex variable.

In all applications in this book, the coefficients $a_n, a_{n-1}, \ldots, a_0$ of a polynomial are real numbers, in which case we speak of a **real polynomial**.

The numbers $\alpha_1, \alpha_2, \ldots, \alpha_n$ in the Theorem are clearly values of x for which $p(x)$ has the value zero; they are called the **zeros** of $p(x)$ or the **roots** or **solutions** of the polynomial equation $p(x) = 0$. The zeros of a real polynomial have the following properties:

(i) multiple or repeated zeros may occur (for example, if $\alpha_1 = \alpha_2$),

(ii) zeros may be either real or complex,

(iii) complex zeros occur in conjugate pairs (if $\alpha = u + iv$ is a zero, so is $\tilde{\alpha} = u - iv$),

(iv) if the degree of $p(x)$ is odd, there is at least one real zero. This follows from property (iii), since the number of complex zeros, if any, must be even, while the total number of zeros is odd.

(v) If $p(x)$ is divided by $(x - \alpha)$ the remainder is $p(\alpha)$ (this is the **remainder theorem**). It follows that if α is a zero of $p(x)$, so that $p(\alpha) = 0$, then $p(x)$ is exactly divisible by $(x - \alpha)$.

Examples

(i) $$x^2 - 2x + 1 = (x - 1)(x - 1) ;$$

$x = 1$ is a zero, repeated twice (or with **multiplicity** 2).

(ii), (iii) $$x^3 - x^2 + 3x + 5 = (x + 1)(x - 1 - 2i)(x - 1 + 2i) ;$$

$x = -1$ is a zero, as are the complex conjugates $1 \pm 2i$.

These properties of polynomials are discussed at greater length in, for instance, Goult *et al* (1973).

A polynomial $p(x)$ has many features which commend it to the mathematician. It is everywhere continuous and differentiable, and can easily be evaluated for any value of x (unlike the function sin x, for example, whose evaluation requires either a set of tables or an approximating formula). The derivative of a polynomial is another polynomial, with the same virtues, and so is its indefinite integral. All derivatives of order greater than n, where n is the degree of $p(x)$, are zero.

Many practical problems involve functions lacking some or all of these desirable properties, and it is often expedient to approximate these using simpler and more well-behaved functions. Polynomials are widely used as approximating functions in this context, particularly when the function to be approximated is single-valued and continuous, when the following result holds:

THEOREM (Weierstrass):

Let $f(x)$ be a function which is single-valued and continuous on the interval $0 \leqslant x \leqslant 1$. Then for any positive number ϵ, *however small*, there exists a polynomial $p(x)$ such that

$$|f(x) - p(x)| < \epsilon$$

throughout $0 \leqslant x \leqslant 1$.

The Theorem may be extended to any other finite interval $a \leqslant \xi \leqslant b$ if we make the change of variable $\xi = (b - a)x + a$. In other words, an arbitrary continuous curve of finite length may be approximated to any desired degree of accuracy by a polynomial. As might be expected, a higher degree of accuracy (that is, a smaller choice of ϵ in the Theorem) will demand a higher degree for the polynomial $p(x)$. Note that the Theorem permits $f(x)$ to exhibit all manner of kinks and spikes, provided only that it is continuous.

One proof of Weierstrass' Theorem (given in Ralston, 1965) employs a set of polynomials defined by

$$B_n(x) = \sum_{k=0}^{n} \frac{n!}{k!(n - k)!} f\left(\frac{k}{n}\right) x^k(1 - x)^{n-k}, \qquad \text{(A3.1)}$$

and known as **Bernstein polynomials.** We see that $B_n(x)$ is a polynomial of degree n, and that its coefficients depend upon $(n + 1)$ equally spaced evaluations of the function $f(x)$ on the interval $0 \leqslant x \leqslant 1$, at the points $x = 0, \frac{1}{n}, \frac{2}{n}, \ldots, \frac{n-1}{n}, 1$. The proof entails a rigorous demonstration that for any choice of ϵ in the Theorem, a number N can be found such that

$$|f(x) - B_n(x)| < \epsilon \ , \qquad (0 \leqslant x \leqslant 1) \ ,$$

for all $n > N$. Thus, by taking n large enough, the discrepancy between $f(x)$ and $B_n(x)$ may be kept within arbitrarily small bounds.

For various reasons the Bernstein method is not used in practice for the approximation of functions. However, its interest in the context of curve design is more than purely theoretical. If the scalars $f(k/n)$ in (A3.1) are replaced by vectors $\mathbf{r}_k$, the resulting vector-valued polynomial represents, for $0 \leqslant x \leqslant 1$, the Bernstein-Bézier curve segment of degree n defined in Section 5.1.4. We may therefore regard the curve segment as a Bernstein approximation to the polygon whose vertices are the $\mathbf{r}_k$. This being so, we may expect that we can modify the curve in a predictable way by altering the positions of the vertices. As we saw earlier, Bézier's UNISURF method is based on this principle.

The scalar functions

$$\frac{n!}{k!(n-k)!} \, x^k(1-x)^{n-k}, \quad k = 0,1,2, \ldots ,n,$$

which occur in (A3.1) constitute what is called the **Bernstein basis** for the set of all polynomials of degree n. This is to say that any polynomial of degree n can be expressed as a linear combination of them. For instance, when $n = 3$ the basis consists of $p_0(x) = 1 - 3x + 3x^2 - x^3$, $p_1(x) = 3(x - 2x^2 + x^3)$, $p_2(x) = 3(x^2 - x^3)$ and $p_3(x) = x^3$, and the general cubic $ax^3 + bx^2 + cx + d$ can be written in terms of it as

$$dp_0(x) + \frac{1}{3}(c + 3d)p_1(x) + \frac{1}{3}(b + 2c + 3d)p_2(x) + (a + b + c + d)p_3(x) \ .$$

The functions comprising the Bernstein basis for all fifth-degree polynomials are depicted in Figure A3.1

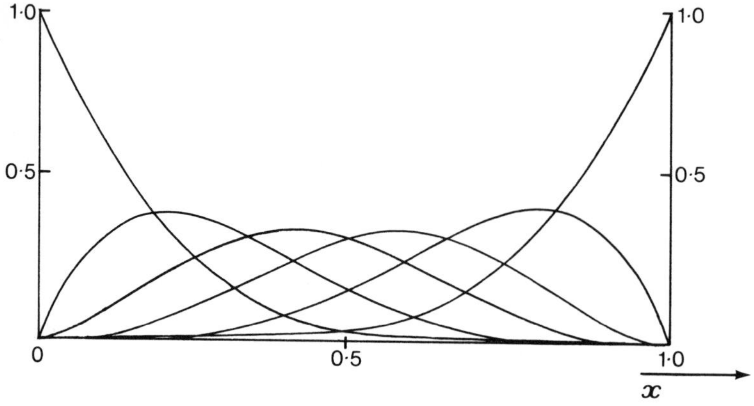

Figure A3.1 Bernstein basis functions for polynomials of degree 5.

Appendix 4

Numerical solution of polynomial and other non-linear equations

A4.1 SOLUTION OF A SINGLE EQUATION

Although polynomials possess the desirable features cited in Appendix 3, they also have their disadvantages. One of these is the difficulty of finding solutions of a polynomial equation $p(x) = 0$ (or zeros of $p(x)$) when the degree of the polynomial is greater than two. In principle, we may calculate exact solutions of cubic (third-degree) and quartic (fourth-degree) equations, but the procedures are lengthy and involved. Polynomial equations of higher degree cannot, in general, be solved exactly. Fortunately, a simple numerical procedure known as the **Newton-Raphson method** usually gives the required solutions to any desired accuracy with great efficiency. The method is not restricted to the solution of polynomial equations, and works also for other kinds of nonlinear equations such as $2 \sin x - \cosh x = 0$.

The Newton-Raphson method generates a sequence of successively improved approximations to the solution of an equation $f(x) = 0$ using the relation

$$x_{r+1} = x_r - \frac{f(x_r)}{f'(x_r)}. \tag{A4.1}$$

This formula has a simple geometrical interpretation. Reference to Figure A4.1 shows that the tangent to the curve $y = f(x)$ at the point $(x_r, f(x_r))$ may be expected to cut the x-axis at a point lying closer than x_r to the true solution X. From the right-angled triangle, we see that

$$\tan\theta = f'(x_r) = \frac{f(x_r)}{x_r - x_{r+1}},$$

which, when solved for x_{r+1}, results in equation (A4.1). The process may now be repeated with x_{r+1} as a new approximation to X, to give a new point x_{r+2} which lies still closer to X, and so on. A reasonable starting approximation x_0 is a prerequisite for the success of the method.

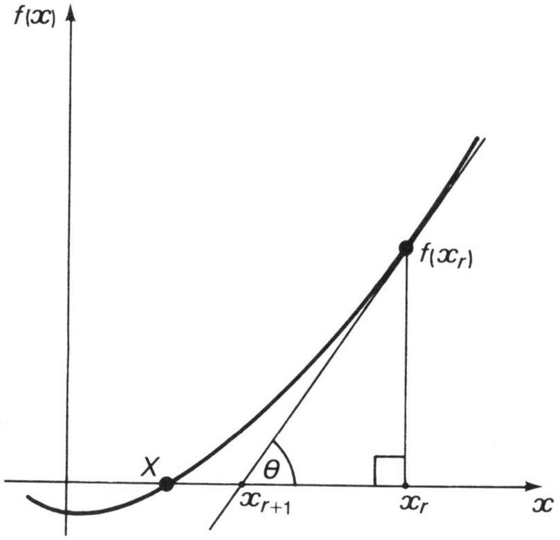

Figure A4.1 – Geometrical basis of the Newton-Raphson method

Example

Find the point of intersection of the straight line $y = x + 2$ with the cubic curve $y = x^3 - x^2 + 3x + 1$.

The intersection occurs at a point whose x-coordinate is given by

$$x + 2 = x^3 - x^2 + 3x + 1$$

or

$$p(x) = x^3 - x^2 + 2x - 1 = 0 .$$

We note that $p(0) = -1$ and $p(1) = 1$. This implies that $p(x)$ changes sign (and hence passes through zero) somewhere in the range $0 < x < 1$. We take $x_0 = 0.5$ as an initial guess at the zero of $p(x)$.

The appropriate Newton-Raphson relation is

$$x_{r+1} = x_r - \frac{x_r^3 - x_r^2 + 2x_r - 1}{3x_r^2 - 2x_r + 2},$$

and with $x_0 = 0.5$ this gives the sequence

$$\begin{aligned} x_1 &= 0.571\,428\,57 \text{ (to 8 decimal places)}, \\ x_2 &= 0.569\,841\,27, \\ x_3 &= 0.569\,840\,30, \\ x_4 &= 0.569\,840\,29, \\ x_5 &= 0.569\,840\,29 = x_4 . \end{aligned}$$

Thus four cycles (or **iterations**) of the process have given us a solution correct to 8 decimal places. A typical feature of this method is that the number of correct significant figures roughly doubles at each iteration, unless a multiple root is being calculated, when convergence is much slower.

The two remaining solutions of $p(x) = 0$ are complex. We may see this by considering $p'(x) = 3x^2 - 2x + 2$, which is not zero for any real value of x. This shows that $p(x)$ is a function which increases continually as x goes from $-\infty$ to $+\infty$, and its graph therefore crosses the x-axis once only. We conclude that $x = 0.569\ 840\ 29$ is the only real solution of $p(x) = 0$, and that the required intersection point is (0.569 840 29, 2.569 840 29).

The Newton-Raphson method is only one of a multitude of methods for solving non-linear equations. It was chosen here because it usually combines the virtues of efficiency and simplicity. An account of some further standard techniques is given in Ralston (1965), where a critical comparison is also made of several procedures restricted to the computation of real and complex zeros of polynomials.

A4.2 NUMERICAL SOLUTION OF A SYSTEM OF NONLINEAR EQUATIONS

We now show how the Newton-Raphson method may be used for solving simultaneous equations. We illustrate for the case of two equations, but larger systems may be solved by an obvious extension of the method.

Let us return for the moment to the one-dimensional situation. An alternative derivation of equation (A4.1) is as follows. Suppose x_r is an approximation to X, the true solution of $f(x) = 0$. Then by Taylor's Theorem we may write

$$f(X) = f(x_r) + h f'(x_r) + \text{ terms of order } h^2 ,$$

where $h = X - x_r$. Assuming that h is small, we retain only the first two terms on the right. Since X is the true solution, $f(X) = 0$, and we are left with

$$f(x_r) + h_r f'(x_r) = 0 ,$$

where h_r is only an approximation to our original h because we have neglected the higher order terms. Then $h_r = -f(x_r)/f'(x_r)$ provided $f'(x_r) \neq 0$, and since $X = x_r + h \simeq x_r + h_r$ we obtain the Newton-Raphson formula

$$X \simeq x_{r+1} = x_r - \frac{f(x_r)}{f'(x_r)} .$$

In the two-dimensional problem, we have two equations to solve simultaneously,

$$\left.\begin{aligned} F(x,y) &= 0 \\ \text{and} \qquad G(x,y) &= 0 \end{aligned}\right\} .$$

Suppose the true solution to be (X,Y), and that we have an approximate solution (x_r,y_r). Set $h = X - x_r$ as previously, and now also $k = Y - y_r$. The two-dimensional Taylor expansions of $F(X,Y)$ and $G(X,Y)$ about (x_r,y_r) are respectively

$$F(X,Y) = F(x_r,y_r) + h\frac{\partial F}{\partial x}(x_r,y_r) + k\frac{\partial F}{\partial y}(x_r,y_r) + \ldots$$

and

$$G(X,Y) = G(x_r,y_r) + h\frac{\partial G}{\partial x}(x_r,y_r) + k\frac{\partial G}{\partial y}(x_r,y_r) + \ldots \quad .$$

As before, we neglect higher order terms on the right and set $F(X,Y)=G(X,Y)=0$ on the left. This gives a pair of linear equations for h_r and k_r, which are approximations to h and k:

$$h_r\frac{\partial F}{\partial x}(x_r,y_r) + k_r\frac{\partial F}{\partial y}(x_r,y_r) = -F(x_r,y_r)$$

$$\text{and} \qquad h_r\frac{\partial G}{\partial x}(x_r,y_r) + k_r\frac{\partial G}{\partial y}(x_r,y_r) = -G(x_r,y_r)\,. \tag{A4.2}$$

The solution is

$$\left.\begin{aligned} h_r &= D^{-1}\left[G\frac{\partial F}{\partial y} - F\frac{\partial G}{\partial y}\right] \\ k_r &= D^{-1}\left[F\frac{\partial G}{\partial x} - G\frac{\partial F}{\partial x}\right] \end{aligned}\right\} , \tag{A4.3}$$

where D is the determinant of the **Jacobian matrix J** defined by

$$\mathbf{J} = \begin{bmatrix} \dfrac{\partial F}{\partial x} & \dfrac{\partial F}{\partial y} \\[2ex] \dfrac{\partial G}{\partial x} & \dfrac{\partial G}{\partial y} \end{bmatrix} . \tag{A4.4}$$

In the last three equations all functions and derivatives are understood to be evaluated at (x_r, y_r). The improved solution is now (x_{r+1}, y_{r+1}), where $x_{r+1} = x_r + h_r$, $y_{r+1} = y_r + k_r$. Note that for the process to succeed **J** must be non-singular at all the approximate solution points. Otherwise, D in (A4.3) will be zero and the iteration will fail.

The extension of this approach to the solution of systems of more than two equations is straightforward. We start by noting that equation (A4.2) can be written as

$$\mathbf{J}(\mathbf{x}_r)\boldsymbol{\delta}_r = -\mathbf{F}(\mathbf{x}_r) \;, \qquad \text{(A4.5)}$$

where $\mathbf{x}_r = \begin{bmatrix} x_r \\ y_r \end{bmatrix}$, $\boldsymbol{\delta}_r = \begin{bmatrix} h_r \\ k_r \end{bmatrix}$ and $\mathbf{F}(\mathbf{x}_r) = \begin{bmatrix} F(x_r) \\ G(x_r) \end{bmatrix}$, while $\mathbf{J}(x_r)$ is defined by (A4.4). If we are solving a system of n equations rather than just two, (A4.5) is still the fundamental relation. In this case however, x_r, $\boldsymbol{\delta}_r$ and $\mathbf{F}(\mathbf{x}_r)$ become $n \times 1$ vectors while $\mathbf{J}(\mathbf{x}_r)$ is an $n \times n$ matrix. The equation corresponding to (A4.3) is $\boldsymbol{\delta}_r = -\mathbf{J}^{-1}(\mathbf{x}_r)\,\mathbf{F}(\mathbf{x}_r)$, and a single iteration of the Newton-Raphson process is completed by setting $\mathbf{x}_{r+1} = \mathbf{x}_r + \boldsymbol{\delta}_r$.

Example

Find a solution, to three significant figures, of the simultaneous equations

$$\left.\begin{aligned} F(x,y) &= x^2 + y^2 - 4 = 0 \\ \text{and} \qquad G(x,y) &= xy - 1 = 0 \end{aligned}\right\} .$$

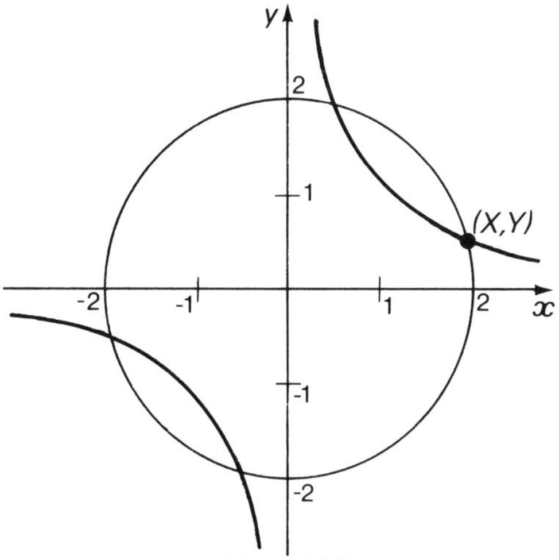

Figure A4.2

We obtain an approximate solution by noting that the equations are those of a circle of radius 2, centred at the origin, and a rectangular hyperbola whose asymptotes are the x- and y-axes as shown in Figure A4.2.

It appears that $(x_0,y_0) = (2,0)$ is a reasonable starting approximation to one of the four solutions. We have

$$F = x^2 + y^2 - 4 \qquad G = xy - 1$$

$$\frac{\partial F}{\partial x} = 2x \qquad \frac{\partial G}{\partial x} = y$$

$$\frac{\partial F}{\partial y} = 2y \qquad \frac{\partial G}{\partial y} = x \ .$$

When these are evaluated at (2,0) and the results substituted into equations (A4.3) and (A4.4), we obtain

$$h_0 = 0, \quad k_0 = \tfrac{1}{2} \ .$$

Then $(x_1,y_1) = (2,\tfrac{1}{2})$.

The functions and derivatives are now evaluated at this new point, and the results substituted into equations (A4.3) and (A4.4), when we find

$$h_1 = -\frac{1}{15}, \quad k_1 = \frac{1}{60} \ .$$

This gives, to three significant figures, $(x_2,y_2) = (1.93,0.517)$.

Further (and subsequent) iterations give the solution of the equations to be $(X,Y) = (1.93,0.518)$, to the accuracy desired.

The numerical values are easily confirmed for this simple problem, because elimination of y between the two original equations leads to an equation which is quadratic in x^2 and can therefore be solved algebraically. One solution proves to be $([2+\sqrt{3}]^{1/2},[2-\sqrt{3}]^{1/2}) = (1.93,0.518)$.

In general, it is less easy to obtain good starting approximations when solving a system of equations than when solving a single equation, though this problem is usually not troublesome in curve and surface applications. For instance, if points are being computed on the line of intersection of two surfaces, the last point computed will be a reasonable starting approximation for the next one provided the spacing is fairly close. Many methods exist in addition to the simple one given here. A standard work on the solution of nonlinear systems is Ortega and Rheinboldt (1970), while a more recent survey is given by Dennis (1976).

The solution of systems of simultaneous linear equations is not dealt with in this book; an elementary treatment is given by Goult *et al* (1974).

Appendix 5

Approximation using polynomials

A5.1 INTRODUCTION

The methods employed in polynomial approximation fall into two fairly distinct classes. In the first the functions to be approximated are known, and can be evaluated, at all points in some range of x. For example, we may wish to find the cubic polynomial which is the 'best' approximation to the function e^x over the interval $-1 \leqslant x \leqslant 1$. There are several possible criteria of 'best fit', and a considerable amount of theory exists on problems of this type. Much of it involves orthogonal polynomials such as Legendre and Chebyshev polynomials (see for instance Chapters 6 and 7 of Ralston, 1965).

Here, however, we confine attention to methods of the second class, which apply when the function $y(x)$ to be approximated is given in a tabulated form, so that y-values are only available for certain discrete values of x. The procedure adopted will depend upon whether

(a) the given points (x_i,y_i) are known to contain statistical or other fluctuations, or
(b) they are known to be reliable.

Typical examples might be (a) points resulting from some experimental procedure which is subject to errors of measurement, and (b) points taken from standard tables such as log tables.

In case (b) it is reasonable to construct an **interpolating function** which actually passes through all the given points (x_i,y_i). In case (a) this could lead to disastrous results; an interpolating function might well have the effect of amplifying the statistical fluctations of the data, when what is really required is that these should be smoothed out. When the data points are unreliable, then, we seek an approximating function which passes close to all the points but not necessarily through any of them. Since some curve and surface fitting systems employ a preliminary stage of this kind to smooth out anomalies in the data supplied, we illustrate a standard method of approach for case (a) in the following section. The same principles apply to least-squares approximation by composite curves and surfaces. We return in Section A5.3 to case (b) and the problem of interpolating reliable data.

A5.2 LEAST-SQUARES POLYNOMIAL CURVE FITTING

Consider the following table of data, the values from which are plotted in Figure A5.1:

x	y
0	0.4
1	1.7
2	2.1
3	1.2
4	0.9
5	1.8

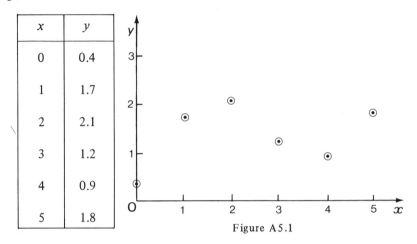

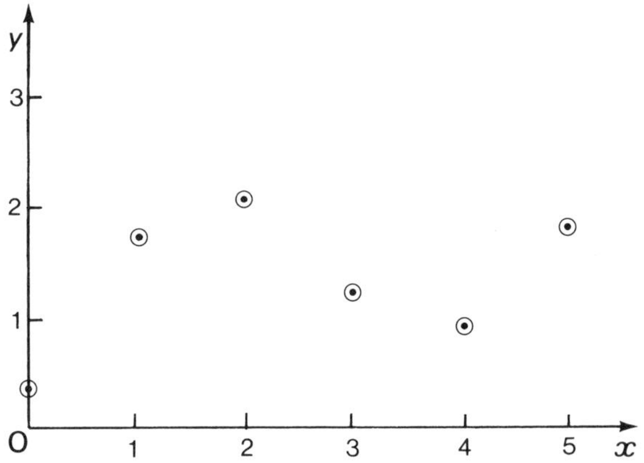

Figure A5.1

We will assume that any errors in the x-values are negligible, but that the y-values are subject to significant inaccuracies.

If we wish to approximate $y(x)$ by a polynomial $p(x)$, we must first decide on the degree of the polynomial. A linear function $a_1x + a_0$, whose graph is a straight line, could clearly not be made to fit the data at all closely. The graph of the quadratic polynomial $a_2x^2 + a_1x + a_0$, is a parabola, and our points do not appear to lie, even approximately, on a parabola. However, a cubic polynomial $a_3x^3 + a_2x^2 + a_1x + a_0$ can have a maximum and a minimum provided that its derivative $3a_3x^2 + 2a_2x + a_1$ has two real zeros. A cubic, then, is a plausible choice for our approximating polynomial.

Having decided that the degree of $p(x)$ is three, we consider next the **deviations** δ_i defined by

$$\delta_i = p(x_i) - y_i = a_3x_i^3 + a_2x_i^2 + a_1x_i + a_0 - y_i \ ,$$

where $i = 0, 1, \ldots, 5$ labels the data points in the table. These deviations are the discrepancies between $p(x)$, evaluated at the tabulation points, and the given y-values at those points. The least-squares method aims to minimise the expression

$$S = \sum_{i=0}^{5} \delta_i^2 = \sum_{i=0}^{5} [a_3 x_i^3 + a_2 x_i^2 + a_1 x_i + a_0 - y_i]^2 ,$$

which is the sum of the squared deviations for all the data points, with respect to the coefficients of $p(x)$. This is achieved by equating to zero all the partial derivatives of S with respect to these coefficients:

$$\frac{\partial S}{\partial a_r} = 2 \sum_{i=0}^{5} [a_3 x_i^3 + a_2 x_i^2 + a_1 x_i + a_0 - y_i] x^r = 0 , \qquad r = 3, 2, 1, 0.$$

On rearranging this and omitting the limits on the summations for clarity, we obtain

$$a_3 \Sigma x_i^{3+r} + a_2 \Sigma x_i^{2+r} + a_1 \Sigma x_i^{1+r} + a_0 \Sigma x_i^r = \Sigma x_i^r y_i , \qquad r = 3, 2, 1, 0.$$

This system of four equations, when written out in full, is

$$\begin{aligned}
r = 3: &\quad a_3 \Sigma x_i^6 + a_2 \Sigma x_i^5 + a_1 \Sigma x_i^4 + a_0 \Sigma x_i^3 = \Sigma x_i^3 y_i , \\
r = 2: &\quad a_3 \Sigma x_i^5 + a_2 \Sigma x_i^4 + a_1 \Sigma x_i^3 + a_0 \Sigma x_i^2 = \Sigma x_i^2 y_i , \\
r = 1: &\quad a_3 \Sigma x_i^4 + a_2 \Sigma x_i^3 + a_1 \Sigma x_i^2 + a_0 \Sigma x_i = \Sigma x_i y_i , \\
r = 0: &\quad a_3 \Sigma x_i^3 + a_2 \Sigma x_i^2 + a_1 \Sigma x_i + a_0 \Sigma 1 = \Sigma y_i .
\end{aligned}$$

This is a symmetric 4X4 system of linear equations, whose coefficients are all summations from $i = 0$ to 5 which may be evaluated from the original data to give, for the present example,

$$\begin{aligned}
20515 a_3 + 4425 a_2 + 979 a_1 + 225 a_0 &= 333.5, \\
4425 a_3 + 979 a_2 + 225 a_1 + 55 a_0 &= 80.3, \\
979 a_3 + 225 a_2 + 55 a_1 + 15 a_0 &= 22.1, \\
225 a_3 + 55 a_2 + 15 a_1 + 6 a_0 &= 8.1 .
\end{aligned}$$

To two decimal places, the solution of this system is $a_3 = 0.15$, $a_2 = -1.21$, $a_1 = 2.59$, $a_0 = 0.35$. These are the values of the coefficients of $p(x)$ which minimise S, the sum of squared deviations. The required least-squares cubic is therefore

$$p(x) = 0.15x^3 - 1.21x^2 + 2.59x + 0.35 .$$

At the tabulation points, $p(x)$ has the values 0.35, 1.88, 1.89, 1.28, 0.95, 1.80. These are all within about 0.2 of the given y-values. If this agreement is considered insufficient, a fourth-degree polynomial should give a better fit, at the expense of some extra computation. As we shall see in the next section, it is possible to find a fifth-degree polynomial which actually passes through all six data points, in which case S assumes its minimum possible value, zero. As pointed out in Section A5.1 however, this is dangerous where unreliable data are concerned, and in practice a polynomial of low degree is usually fitted to a comparatively large number of data points.

With no change in principle, the method shown may be used for fitting any approximating function of the form $\sum_{r=0}^{n} a_r g_r(x)$ to a given set of data. Here the $g_r(x)$ are specified functions and the a_r are coefficients to be determined. Polynomial approximation corresponds to the choice $g_r(x) = x^r$.

The least-squares method is also suitable for fitting curves defined in a piecewise manner, as in Chapter 6, to given sets of data in the plane. Correspondingly, surfaces defined in a patchwise manner, as in Chapter 7, may be constructed to approximate sets of data in three dimensions. For further information on applications of this kind the reader may profitably consult Hayes (1973), Hayes and Halliday (1974), Powell (1977), Schumaker (1976) and the references therein.

A5.3 POLYNOMIAL INTERPOLATION: LAGRANGE'S METHOD

We now turn to the problem of interpolation, which concerns the exact fitting of reliable data. First we show how to determine a polynomial which passes through just three given points (x_0, y_0), (x_1, y_1) and (x_2, y_2), where $x_0 < x_1 < x_2$.

Consider the expression

$$L_0(x) = \frac{(x - x_1)(x - x_2)}{(x_0 - x_1)(x_0 - x_2)}.$$

If we set $x = x_0$ the numerator becomes identical with the denominator, and hence $L_0(x_0) = 1$. On the other hand, if we set $x = x_1$ or $x = x_2$ the numerator is zero, so that $L_0(x_1) = L_0(x_2) = 0$. Similarly,

$$L_1(x) = \frac{(x - x_0)(x - x_2)}{(x_1 - x_0)(x_1 - x_2)} = \begin{cases} 1, & x = x_1 \\ 0, & x = x_0 \text{ or } x = x_2 \end{cases}$$

and $$L_2(x) = \frac{(x - x_0)(x - x_1)}{(x_2 - x_0)(x_2 - x_1)} = \begin{cases} 1, & x = x_2 \\ 0, & x = x_0 \text{ or } x = x_1 \end{cases}.$$

Now let us use these three expressions to construct a function involving the given y-values:

$$p_L(x) = L_0(x)y_0 + L_1(x)y_1 + L_2(x)y_2 .$$

We note that when $x = x_0$ only the first term on the right-hand side is non-zero. Since $L_0(x_0) = 1$, we obtain $p_L(x_0) = y_0$. When $x = x_1$ the second term alone is non-zero, and $p_L(x_1) = y_1$; similarly, $p_L(x_2) = y_2$. The function $p_L(x)$ therefore agrees in value with $y(x)$ at all the tabulation points, and we have found a function which interpolates (that is fits exactly) the three given points. Since, from their definitions, $L_0(x)$, $L_1(x)$ and $L_2(x)$ are all quadratic polynomials in x, it follows that our interpolating function $p_L(x)$ is also a quadratic polynomial. This is not surprising; the three given items of information (function values) are sufficient to determine the three coefficients of the interpolating quadratic.

The procedure outlined here is easily generalised. We find that in order to interpolate $(n + 1)$ given points (x_i, y_i), $i = 0, 1, 2, \ldots, n$, such that $x_0 < x_1 < \ldots < x_n$, we generally need a polynomial of degree n, which involves $(n + 1)$ coefficients and which may be expressed as

$$p_L(x) = \sum_{i=0}^{n} L_i(x)y_i , \tag{A5.1}$$

where

$$L_i(x) = \frac{(x - x_0)(x - x_1) \ldots . (x - x_{i-1})(x - x_{i+1}) \ldots . (x - x_n)}{(x_i - x_0)(x_i - x_1) \ldots . (x_i - x_{i-1})(x_i - x_{i+1}) \ldots . (x_i - x_n)}$$
$$= \prod_{\substack{j=0 \\ j \neq i}}^{n} \frac{(x - x_j)}{(x_i - x_j)} .$$

This is **Lagrange's interpolating formula,** and the expressions $L_i(x)$ are known as **Lagrangean interpolation coefficients.** As can be seen, they are all polynomials of degree n in x, as in general will be $p_L(x)$ itself. In some cases, however, when the $L_i(x)$ are put into equation (A5.1) the terms of highest degree in x may cancel, so that the degree of $p_L(x)$ is $< n$. This would happen, for instance, if the $(n + 1)$ data points happened to lie on a straight line, in which event $p_L(x)$ would boil down to a linear (first-degree) polynomial.

The uniqueness of the interpolating polynomial of degree $\leqslant n$ given by equation (A5.1) is easily proved. Suppose that $q(x)$ also has degree $\leqslant n$ and interpolates the data set. Then $d(x) = p_L(x) - q(x)$ is a further polynomial of degree $\leqslant n$, which has the value zero at the $(n+1)$ points x_i, because $p_L(x)$ and $q(x)$ coincide in value there. But a polynomial of degree $\leqslant n$ can have at most n zeros, as shown in Appendix 3. Then $d(x)$ must be identically zero, so that $p_L(x) = q(x)$.

Example

Find an interpolated value of y at $x = 6$ from the following table of data:

i	x_i	y_i
0	2	0.6931
1	4	1.3863
2	5	1.6094
3	7	1.9459

The Langrangean formula gives the (cubic) interpolating polynomial as

$$p_L(x) = \frac{(x-4)(x-5)(x-7)}{(2-4)(2-5)(2-7)}\,0.6931 + \frac{(x-2)(x-5)(x-7)}{(4-2)(4-5)(4-7)}\,1.3863$$

$$+ \frac{(x-2)(x-4)(x-7)}{(5-2)(5-4)(5-7)}\,1.6094 + \frac{(x-2)(x-4)(x-5)}{(7-2)(7-4)(7-5)}\,1.9459\,.$$

We could now expand the products, collect like terms and put $p_L(x)$ explicitly in the form $a_3x^3 + a_2x^2 + a_1x + a_0$. For present purposes, however, this is unnecessary. Since all we want is the value of $p_L(x)$ at $x = 6$, we may substitute this value directly into the last equation, to obtain

$$p_L(6) = \frac{2}{30}\,0.6931 - \frac{2}{3}\,1.3863 + \frac{4}{3}\,1.6094 + \frac{8}{30}\,1.9459$$

$$= 1.7868\,.$$

In fact the given data were taken from a four-figure table of Napierian logarithms, and our interpolated value differs by only about 0.3% from the value $y(6) = \log_e 6 = 1.7918$ given by that table.

We are not always so fortunate; experience has shown that Lagrangean interpolation can sometimes give quite misleading results. A standard cautionary example is the cubic polynomial $p_L(x)$ which interpolates the points

x	0	1	8	27
y	0	1	2	3

taken from the curve $y = x^{1/3}$. Both functions are plotted in Figure A5.2, and any interpolated values are clearly not likely to be very satisfactory.

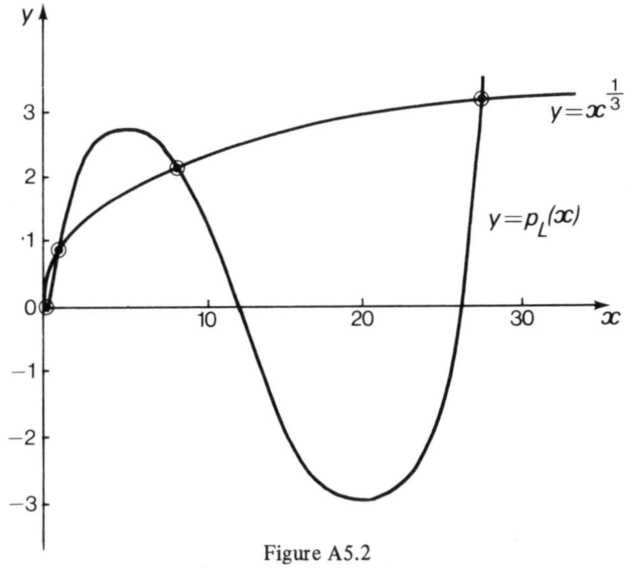

Figure A5.2

The disparity arises because we are using a polynomial to approximate a function which does not behave like a polynomial. Clearly, more accurate results would be achieved by using linear (straight-line) interpolation between successive pairs of data points in this particular case.

The example given, though admittedly extreme, leads us to question the accuracy which we can expect from polynomial interpolation. Analysis shows that if the points (x_i,y_i) are derived from a single-valued function $y(x)$, then a bound on the error is given by

$$|y(x) - p_L(x)| < (x - x_0)(x - x_1)\ldots.(x - x_n)\frac{M}{(n + 1)!},$$

where M is the maximum modulus of $y^{(n+1)}(x)$ on the interval $x_0 \leqslant x \leqslant x_n$. Error bounds such as this, involving values of high-order derivatives, are typical of numerical analysis, but are often unduly pessimistic and of limited practical use.

However, in the synthesis of curves for engineering purposes we are not usually seeking to approximate a mathematical function, but to represent a curve on a drawing or model. In this context error formulae such as that quoted above have no relevance; there is no way of finding M. But if a function is interpolated through a set of points measured from the original curve, it is only necessary to plot out the interpolated curve and compare the two to see whether the representation is satisfactory. Indeed, the mathematical curve may be judged 'fairer' than an original draughted curve and accepted in preference to it, even though significant deviations exist between them. In view of these considerations we will not go further into the subject of errors. The interested reader may consult Ralston (1965) for further details.

A5.4 POLYNOMIAL INTERPOLATION: HERMITE'S METHOD

Suppose now that at the $(n + 1)$ points x_i, $i = 1, 2 \ldots, n$, we are given not only the function values y_i but also the gradients y_i'. We may determine the unique polynomial of lowest degree having these same properties at all the points x_i by using the formula

$$p_H(x) = \sum_{i=0}^{n} H_i(x)y_i + \sum_{i=0}^{n} H_i^*(x)y_i' ,$$

in which

$$H_i(x) = [1 - 2L_i'(x_i)(x - x_i)]L_i^2(x)$$

and

$$H_i^*(x) = (x - x_i)L_i^2(x) ,$$

with the functions $L_i(x)$ as defined on p. 305. Since the $L_i(x)$ are polynomials of degree n, it follows that $p_H(x)$ is a polynomial of degree $2n + 1$. This is to be expected, because such a polynomial requires $(2n + 2)$ items of information to determine all its coefficients, and we use known values of both y and y' at the $(n + 1)$ points x_i.

The formula quoted above is **Hermite's** interpolation formula.

A5.5 POLYNOMIAL INTERPOLATION: DIVIDED DIFFERENCES

We return now to the situation where only the points (x_i, y_i) are specified. Instead of using Lagrangean interpolation as in Section A5.3 we may alternatively construct an interpolating polynomial in the following form, originally due to Newton:

$$p(x) = \alpha_0 + (x - x_0)\alpha_1 + (x - x_0)(x - x_1)\alpha_2 + \ldots$$
$$+ (x - x_0)(x - x_1)\ldots(x - x_{n-1})\alpha_n, \qquad \text{(A5.2)}$$

so that $p(x)$ is a polynomial of degree n in x and the α's are to be determined. We require that $p(x_i) = y_i$ at each data point. On setting $x = x_0, x_1, \ldots, x_n$ respectively in equation (A5.2), this requirement leads to a system of $(n + 1)$ equations:

$$\begin{aligned}
y_0 &= \alpha_0 \\
y_1 &= \alpha_0 + (x_1 - x_0)\alpha_1 \\
y_2 &= \alpha_0 + (x_2 - x_0)\alpha_1 + (x_2 - x_0)(x_2 - x_1)\alpha_2 \\
&\ldots \\
y_n &= \alpha_0 + (x_n - x_0)\alpha_1 + (x_n - x_0)(x_n - x_1)\alpha_2 + \ldots \\
&\ldots + (x_n - x_0)(x_n - x_1)\ldots(x_n - x_{n-1})\alpha_n .
\end{aligned} \qquad \text{(A5.3)}$$

These equations may be solved uniquely for $\alpha_0, \alpha_1, \ldots, \alpha_n$. The first gives $\alpha_0 = y_0$. Putting this into the second, we obtain $\alpha_1 = (y_1 - y_0)/(x_1 - x_0)$, since $(x_1 - x_0) \neq 0$. Then α_2 may be found from the third equation, and so on.

Note that the introduction of a further data point (x_{n+1}, y_{n+1}) has the effect of adding a further equation, involving α_{n+1}, to the end of this system. The earlier equations remain unaltered, however, and consequently the values of $\alpha_0, \alpha_1, \ldots, \alpha_n$ are independent of the actual number of data points.

It is fairly clear from equations (A5.3) that α_0 depends only upon y_0, α_1 depends upon y_0 and y_1, α_2 depends upon y_0, y_1 and y_2, and so on. To emphasise these relationships, we use the notation

$$\alpha_i = y[x_0, x_1, \ldots, x_i] .$$

A number of alternative notations are also current.

We can also construct similar entities from any consecutive subset of the data points. For example, we can compute $y[x_1, x_2, \ldots, x_{i+1}]$ from the subset obtained by omitting the point (x_0, y_0) and adding the point $x_i, y_i)$. It may be shown (see, for example, Hildebrand, 1956) that such expressions are related by

$$y[x_s, \ldots, x_t] = \frac{y[x_{s+1}, \ldots, x_t] - y[x_s, \ldots, x_{t-1}]}{x_t - x_s} \qquad \text{(A5.4)}$$

for which reason they are known as **divided differences.** Equation (A5.4) enables us to build up a table of divided differences according to a scheme such as the following, in which we have written $y_i = y[x_i]$ for uniformity of notation:

x-values	y-values	1st differences	2nd differences	3rd differences
x_0	$y[x_0]$			
		$y[x_0,x_1]$		
x_1	$y[x_1]$		$y[x_0,x_1,x_2]$	
		$y[x_1,x_2]$		$y[x_0,x_1,x_2,x_3]$
x_2	$y[x_2]$		$y[x_1,x_2,x_3]$	
		$y[x_2,x_3]$		
x_3	$y[x_3]$			

In terms of this notation, equation (A5.4) becomes

$$p(x) = y[x_0] + (x-x_0)y[x_0,x_1] + (x-x_0)(x-x_1)y[x_0,x_1,x_2] + \ldots$$

$$\ldots + (x-x_0)(x-x_1)\ldots(x-x_{n-1})y[x_0,x_1,\ldots,x_n] \; . \qquad \text{(A5.5)}$$

Example

We once again calculate an interpolated value for $y(6)$ from the table of data on p. 306.

The numerical entries in the difference table are calculated using equation (A5.4), with the following results:

Divided Difference Table for $y = \log_e x$

x	y	1st differences	2nd differences	3rd differences
2	0.6931			
		$\dfrac{1.3863 - 0.6931}{4 - 2} = 0.3466$		
4	1.3863		$\dfrac{0.2231 - 0.3466}{5 - 2} = -0.0412$	
		$\dfrac{1.6094 - 1.3863}{5 - 4} = 0.2231$		$\dfrac{-0.0183 + 0.0412}{7 - 2} = 0.0046$
5	1.6094		$\dfrac{0.1682 - 0.2231}{7 - 4} = -0.0183$	
		$\dfrac{1.9459 - 1.6094}{7 - 5} = 0.1682$		
7	1.9459			

Equation (A5.5) then gives the interpolating polynomial to be

$$p(x) = 0.6931 + (x-2)0.3466 - (x-2)(x-4)0.0412 + (x-2)(x-4)(x-5)0.0046$$

$$= 0.6931 + (x-2)\,[0.3466 + (x-4)\,[-0.0412 + (x-5)0.0046]\,] .$$

In this latter ('nested') form the polynomial may be evaluated efficiently, with a minimum number of multiplications, by working outwards from the innermost square bracket. With $x = 6$ we obtain

$$y(6) \simeq p(6) = 1.7867 .$$

Apart from a discrepancy in the last digit, due to differences in the effects of rounding errors, we have agreement with the result from Lagrangean interpolation. This we expect, since the two methods are equivalent.

When compared with the Lagrangean method, interpolation using divided differences has a distinct advantage, in that the evaluation of the interpolating polynomial for a particular value of x is very much simpler computationally. Once the necessary preliminary work has been done in setting up the divided difference table, any number of interpolated values may be obtained, as shown in the foregoing example, with a minimal amount of arithmetic.

Other related methods exist, notably those of Neville and Aitken, which compute interpolated values by iterative means. If a y-value is required at some point x lying between x_0 and x_1, the first iteration gives the value corresponding to linear (straight-line) interpolation between (x_0, y_0) and (x_1, y_1). Subsequent iterations give results corresponding to quadratic interpolation over three points, cubic interpolation over four points, and so on. Smooth convergence of the sequence of successive approximations usually implies that the final error is small, but the entire fairly lengthy process must be repeated for each new value of x. Further details may be found in Hildebrand (1956).

Before leaving the subject of divided differences we prove a fundamental result of relevance in the theory of B-splines (Section 6.2.4):

Theorem

The nth divided differences of a polynomial $p(x)$ of degree n are all equal, and the $(n+1)$th divided differences are consequently all zero.

Proof

Set $p(x) = \sum_{i=0}^{n} a_i x^i$. The first divided difference of $p(x)$ relative to x_0 and a general point x is

$$p[x_0,x] = \frac{p[x] - p[x_0]}{x - x_0} = \frac{1}{x - x_0}\sum_{i=0}^{n} a_i(x^i - x_0^i)$$

$$= \sum_{i=0}^{n} a_i(x^{i-1} + x_0x^{i-2} + x_0^2x^{i-3} + \ldots . + x_0^{i-1}).$$

This is a polynomial of degree $(n - 1)$ in x. Similarly the second divided difference $p[x_0,x_1,x]$ will be a polynomial of degree $(n - 2)$ in x. Repetition of this argument shows that the nth divided difference $p[x_0,x_1,\ldots,x_{n-1},x]$ is a polynomial of degree zero in x, and therefore has a constant value for all x. It follows immediately that any $(n + 1)$th difference of $p(x)$ is zero.

The interpolation formulae discussed in this Appendix may all be applied to functions which are tabulated at unequal intervals. They simplify considerably when the interval of tabulation is constant, and the rather cumbersome divided differences may then be replaced by **finite differences**, in terms of which a considerable number of interpolation formulae exist. We do not deal with finite differences here; the reader may consult Chapters 4 and 5 of Hildebrand (1956) for a thorough treatment.

A5.6 NUMERICAL INTEGRATION AND DIFFERENTIATION

It is clear that approximations may be made to integrals and derivatives of tabulated functions by integrating of differentiating interpolating polynomials. Several well-known and reliable formulae for numerical integration may be derived in this way, notably the Trapezium Rule and Simpson's Rule, which correspond to integration of a linear interpolant over a single interval and a quadratic interpolant over a pair of intervals respectively. In the context of curve and surface fitting we are unfortunately more often concerned with the approximation of derivatives, which is an inherently inaccurate process owing to sensitivity of the relevant formulae to rounding errors in the function values. For this reason, numerical differentiation should be avoided whenever possible.

The difficulty of accurate derivative approximation may be illustrated by a simple example based on the approximation (in which $h = x_1 - x_0$)

$$y_0' = \frac{y_1 - y_0}{h},$$

which estimates y_0' as the slope of the linear interpolating function fitting (x_0,y_0) and (x_1,y_1). If h is large, this is clearly a poor approximation. If we let

$h \to 0$ the approximation becomes exact in the limit, but only in principle. In practice, when h is small, y_1 and y_0 are very nearly equal. In forming the difference $y_1 - y_0$ we are therefore likely to incur a loss of significant figures, so that rounding errors present in the function values may be amplified in their effect. Division by h, which is small, may then lead to a large absolute error in $y_0{}'$. Unfortunately the problem of choosing an optimum value of h so as to avoid both these extremes of behaviour is no simple matter.

For further information on numerical integration and differentiation the reader may once again consult Hildebrand (1956).

A5.7 FURTHER READING ON NUMERICAL ANALYSIS

The references given in this Appendix and the last are all authoritative and informative, but for further reading in the area of numerical analysis two other books may also be recommended. The first is Stark (1970), which expands upon most of the topics covered here, and the second is Acton (1970), which also covers many other topics in numerical analysis. Both are readable and stimulating, and are oriented towards the computing practitioner rather than the potential numerical analyst.

Bibliography

This bibliography contains a fair number of references in addition to those cited in the text, and has been brought up to date at the proof stage by the inclusion of relevant papers which have come to our notice up to mid-1978. Some of the references given themselves contain extensive bibliographies, notably de Boor (1976), Forrest (1972a), Griffiths (1978), Kuo (1971), Newman and Sproull (1973), Schoenberg (1973) and Schumaker (1976).

Abramowitz, M. and Stegun, I. A. (1964), *Handbook of Mathematical Functions,* U.S. National Bureau of Standards, (also published by Dover Publications Inc., 1965).

Acton, F. S. (1970), *Numerical Methods that Work*, Harper and Row.

Adams, J. A. (1975), 'The Intrinsic Method for Curve Definition', *Computer Aided Design,* **7**, 4, 243-249.

Ahlberg, J. H., Nilson, E. N. and Walsh, J. L. (1967), *The Theory of Splines and their Applications*, Academic Press.

Ahuja, D. V. (1968), 'An Algorithm for generating Spline-like Curves', *IBM Syst. J.*, 3 & 4, 206-217.

Ahuja, D. V. and Coons, S. A. (1968), 'Geometry for Construction and Display', *IBM Syst. J.,* 3 & 4, 188-205.

Akima, H. (1970), 'A New Method of Interpolation and Smooth Curve Fitting based on Local Procedures', *Journal ACM* **17**, 4, 589-602.

Akima, H. (1974), 'A Method of Bivariate Interpolation and Smooth Surface Fitting based on Local Procedures', *Comm. ACM* **17,** 1, 18-20.

Akima, H. (1978), 'A Method of Bivariate Interpolation and Smooth Surface Fitting for Irregularly Distributed Data Points', *ACM Trans. Math. Software* **4,** 2, 148-159.

Armit, A. P. (1971), 'Example of an Existing System in University Research. Multipatch and Multiobject Design Systems', *Proc. Roy. Soc. Lond.* **A 321,** 235-242.

Armit, A. P. (1972), 'Interactive 3D Shape Design – MULTIPATCH and MULTIOBJECT'. In *Curved Surfaces in Engineering* (proc. Conference at Churchill College, Cambridge, 1972), IPC Science and Technology Press Ltd.

Ball, A. A. (1974), 'CONSURF. Part 1: Introduction of the Conic Lofting Tile', *Computer Aided Design*, **6**, 4, 243-249.

Ball, A. A. (1975), 'CONSURF. Part 2: Description of the Algorithms', *Computer Aided Design*, **7**, 4, 237-242.

Ball, A. A. (1977), 'CONSURF. Part 3: How the Program is used', *Computer Aided Design*, **9**, 1, 9-12.

Ball, A. A. (1978), 'A Simple Specification of the Parametric Cubic Segment', *Computer Aided Design*, **10**, 3, 181-182.

Bär, G. (1977), 'Parametrische Interpolation empirischer Raumkurven', *ZAMM*, **57**, 305-314.

Barnhill, R. E. (1974), 'Smooth Interpolation over Triangles'. In *Computer-Aid Geometric Design*, (R. E. Barnhill and R. F. Riesenfeld, eds.), Academic Press.

Barnhill, R. E., Birkhoff, G. and Gordon, W. J. (1973), 'Smooth Interpolation in Triangles', *J. Approx. Th.* **8**, 114-128.

Bates, K. J. (1972), 'The AUTOKON AUTOMOTIVE and AEROSPACE Packages'. In *Curved Surfaces in Engineering* (proc. Conference at Churchill College, Cambridge, 1972), IPC Science and Technology Press Ltd.

Ben-Israel, A. (1966), 'A Newton-Raphson Method for the Solution of Systems of Equations', *J. Math. Anal. Appl.*, **15**, 243-252.

Bézier, P. (1968), 'How Renault uses Numerical Control for Car Body Design and Tooling', SAE Paper 680010.

Bézier, P. (1971), 'Example of an Existing System in the Motor Industry: The UNISURF System' *Proc. Roy. Soc. Lond.* **A 321**, 207-218.

Bézier, P. (1972), *Numerical Control: Mathematics and Applications*, Wiley.

Bézier, P. (1974a), 'Mathematical and Practical Possibilities of UNISURF'. In *Computer-Aided Geometric Design*, (R. E. Barnhill and R. F. Reisenfeld, eds.), Academic Press.

Bézier, P. (1974b), 'UNISURF System: Principles, Program, Language', *Proc. 1973 PROLAMAT Conference, Budapest* (J. Hatvany, ed.), North Holland Publ. Co., Amsterdam.

Bloor, M. S., de Pennington, A. and Woodwark, J. R. (1978), RISP: Bridging the Gap between Conventional Surface Elements', *Proc. CAD78 Conference, Brighton, 1978*, IPC Science and Technology Press Ltd.

Böhm, W. (1977), 'Über die Konstruktion von B-Spline-Kurven', *Computing*, **18**, 161-166.

Bolton, K. M. (1975), 'Biarc Curves', *Computer-Aided Design* **7**, 2, 89-92.

Bradley, D. M. and Miller, C. P. (1978), 'The Implementation, Testing and Use of CASPA', *Proc. CAD78 Conference, Brighton, 1978*, IPC Science and Technology Press Ltd.

Braid, I. C. (1973), *Designing with Volumes*, Cantab Press, Cambridge.

Braid, I. C. (1975), 'The Synthesis of Solids bounded by Many Faces', *Comm. ACM* **18**, 4, 209-216.

Braid, I. C. (1976), 'A New Shape Design System', CAD Group Document No. 89, Computer Laboratory, Cambridge University.

Butterfield, K. R. (1976), 'The Computation of all the Derivatives of a B-spline Basis', *J. Inst. Maths. Applics.* **17**, 15–25.

Butterfield, K. R. (1978), Ph.D. Thesis, Brunel University, Uxbridge, Middlesex.

C.A.D. Centre (1972), *'An Introduction to Numerical Master Geometry'*, Computer Aided Design Centre, Madingley Road, Cambridge.

CAM-I (1976), 'User Documentation for Sculptured Surfaces Releases SSX5 and SSX5A', Publication No. PS-76-SS-02, Computer Aided Manufacturing International, Inc., Arlington, Texas.

CAM-I (1977), 'Sculptured Surfaces Users Course', Publication No. TM-77-SS-01, Computer Aided Manufacturing International, Inc., Arlington, Texas.

Cline, A. K. (1974), 'Scalar and Planar valued Curve Fitting using Splines under Tension', *Comm. ACM* **17**, 4, 218-220.

Coles, W. A. (1977), 'Use of Graphics in an Aircraft Design Office', *Computer-Aided Design* **9**, 23–28.

Collins, P. S. and Gould, S. S. (1974), 'Computer-Aided Design and Manufacture of Surfaces for Bottle Moulds, *Proc. CAM74 Conference on Computer Aided Manufacture and Numerical Control*, Strathclyde University, 1974.

Coons, S. A. (1967), 'Surfaces for Computer Aided Design of Space Forms', Report MAC-TR-41, Project MAC, M.I.T.

Coons, S. A. (1977), 'Modification of the Shape of Piecewise Curves', *Computer-Aided Design,* **9**, 3, 178–180.

Cox, M. G. (1971), 'An Algorithm for approximating Convex Functions by means of First-Degree Splines', *Comput. J.* **14**, 3, 272–275.

Cox, M. G. (1972), 'The Numerical Evaluation of B-Splines', *J. Inst. Maths. Applics.* **10**, 134–149.

Cox, M. G. (1975), 'An Algorithm for Spline Interpolation', *J. Inst. Maths. Applics.* **15**, 95-108.

Cox, M. G. (1976), 'The Numerical Evaluation of a Spline from its B-Spline Representation', NPL Report NAC 68, National Physical Laboratory, Teddington, Middlesex.

Cox, M. G. and Hayes, J. G. (1973), 'Curve Fitting: A Guide and Suite of Programs for the Non-specialist User', NPL Report NAC 26, National Physical Laboratory, Teddington, Middlesex.

Curry, H. B. and Schoenberg, I. J. (1966), 'On Polya Frequency Functions IV: The Fundamental Spline Functions and their Limits', *J. Analyse Math.* **17**, 71–107.

De Boor, C. (1972), 'On Calculating with B-splines', *J. Approx. Th.* **6**, 50-62.

De Boor, C. (1976), 'Splines as Linear Combinations of B-splines. A Survey', In *Approximation Theory II* (G. G. Lorentz, C. K. Chui and L. L. Schumaker, eds.), Academic Press.

De Boor, C. (1977), 'Package for Calculating with B-splines' *SIAM J. Numer.*

Anal. **14**, 3, 441–472.

Dennis, J. E. (1976), 'Non-linear Least Squares and Equations', Report CSS 32, Atomic Energy Research Establishment, Harwell. Also in *The State of the Art in Numerical Analysis* (D. A. H. Jacobs, ed.), Academic Press (1977).

Deuflhard, P. (1974), 'A Modified Newton Method for the Solution of Ill-conditioned Systems of Non-linear equations with Applications to Multiple Shooting', *Numer. Math.* **22**, 289–315.

Dimsdale, B. (1977), 'Bicubic Patch Bounds', *Comp. & Maths. with Appls.* **3**, 2, 95–104.

Dimsdale, B. (1978), 'Convex Cubic Splines', *IBM J. Res. Develop.* **22**, 168–178.

Dimsdale, B. and Burkley, R. M. (1976), 'Bicubic Patch Surfaces for High-speed Numerical Control Processing', *IBM J. Res. Develop.* 358–367.

Dimsdale, B. and Johnson, K. (1975), 'Multiconic Surfaces', *IBM J. Res. Develop.* **19**, 6, 523–529.

Dollries, J. F. (1963), 'Three-dimensional Surface Fit and Numerically Controlled Machining from a Mesh of Points', Technical Information Series Report No. R63FPD319, General Electric Co.

Einar, H. and Skappel, E. (1973), 'FORMELA: A general Design and Production Data System for Sculptured Products', *Computer-Aided Design* **5**, 2, 68–76.

Ellis, T. M. R. and McLAIN, D. H. (1977), 'A New Method of Cubic Curve Fitting using Local Data', *ACM Trans. Math. Software* **3**, 2, 175–178.

Emmerson, W. C. (1976), 'CAD in the Motor Industry', *Computer-Aided Design* **8**, 3, 193–197.

Epstein, M. P. (1976), 'On the Influence of Parametrisation in Parametric Interpolation', *SIAM J. Numer. Anal.* **13**, 2, 261–268.

Faux, I. D. (1978), 'Simple Cross-Sectional Designs', *Proc. CAM78 Conference, National Engineering Laboratory, East Kilbride, 1978.*

Ferguson, J. C. (1963), 'Multivariable Curve Interpolation', Report No. D2-22504, The Boeing Co., Seattle, Washington.

Ferguson, J. C. (1964), 'Multivariate Curve Interpolation', *Journal ACM,* **11**, 2, 221–228.

Flanagan, D. L. and Hefner, O. V. (1967), 'Surface Moulding – New Tool for the Engineer', *Aeronautics and Astronautics,* April 1967, 58–62.

Flutter, A. G. (1976), 'The POLYSURF System', *Proc. 1973 PROLAMAT Conference, Budapest* (J. Hatvany, ed.), North Holland Publ. Co., Amsterdam.

Flutter, A. G. and Rolph, R. N. (1976), 'POLYSURF: An Interactive System for Computer-Aided Design and Manufacture of Components', *CAD 76 Proceedings,* 150–158; CAD Centre, Madingley Road, Cambridge.

Forrest, A. R. (1968), 'Curves and Surfaces for Computer-Aided Design', Ph.D Thesis, University of Cambridge.

Forrest, A. R. (1971), 'Computational Geometry', *Proc. Roy. Soc. Lond.*

A 321, 187--195.

Forrest, A. R. (1972a), 'On Coons' and other Methods for the Representation of Curved Surfaces', *Computer Graphics and Image Processing* **1**, 341-359.

Forrest, A. R. (1972b), 'Interactive Interpolation and Approximation by Bézier Polynomials', *Comput. J.* **15**, 1, 71-79.

Forrest, A. R. (1972c), 'Mathematical Principles for Curve and Surface Representation'. In *Curved Surfaces in Engineering* (proc. Conference at Churchill College, Cambridge, 1972), IPC Science and Technology Press Ltd.

Forrest, A. R. (1974), 'Computational Geometry – Achievements and Problems'. In *Computer-Aided Geometric Design* (R. E. Barnhill and R. F. Riesenfeld, eds.), Academic Press.

Freemantle, A. C. and Freeman, P. L. (1972), 'The Evolution and Application of Lofting Techniques at Hawker Siddeley Aviation'. In *Curved Surfaces in Engineering* (proc. Conference at Churchill College, Cambridge, 1972), IPC Science and Technology Press Ltd.

Gaffney, P. W. (1977), 'To Compute the Optimum Interpolation Fromula', Report CSS 52, Atomic Energy Research Establishment, Harwell.

Ghezzi, C. and Tisato, F. (1973), 'Interactive Computer-Aided Design for Sculptured Surfaces', *Proc. 1973 PROLAMAT Conference, Budapest* (J. Hatvany, ed.), North Holland Publishing Co., Amsterdam.

Gill, P. E. and Murray, W. (1976), 'Algorithms for the Solution of the Nonlinear Least Squares Problem', Report NAC 71, National Physical Laboratory, Teddington, Middlesex.

Gordon, W. J. (1969), 'Spline-blended Surface interpolation through Curve Networks', *Journal of Mathematics and Mechanics* **18**, 10, 931-952.

Gordon, W. J. and Riesenfeld, R. F, (1974a), 'B-spline Curves and Surfaces'. In *Computer Aided Geometric Design* (R. E. Barnhill and R. F. Riesenfeld, eds.), Academic Press.

Gordon, W. J. and Riesenfeld, R. F. (1974b), 'Bernstein-Bézier Methods for the Computer-Aided Design of Free Form Curves and Surfaces', *Journal ACM* **21**, 2, 293-310.

Gordon, W. J. and Wixom, J. A. (1978), 'Shephard's Method of "Metric Interpolation" to Bivariate and Multivariate Interpolation', *Mathematics of Computation,* **32**, 141, 253-264.

Gossling, T. H. (1976), 'The 'Duct' System of Design for Practical Objects', *Proc. World Congress on the Theory of Machines and Mechanisms, Milan.*

Gould, S. S. (1972), 'Surface Programs for Numerical Control'. In *Curved Surfaces in Engineering* (Proc. Conference at Churchill College, Cambridge, 1972), IPC Science and Technology Press Ltd.

Goult, R. J., Hoskins, R. F., Milner, J. A. and Pratt, M. J. (1973), *Applicable Mathematics,* MacMillan.

Goult, R. J., Hoskins, R. F., Milner, J. A. and Pratt, M. J. (1974), *Computational Methods in Linear Algebra,* Stanley Thornes (Publishers) Ltd., London and Wiley Interscience, New York.

Gregory, J. A., (1974), 'Smooth Interpolation without Twist Constraints', In *Computer-Aided Geometric Design* (R. E. Barnhill and R. F. Riesenfeld, eds.), Academic Press.

Greville, T. N. E. (1967), 'On the Normalisation of the B-splines and the Location of the Nodes for the case of Unequally Spaced Knots'. Supplement to the paper 'On Spline Functions' by I. J. Schoenberg in *Inequalities* (O. Shisha, ed.), Academic Press.

Greville, T. N. E. (ed.) (1969), *Theory and Applications of Spline Functions*, Academic Press.

Griffiths, J. G. (1978), 'Bibliography of Hidden-line and Hidden-surface Algorithms', *Computer-Aided Design* **10**, 3, 203-206.

Hart, W. B. (1971), 'Glider Fuselage Design with the Aid of Computer Graphics', *Computer-Aided Design* **3**, 2, 3-8.

Hart, W. B. (1972), 'Current and Potential Applications to Industrial Design and Manufacture'. In *Curved Surfaces in Engineering* (proc. Conference at Churchill College, Cambridge, 1972) IPC Science and Technology Press Ltd.

Hart, W. B. (1972), 'The Application of Computer-Aided Design Techniques to Glassware and Mould Design', *Computer-Aided Design* **4**, 2, 57-66.

Hartley, P. J. and Judd, C. J. (1978), 'Parametrisation of Bézier-type B-spline Curves and Surfaces', *Computer-Aided Design* **10**, 2, 130-134.

Hayes, J. G. (1973), 'Available Algorithms for Curve and Surface Fitting', NPL Report NAC 39, National Physical Laboratory, Teddington, Middlesex.

Hayes, J. G. (1974), 'New Shapes from Bicubic Splines', NPL Report NAC 58, National Physical Laboratory, Teddington, Middlesex.

Hayes, J. G. and Halliday, J. (1974), 'The Least-squares Fitting of Cubic Spline Surfaces to General Data Sets', *J. Inst. Maths. Applics.* **14**, 89-103.

Hildebrand, F. B. (1956), *Introduction to Numerical Analysis*, MacGraw-Hill.

Hyodo, Y. (1973), 'HAPT-3D. A Programming System for Numerical Control', *Proc. 1973 PROLAMAT Conference, Budapest* (J. Hatvany, ed.), North Holland Publishing Co., Amsterdam.

IITRI (1967), *APT Part Programming*, McGraw-Hill.

Inselberg, A. (1976), 'Cubic Splines with Infinite Derivatives at Some Knots', *IBM J. Res. Develop.*, September 1976, 430-436.

Kuo, C. (1971), *Computer Methods for Ship Surface Design*, Longman.

Lavick, J. J. (1971), 'Computer-Aided Design at McDonnell Douglas'. In *Advanced Computer Graphics* (R. D. Parslow and R. Elliott Green, eds.), Plenum Press, London and New York.

Lee, T. M. P. (1971), 'Analysis of an Efficient Homogeneous Tensor Representation of Surfaces for Computer Display'. In *Advanced Computer Graphics* (R. D. Parslow and R. Elliott Green, eds.), Plenum Press, London and New York.

Levin, J. (1976), 'A Parametric Algorithm for drawing Pictures of Solid Objects composed of Quadric Surfaces', *Comm. ACM* **19**, 10, 555-563.

Liming, R. A. (1944), *Practical Analytical Geometry with Applications to Aircraft,* Macmillan, New York.

MacCallum, K. J. (1970), 'Surfaces for Interactive Graphical Design', *Comput. J.* **13,** 4, 352-358.

MacCallum, K. J. (1972), 'Mathematical Design of Hull Surfaces', *The Naval Architect,* July 1972, 359-373.

Mair, S. G. and Duncan, J. P. (1975), 'Polyhedral NC Program Documentation', Report, Department of Mechanical Engineering, University of British Columbia, Vancouver.

Malcolm, M. A. (1977), 'On the Computation of Nonlinear Spline Functions', *SIAM J. Numer. Anal.* **14,** 2, 254-282.

Manning, J. R. (1974), 'Continuity Conditions for Spline Curves', *Comput. J.* **17,** 181-186.

Maxwell, E. A. (1958), *Co-ordinate Geometry with Vectors and Tensors*, Oxford University Press.

Mehlum, E. (1964), 'A Curve-Fitting Method based on a Variational Criterion', *BIT* **4,** 213-223.

Mehlum, E. (1969), *Curve and Surface Fitting based on a Variational Criterion for Smoothness*, Central Institute of Industrial Research, Oslo.

Mehlum, E. (1974), 'Non-linear Splines' In *Computer-Aided Geometric Design* (R. E. Barnhill and R. F. Riesenfeld, eds.), Academic Press.

Mehlum, E. and Sφrenson, P. F. (1971), 'Example of an Existing System in the Shipbuilding Industry: the AUTOKON System', *Proc. Roy. Soc. Lond.* **A. 321,** 219-233.

Meinardus, G. (1976), 'Algebräische Formulierung von Spline-Interpolationen', International Series in Numerical Mathematics, Vol. 32, *Moderne Methoden der numerischen Mathematik*, 125-138. Birkhäuser Verlag, Basel & Stuttgart.

Mitchell, A. R. and Wait, R. (1977), *The Finite Element Method in Partial Differential Equations*, Wiley Interscience.

Moore, C. L. (1959), 'Method of Fitting a Smooth Surface to a Mesh of Points, Technical Information Series Report No. R59FPD927, General Electric Co.

Newman, W. M. and Sproull, R. F. (1973), *Principles of Interactive Computer Graphics,* McGraw-Hill.

Nielson, G. M. (1974), 'Some Piecewise Polynomial Alternatives to Splines in Tension'. In *Computer-Aided Geometric Design* (R. E. Barnhill and R. F. Riesenfeld, eds.), Academic Press.

Nutbourne, A. W. (1973), 'A Cubic Spline Package. Part 2 – The Mathematics'. *Computer-Aided Design* **5**, 1, 7-13.

Nutbourne, A. W., McLellan, P. M. and Kensit, R. M. L. (1972), 'Curvature Profiles for Plane Curves', *Computer-Aided Design,* **4,** 4, 176-184.

Ortega, J. M. and Rheinboldt, W. C. (1970), *Iterative Solution of Non-linear*

Equations in Several Variables, Academic Press.

Pal, T. K. (1978a), 'Intrinsic Spline Curve with Local Control', *Computer-Aided Design* **10,** 1, 19–29.

Pal, T. K. (1978b), 'Mean Tangent Rotational Angles and Curvature Integration', *Computer-Aided Design* **10**, 1, 30–34.

Pal, T. K. and Nutbourne, A. W. (1977), 'Two-dimensional Curve Synthesis using Linear Curvature Elements', *Computer-Aided Design* **9**, 2, 121–134.

Penrose, R. (1955), 'A Generalised Inverse for Matrices', *Proc. Camb. Phil. Soc.* **51**, 406–413.

Peters, G. J. (1974), 'Interactive Computer Graphics Application of the Parametric Bicubic Surface to Engineering Design Problems', In *Computer-Aided Geometric Design* (R. E. Barnhill and R. F. Riesenfeld, eds.) Academic Press.

Powell, M. J. D. (1972), 'Problems related to Unconstrained Optimisation'. In *Numerical Methods for Unconstrained Optimisation* (W. Murray, ed.), Academic Press.

Powell, M. J. D. (1977), 'Numerical Methods for Fitting Functions of Two Variables'. In *The State of the Art in Numerical Analysis* (D. A. H. Jacobs, ed.), Academic Press.

Protter, M. H. and Morrey, C. B. (1964), *Modern Mathematical Analysis,* Addison-Wesley.

Pruess, S. (1976), 'Properties of Splines in Tension', *J. Approx. Th.* **17**, 86–96.

Rabinowitz, P. (ed.) (1970), *Numerical Methods for Non-Linear Equations,* Gordon and Breach, London.

Ralston, A. (1965), *A First Course in Numerical Analysis,* McGraw-Hill.

Riesenfeld, R. F. (1975), 'Nonuniform B-spline Curves', Proc. 2nd USA–Japan Computer Conference, 1975, 551–555.

Roberts, L. G. (1963), 'Machine Perception of Three-Dimensional Solids', Technical Report TR315, MIT Lincoln Laboratory, Lexington, Mass.

Roberts, L. G. (1965), 'Homogeneous Matrix Representation and Manipulation of N-dimensional Constructs', *The Computer Display Review*, Adams Associates, May 1965, 1–16.

Robson, A. (1940), *An Introduction to Analytic Geometry,* Cambridge University Press.

Rogers, D. F. and Adams, J. A, (1976), *Mathematical Elements for Computer Graphics*, McGraw-Hill.

Sabin, M. A. (1968), 'Parametric Surface Equations for Non-rectangular Regions', Report No. VTO/MS/147, Dynamics and Mathematical Services Dept., British Aircraft Corporation, Weybridge.

Sabin, M. A. (1971), 'An existing System in the Aircraft Industry. The British Aircraft Corporation Numerical Master Geometry System', *Proc. Roy. Soc. Lond.* **A 321**, 197–205.

Sabin, M. A. (1972), 'Comments on some Algorithms for the Representation of

Curves by Straight Line Segments', letter in *Comput J.* **15**, 2, 104.

Sabin, M. A. (1976), 'A Method for displaying the Intersection Curve of two Quadric Surfaces, *Comput. J.* **19**, 336–338.

Schechter, A. (1978a), 'Synthesis of 2D Curves by Blending Piecewise Linear Curvature Profiles', *Computer-Aided Design,* **10**, 1, 8–18.

Schechter, A. (1978b), 'Linear Blending of Curvature Profiles', *Computer-Aided Design* **10**, 2, 101–109.

Schoenberg, I. J. (1973), *Cardinal Spline Interpolation,* SIAM, Philadelphia.

Schoenberg, I. J. and Whitney, A. (1953), 'On Polya Frequency Functions III: The Positivity of Translation Determinants with an Application to the Interpolation Problem by Spline Curves', *Trans. Amer. Math. Soc.* **74**, 246–259.

Schumaker, L. L. (1969), 'Some Algorithms for the Computation of Interpolating and Approximating Spline Functions'. In *Theory and Applications of Spline Functions,* (T. N. E. Greville, ed.), Academic Press.

Schumaker, L. L. (1976), 'Fitting Surfaces to Scattered Data'. In *Approximation Theory II* (G. G. Lorentz, C. K. Chui and L. L. Schumaker, eds.), Academic Press.

Schweikert, D. G. (1966), 'An Interpolation Curve using a Spline in Tension', *J. Math. & Phys.* **45**, 312–317.

Semple, J. G. and Kneebone, G. T. (1952), *Algebraic Projective Geometry,* Oxford University Press.

Shephard, D. (1968), 'A Two-dimensional Interpolation Function for Irregularly Spaced Data', *Proc. ACM National Conference*, 1968, 517–524.

Shu, H., Hori, S., Mann, W. R. and Little, R. N. (1970), 'The Synthesis of Sculptured Surfaces', In *Numerical Control Programming Languages* (W. H. P. Leslie, ed.), North Holland Publishing Co., Amsterdam.

Smith, D. J. L. and Merryweather, H. (1973), 'The Use of Analytic Surfaces for the Design of Centrifugal Impellers by Computer Graphics', *Int. J. Num. Meth. Eng.* **17**, 137–154.

Smith, L. B. (1971), 'Drawing Ellipses, Hyperbolas or Parabolas with a Fixed Number of Points and Maximum Inscribed Area', *Comput. J.* **14**, 81–86.

Sommerville, D. M. Y. (1934), *Analytic Geometry of Three Dimensions,* Cambridge University Press.

Späth, H. (1974), *Spline Algorithms for Curves and Surfaces,* Utilitas Mathematica Publishing Inc., Winnipeg, Canada.

Stark, P. A. (1970), *Introduction to Numerical Methods,* Collier-Macmillan.

Thomas, D. H. (1976), 'Pseudospline Interpolation for Space Curves' *Mathematics of Computation* **30**, 133, 58–67.

Varah, J M. (1977), 'On the Condition Number of Local Bases for Piecewise Cubic Polynomials', *Mathematics of Computation* **31**, 137, 37–44.

Veron, M., Ris, G. and Musse, J.-P. (1976), 'Continuity of Biparametric Surface Patches', *Computer–aided Design,* **8**, 4, 267–273.

Walker, L. F. (1972), 'Curved Surfaces in Shipbuilding Design and Production'. In *Curved Surfaces in Engineering* (proc. Conference at Churchill College, Cambridge, 1972), IPC Science and Technology Press.

Walter, H. (1973), 'Computer-Aided Design in the Aircraft Industry'. In *Computer-Aided Design* (J. Vlietstra and R. F. Wielinga, eds.), North Holland Publishing Co.

Warnock, J. E. (1969). 'A Hidden Surface Algorithm for Computer Generated Half-tone Pictures', Report AD-753-671, National Technical Information Service, U.S. Dept. of Commerce, Springfield, Va.

Weatherburn, C. E. (1961), *Differential Geometry of Three Dimensions*, Cambridge University Press.

Wellman, B. L. (1957), *Technical Descriptive Geometry,* McGraw-Hill.

Wielinga, R. F. (1974), 'Constrained Interpolation using Bézier Curves as a New Tool in Computer-Aided Geometric Design'. In *Computer-Aided Geometric Design* (R. E. Barnhill and R. F. Riesenfeld, eds.), Academic Press.

Willmore, T. J. (1958), *Differential Geometry*, Oxford University Press.

Wilson, H. B. and Farrior, D. S. (1976), 'Computation of Geometrical and Inertial Properties for General Areas and Volumes of Revolution', *Computer-Aided Design* **8**, 4, 257-263.

Woo, T. C. (1977a), 'Computer-Aided Recognition of Volumetric Designs'. In *Advances in Computer-Aided Manufacture,* (D. McPherson, ed.) North Holland Publishing Co.

Woo, T. (1977b), 'Progress in Shape Modelling', *Computer,* Dec. 1977, 40-46.

Yuille, I. M. (1972), 'Ship Design'. In *Curved Surfaces in Engineering* (proc. Conference at Churchill College, Cambridge, 1972), IPC Science and Technology Press Ltd.

Index

O

P

Q

R

NOTES

NOTES

NOTES